P. K. Singh, PhD
S. K. Dasgupta, PhD
S. K. Tripathi, PhD
Editors

Hybrid Vegetable Development

Hybrid Vegetable Development has been co-published simultaneously as *Journal of New Seeds*, Volume 6, Numbers 2/3 and Number 4 2004.

Pre-publication REVIEWS, COMMENTARIES, EVALUATIONS . . .

"Covers major tropical and temperate vegetables and PROVIDES USEFUL INFORMATION FOR THOSE LOOKING FOR AN OVERVIEW OF THE SUBJECT. This book can be used as a reference and guidebook for those engaged in the genetic improvement of vegetable crops. I believe the book WILL BE VERY USEFUL FOR GRADUATE STUDENTS AND RESEARCHERS and should be made available on the shelves of university libraries."

Jagveer S. Sindhu, PhD
Director
Asia & Pacific Seed Association (APSA)
Bangkok, Thailand

Hybrid Vegetable Development

Hybrid Vegetable Development has been co-published simultaneously as *Journal of New Seeds*, Volume 6, Numbers 2/3 and Number 4 2004.

The *Journal of New Seeds* Monographic "Separates"

Below is a list of " separates," which in serials librarianship means a special issue simultaneously published as a special journal issue or double-issue *and* as a "separate" hardbound monograph. (This is a format which we also call a "DocuSerial.")

"Separates" are published because specialized libraries or professionals may wish to purchase a specific thematic issue by itself in a format which can be separately cataloged and shelved. as opposed to purchas - ing the journal on an on-going basis. Faculty members may also more easily consider a "separate" for classroom adoption.

"Separates" are carefully classified separately with the major book jobbers so that the journal tie-in can be noted on new book order slips to avoid duplicate purchasing.

You may wish to visit Haworth's website at . . .

http://www.HaworthPress.com

. . . to search our online catalog for complete tables of contents of these separates and related publications.

You may also call 1-800-HAWORTH (outside US/Canada: 607-722-5857). or Fax 1-800-895-0582 (outside US/Canada: 607-771-0012), or e-mail at:

docdelivery@haworthpress.com

Hybrid Vegetable Development. edited by P. K. Singh, PhD, S. K. Dasgupta. PhD, and S K. Tripathi. PhD (Vol. 6, No. 2/3 and No. 4, 2004). *A comprehensive sourcebook that includes detailed information. trends, and techniques on all aspects of hybrid vegetable development.*

Bacillus thuringiensis: A Cornerstone of Modern Agriculture, edited by Matthew Metz. PhD (Vol. 5, No. 1 and No. 2/3. 2003). *"A well-rounded discussion of the most pressing issues surrounding Bt technology, complemented by RELEVANT AND SIGNIFICANT research paper" (James Pearse, PhD. BSc, Post-Doctoral Researcher, Syngenta Ltd.)*

Seed Policy, Legislation and Law: Widening a Narrow Focus. edited by Niels P. Louwaars (Vol. 4, No. 1/2. 2002). *International experts focus on national/international seed policy and law.*

Hybrid Seed Production in Vegetables: Rationale and Methods in Selected Crops. edited by Amarjit S. Basra, PhD (Vol. 1, No. 3/4. 1999). *This essential guide will help crop scientists and growers increase the quality and yield of vegetables such as cucumbers, pumpkins, squash, peppers, onions, gourds, and the fruits watermelon and winter melon. Containing suggestions and methods for overcoming male plant sterility, inbreeding, and challenges to pollination. This book will help you successfully breed hybrid plants to produce bountiful and healthy crops.*

Hybrid Vegetable Development

P. K. Singh, PhD
S. K. Dasgupta, PhD
S. K. Tripathi, PhD
Editors

Hybrid Vegetable Development has been co-published simultaneously as *Journal of New Seeds*, Volume 6, Numbers 2/3 and Number 4 2004.

CRC Press
Taylor & Francis Group
Boca Raton London New York

CRC Press is an imprint of the
Taylor & Francis Group, an **Informa** business

Indexing, Abstracting & Website/Internet Coverage

This section provides you with a list of major indexing & abstracting services and other tools for bibliographic access. That is to say, each service began covering this periodical during the year noted in the right column. Most Websites which are listed below have indicated that they will either post, disseminate, compile, archive, cite or alert their own Website users with research-based content from this work. (This list is as current as the copyright date of this publication.)

Abstracting, Website/Indexing Coverage ········· Year When Coverage Began

- *AGRICOLA Database <http://www.natl.usda.gov/ag98>*............1999

- *AGRIS <http://www.fao.org/agris/>*..............................1999

- *BIOBASE (Current Awareness in Biological Sciences)*
  <http://www.elsevier.nl> 2000

- *Biology Digest (in print & online) <http://www.infotoday.com>*.......1999

- *Bioscience Information Service of Biological Abstracts (BIOSIS)*
  <http://www.biosis.org>.....................................2001

- *CAB ABSTRACTS c/o CAB International/CAB ACCESS ...*
  available in print, diskettes updated weekly, and on INTERNET.
  Providing full bibliographic listings, author affiliation,
  augmented keyword searching <http://www.cabi.org> 2002

- *Cambridge Scientific Abstracts (Agricultural & Environmental*
  Biotechnology Abstracts) <http://www.csa.com>.................2001

- *Crop Physiology Abstracts (c/o CAB Intl/CAB ACCESS)*
  <http://www.cabfocus.com>....................................*

- *EBSCOhost Electronic Journals Service (EJS)*
  <http://ejournals.ebsco.com>................................2001

- *Field Crop Abstracts (c/o CAB Intl/CAB ACCESS)*
  <http://www.cabi.org> 2002

(continued)

- *Food Science & Technology Abstracts (FSTA)*
  <http://www.foodsciencecentral.com> 1999

- *Foods Adlibra* 1999

- *Google <http://www.google.com>*...........................2004

- *Google Scholar <http://scholar.google.com>* 2004

- *Helminthological Abstracts (c/o CAB Intl/CAB ACCESS)*
  <http://www.cabi.org/>......................................*

- *HORT CABWeb (c/o CAB Intl/CAB ACCESS)* 2001

- *Horticultural Abstracts (c/o CAB Intl/CAB ACCESS)*
  <http://www.cabi-publishing.org/AtoZ.asp>.....................*

- *Journal of Medicinal Food*...................................1999

- *Nutrition Abstracts & Reviews, Series B, Livestock Feeds &*
  Feeding (c/o CAB Intl/CAB ACCESS) <http://www.cabi.org>......2002

- *Plant Breeding Abstracts (c/o CAB Intl/CAB ACCESS)*
  <http://www.cabi.org> 2002

- *Plant Genetic Resources Abstracts (c/o CAB Intl/CAB ACCESS)*
  <http://www.cabi.org/>......................................*

- *Postharvest News and Information (c/o CAB Intl/CAB ACCESS)*
  <http://www.cabi.org> 2002

- *Rural Development Abstracts (CAB Intl)*.........................*

- *Seed Abstracts (c/o CAB Intl/CAB ACCESS)*
  <http://www.cabi.org/>.....................................1999

- *Seed Pathology & Microbiology (c/o CAB Intl/CAB ACCESS)*
  <http://www.cabi.org> 2001

- *Soyabean Abstracts (c/o CAB Intl/CAB ACCESS)*
  <http://www.cabi.org> ....*

- *TROPAG & RURAL (Agriculture and Environment for Developing*
  Regions), available from SilverPlatter, both via the Internet and
  on CD-ROM <http://www.kit.nl> 2003

- *World Agricultural Economics & Rural Sociology Abstracts (c/o CAB*
  Intl/CAB ACCESS) <http://www.cabi-publishing.org/AtoZ.asp>.......*

- *zetoc <http://zetoc.mimas.ac.uk/>*............................2004

*Exact start date to come.

(continued)

*Special Bibliographic Notes related to special journal issues (separates)
and indexing/abstracting:*

- indexing/abstracting services in this list will also cover material in any "separate" that is co-published simultaneously with Haworth's special thematic journal issue or DocuSerial. Indexing/abstracting usually covers material at the article/chapter level.
- monographic co-editions are intended for either non-subscribers or libraries which intend to purchase a second copy for their circulating collections.
- monographic co-editions are reported to all jobbers/wholesalers/approval plans. The source journal is listed as the "series" to assist the prevention of duplicate purchasing in the same manner utilized for books-in-series.
- to facilitate user/access services all indexing/abstracting services are encouraged to utilize the co-indexing entry note indicated at the bottom of the first page of each article/chapter/contribution.
- this is intended to assist a library user of any reference tool (whether print, electronic, online, or CD-ROM) to locate the monographic version if the library has purchased this version but not a subscription to the source journal.
- individual articles/chapters in any Haworth publication are also available through the Haworth Document Delivery Service (HDDS).

Hybrid Vegetable Development

CONTENTS

ABOUT THE EDITORS

P. K. Singh, PhD, is a research officer at Sungro Seeds Ltd. Delhi, a leading research-based company in the development of vegetable seeds in India. He has several years' experience in vegetable breeding and hybrid seed production (crucifer vegetables). He has published numerous research papers in various national and international journals. His present research focuses on vegetable breeding in both fundamental and applied aspects. Dr. Singh has participated in several national and international symposiums.

S. K. Dasgupta, PhD, is a senior research officer at Sungro Seeds Ltd. Delhi. He completed his studies in genetics and plant breeding from BCKV, West Bengal, India, and has worked in the Division of Genetics and Plant Breeding at the Indian Agriculture Research Institute. His principle area of work includes genetics and mapping of disease resistant genes in peas, marker-assisted selection through molecular means, and transgenic vegetables. Dr. Dasgupta has published work in several international journals and has participated in numerous international conferences.

S. K. Tripathi, PhD, is the general manager of research and development at Sungro Seeds Ltd. Delhi. He completed his studies at the University of Rajasthan, then joined Mahyco, the largest seed company in India in its research and development department as chief scientist before moving to his present position at Mahyco's sister organization Sungro Seeds Ltd. His present research centers on breeding and commercialization of cole crops, solanaceous crops, and different types of cucurbits. He has numerous released commercial hybrids and varieties to his credit and has traveled internationally for his research work. Dr. Tripathi has numerous publications in international journals and has participated in several international conferences.

Dedicated to

Dr. B. R. Barwale
World Food Prize Laurate
Father of Indian Seed Industry

Acknowledgments

We feel a deep sense of gratefulness for Dr. Amarjit S. Basra for his constant encouragement. Without his efforts this book probably would not have been possible in this busy schedule of our professional world.

Deep sense of gratitude is also due for our honourable Managing Director, Deputy Managing Director, Director Research, and Joint Director Research of Maharashtra Hybrid Seed Company (Mahyco) and Executive Director of Sungro Seeds Ltd. who had always encouraged us for professional and academic excellence. Under their guidance we have achieved much scientific and professional experience.

Thanks are also due to Dr. S. Shanmugasundaram, Deputy Director General, AVRDC, Taiwan, for providing us with important contacts and connections related to authors and other inputs.

We also congratulate all the contributors who are international experts in their respective fields, in agreeing to give their valuable time to write their respective chapters in spite of such paucity of time.

Finally we would like to thank our colleagues, friends, R&D team members, our families in providing the encouragement all the time in production of this book.

P. K. Singh
S. K. Dasgupta
S. K. Tripathi

[Haworth co-indexing entry note]: "Acknowledgments." Singh. P. K., S. K. Dasgupta, and S. K. Tripathi. Co-published simultaneously in *Journal of New Seeds* (Food Products Press, an imprint of The Haworth Press, Inc.) Vol. 6. No. 2/3, 2004, p. xix; and: *Hybrid Vegetable Development* (ed: P. K. Singh, S. K. Dasgupta, and S. K. Tripathi) Food Products Press, an imprint of The Haworth Press, Inc., 2004, p. xv. Single or multiple copies of this article are available for a fee from The Haworth Document Delivery Service [1-800-HAWORTH. 9:00 a.m. - 5:00 p.m. (EST). E-mail address: docdelivery@haworthpress.com].

xv

Hybrid Tomato Breeding

D. S. Cheema
M. S. Dhaliwal

SUMMARY. Tomato is one of the most popular vegetable crops grown across the world. Its genetics is much studied among vegetable crops, resulted in the reorganization of its commercial exploitation of hybrid vigor since last hundred years. Tomato has tremendous potential of heterosis for earliness, total yield, resistance attributes and uniformity. Hybrid tomato varieties will continue to predominate high input agricultural systems and may expand under some lower input systems where benefits can be demonstrated. Increasing demand for hybrid seeds can stress commercial hybrid seed production abilities. At present most of the hybrid seeds of tomato is being produced by hand emasculation and hand pollination. The advances in genetic emasculation methods can prove very useful for making the hybrid seed production profitable and ensure the supply of hybrid seed at cheaper rates to the growers. *[Article copies available for a fee from The Haworth Document Delivery Service: 1-800-HAWORTH. E-mail address: <docdelivery@haworthpress.com> Website: <http://www.HaworthPress.com> © 2004 by The Haworth Press, Inc. All rights reserved.]*

KEYWORDS. Tomato, floral biology, heterosis, male sterility, emasculation, pollination, seed production, isolation, hybrid seed production

D. S. Cheema and M. S. Dhaliwal are affiliated with the Department of Vegetable Crops, Punjab Agricultural University, Ludhiana 141004, India.

[Haworth co-indexing entry note]: "Hybrid Tomato Brreding." Cheema, D. S., and M. S. Dhaliwal. Co-published simultaneously in *Journal of New Seeds* (Food Products Press, an imprint of The Haworth Press, Inc.) Vol. 6, No. 2/3, 2004. pp. 1-14; and: *Hybrid Vegetable Development* (ed: P. K. Singh, S. K. Dasgupta, and S. K. Tripathi) Food Products Press, an imprint of The Haworth Press, Inc., 2004, pp. 1-14. Single or multiple copies of this article are available for a fee from The Haworth Document Delivery Service [1-800-HAWORTH. 9:00 a.m. - 5:00 p.m. (EST). E-mail address: docdelivery@haworthpress.com].

http://www.haworthpress.com/web/JNS
© 2004 by The Haworth Press, Inc. All rights reserved.
Digital Object Identifier: 10.1300/J153v06n02_01

INTRODUCTION

Tomato (*Lycopersicon esculentum* Mill.), a member of family Solanaceae is one of the most popular vegetable crops grown all over the world. It is grown in a wide range of climates in the open, under protection in plastic green houses, and heated green houses. Its fruit are rich source of vitamins C, A, B, and minerals like calcium, phosphorous and iron.

It is used as raw in sandwiches, salads, etc. It ranks first in the processed vegetables and its main processed products are paste, sauce, and ketchup.

The annual production of tomato at the world level is estimated to be 95,127,000 metric tons from an area of 3542,000 ha. In India, it is grown in an area of 360,000 ha producing 5,450,000 metric tons. Average yield in India is about 15,139 kg/ha as against 26,860 kg/ha at the world level (Anon, 1999).

The Andean zone is likely to be the centre of origin of wild tomato. The most likely ancestor of cultivated tomato is the cherry tomato (*Lycopersicon esculentum* var. *cerasiforme*).

The domestication of tomato took place in Mexico. The name probably has been derived from 'tomatl' in Nanhua tongue of Mexico. The cultivars introduced from Latin America generally had exposed stigma facilitating cross pollination. In the latter stages, plants with a high percentage of fruit set and with short style at the mouth of the anther tube were selected. Rick (1976) also reported that domestication of *Lycopersicon esculentum* has taken place with the transition of exerted to inserted stigma, as a result of the change of allogamy to autogamy. Boswell (1949) had reported that prior to 1800, the European cultivars were introduced to the United States. Since 1800, tomato plants are being grown in most parts of the world. The English traders of East Indian Company introduced it into India in 1822.

PLANT MORPHOLOGY

The tomato plant, though perennial by nature, is almost universally cultivated as annual. The cultivated species has an herbaceous annual to perennial growth habit. It has a strong tap root, though later adventitious roots develop rapidly, if the tap root is damaged. The stem is soft, brittle, and hairy when young and hard, woody and copiously branched when mature. Its organization is in the sequence of two or three lobed sympodia. It is erect to semierect. The plants can be classified into three groups on the basis of their growth habit, i.e., determinate (self-topping bush), determinate (self-topping, dwarf) and indeterminate. The branching pattern is sympodial, but the first few nodes, which do not produce inflorescence, may be monopodial. The side shoots are produced from the axil of the leaves. The leaf is alternate, petiolate; leaflets are unequal odd pinnate; the apex is acuminate, acute, and irregularly serrate.

FLORAL BIOLOGY

The inflorescence is extra axillary cymes with dichotomous or polycotomous branching. It is lateral and arises from the stem opposite to the leaf. It may be terminal or may continue with the vegetative shoot. The plant produces bright yellow flowers. Studies by Kanahama and Saito (1988) on inflorescence type in tomato have revealed that the first flower bud in the first inflorescence was initiated on the terminal meristem of the main stem, while the second and third flower buds were initiated on the lateral side of the pedicels of the first and second flower, respectively. Flowers were arranged alternately on the leaf and right sides of the inflorescence peduncle. The peduncle was helical at the time of opening of the first flower but unfolded during fruit growth. The flowers are pentamerous, bisexual, regular, complete, ebracteate, and hypogynous. The pistil has two to several carpels. The anthers connate appearing in the throat of the corolla and their dehiscence is introrsely longitudinal and occurs soon after or during the corolla opening. Usually the style is shorter than anther; therefore a high degree of self pollination takes place.

FLORAL ANTHESIS

The anthesis of flowers starts around 6 a.m. and the flower continues to open until 11 a.m. The peak period of anther dehiscence is between 8 to 11 a.m. depending upon the initiation of sunshine, atmospheric temperature, and humidity. At temperatures ranging between 18° to 25°C, the pollen remain viable from 2 to 5 days, the stigma becomes receptive 16 to 18 hours before anthesis and retains the receptivity up to 6 days after anthesis, i.e., shortly before the flower withers. This long duration of stigma receptivity from nearly one day prior to 6 days after anthesis permits controlled pollination. These are further facilitated by the long duration viability of pollen as the pollen retains its viability for 2 to 5 days at temperature ranging from 18° to 25° C. At 5°C in a dessicator, it retains viability for six months (McGuire, 1952). Some important aspects of floral biology of tomato have been reviewed by Sidhu et al. (1980). The maximum fertilization and fruit set takes place where pollination takes place at the time of anthesis. Normally fertilization takes place 24-50 hours after pollination.

HETEROSIS

Commercial hybrids of tomato are products of mid-twentieth century though their valued performance was recognized a century and a quarter ago. Tomato has tremendous potential of heterosis for earliness, total yield, resistance attributes and uniformity. The first step towards a large scale production and extension of tomato hybrids was taken in Bulgaria. As early as 1912, Doskellaf began tomato heterosis breeding and the first tomato F_1 cultivar which he developed during that time was quickly extended in production because it was earlier and produced higher total yield than the parental cultivars and remaining standard cultivars of the area. High hybrid seed price was repaid many times over by the realized higher profits. During the years 1960-70, the production of hybrid tomato seed amounted to 10-14 tonnes/year in Bulgaria alone (Yordanov, 1983). Heterosis breeding was quickly extended to the Netherlands, Japan, USA, France, England, etc., after world war-II. In India, studies on heterosis breeding in tomato had been going on since long. However, the Indo-American Hybrid Seeds Company, Bangalore, released the first commercial tomato hybrid in 1975. From 1975 onwards, public sector organizations have also released some F_1 hybrids like TH 2312, TH 802 (both from PAU, Ludhiana), Pusa Hybrid-1, Pusa Hybrid-2 (both from IARI, New Delhi). Although the hybrids are very good for local marketing, main drawback of these hybrids is their short shelf life.

If a hybrid is to be accepted commercially, it must be superior to the cultivars presently grown. This superiority may be expressed in terms of total yield, early yield, nutritional quality, post harvest life, insect pest and disease resistance, adaptability, external appearance, etc. Increased vigor in total yield in tomato to the extent of 300 percent over the commercial variety has been reported. In India, the extent of the cross over the better parent is reported to be up to 263 percent (Dhaliwal, 1986).

In addition to superior performance of hybrids, due to heterosis, intermediate state of gene expression in heterozygotes could be of potential value. The situation could be illustrated by the potential utilization of *nor* (non-ripening) gene in hybrids exhibiting normal pigment production but slowed down ripening of fruit, which ultimately prolong shelf life of fruit (Buescher et al. 1976). In some cases, the heterozygous state of some genes (*Mi, Tm, Tm2*) is accompanied by an undesirable effect such as low pollen fertility, poor fruit set or necrosis under high temperature (Soost, 1959; Laterrot, 1973; Philouze, 1976; Lapushner and Frankel, 1979). This drawback is corrected in F_1 hybrids when the named genes are in heterozygous state. As regards economics of acceptability, the difference in cash value between a hybrid and conventional crop could be much higher than difference between cost of hybrid seed and cost of conventional seed.

METHODS OF HYBRID SEED PRODUCTION

Hybrid seed of tomato can be produced by the following three methods.

- Hand emasculation and hand pollination,
- Using male sterility and hand pollination, and
- Using male sterility and natural pollination

Hand Emasculation and Hand Pollination

Although this procedure is laborios but is commercially viable in crops like tomato where each fruit gives reasonably large number of seeds and seed rate per acre is as low as 50 g. East and Heyes, the pioneers of heterosis breeding, pointed out that tomato hybrids could have considerable practical value because crossing is relatively easy in this Solanaceous vegetable. The growing of the crop has to be monitored carefully. Especially important are the irrigation and fertilization practices, as moisture and nutrient stress can result in reduced fruit and seed yields. Seed yield vary according to plant and fruit type. Yield ranges for compact plants with blocky fruit, to 85-125 kg/ha, for compact plants with round fruit to 130-160 kg/ha, in medium large plants with round fruit to 145-180 kg/ha and in indeterminate plants with round fruits the range is from 140-180 kg/ha. Seed companies have made a commitment to support research programs for he development of hybrid tomato varieties, and in addition, they are supporting research with regards to hybrid seed production.

Emasculation is done in the afternoon and the most suitable stage is when corolla leaves have just opened and form an angle up to 45° with flower axis. At this stage no selfing has yet occurred. Emasculated flowers are pollinated next morning. Pollination is not done on the day of emasculation, as the stigma is not yet receptive by then and such pollination if at all successful, the resultant fruits will produce few seeds. Delayed pollination should also be avoided as the stigma surface dries up particularly in dry and hot windy weather. For large scale hybrid seed production, the process of pollen collection and its application to stigma are mechanized.

In tomato, it is observed that 51 minutes are required to produce 4000 hybrid seed and using pollen sterility of any kind, viz., pollen sterility, positional sterility, and functional sterility can considerably reduce this time. It is reported that emasculation constitutes about 40 percent of the total labor involved in hybrid seed production of tomato (Bullard and Stevenson, 1952).

Using Male Sterility and Hand Pollination

There are four types of male sterility in tomato. Each one is governed by a single recessive gene (Table 1). Stamenless type produces misshapen fruit in

TABLE 1. Description of different male sterile mutants in tomato

Mutant	Description	Inheritance	Governing genes
Pollen sterile	Pollen abortive	Monogenic recessive (Except MS-48, monogenic dominant)	ms series (-49 independent genes)
Stamenless	Stamens absent	Monogenic recessive	sl
Positional sterility	Stigma exerted	Monogenic recessive	ps
Functional sterility	Anthers do not dehisce	Monogenic recessive	ps-2

the F_1 hybrid generation. Where as positional sterility is not stable. Pollen abortive type and functional sterilities are often used in F_1 hybrids production. Pollen abortive type is maintained through back crossing (ms ms × Ms ms), while functional male sterility is maintained by manual selfing. The efficiency of hybrid seed production by involving pollen abortive type male sterile lines are further improved by incorporating morphological markers, e.g., anthocyaninless gene '*aa*.' The gene is reported to be linked with $ms10^{35}$ and male sterile plants are identified from the mixed population of fertile and sterile plants in nursery itself (Philouze, 1974). Consequently only male sterile plants are transplanted in hybrid seed production block economizing area and labor. However, the efficiency of this method depends upon the power of the linkage between the two factors, viz., male sterility and anthocyaninless. The male sterility genes can be exploited advantageously only if the style is accessible to cross pollination. Otherwise disturbing anther cone to expose stigma for pollination will neutralize advantage of male sterility. According to Georgiev, (1991) utilization of male sterility in combination with exerted stigma and with anthocyanin less gene, faces some inherent problems. These include appearance of 2-3% reciprocal recombinants, i.e., absence of anthocyanin in fertile plants, low seed yield due to poor receptivity of stigma and loss of exerted style in 20-60% of the flowers depending upon the genotype and the environmental factors. The exerted style character is more stable in small fruited genotypes than in large fruited ones.

The efficiency of male sterility (pollen abortive types) genes can further be improved if the homogenous stand of the seed parent is available. This should be made possible by regenerating male sterile plants from *in vitro* cultured explants. In such a case, there is no need of genetic markers. Phenotypic fertility of genetically sterile plants can also be restored by application of growth regulators. Progeny of the self seed thus produced will be all male sterile. Partial success with application of gibberellic acids A_3 + or A_4 + A_7 was reported by Yordanov et al. (1971).

The functional male sterility governed by *ps*-2 has indisputable advantage over other types. These include maintenance of male sterility under homozygous state by self pollination, possibility for simultaneous emasculation and pollination of blossomed flowers when the stigma is most receptive and high yield of hybrid seed with low cost. The main disadvantage of this sterility is that up to 5% selfing is possible which is not tolerated in hybrid varieties known for their extreme uniformity. This has been overcome to great extent by the development of lines combining functional sterility (*ps*-2) with short style (*shst*), which decreased chance selfing to less than 0.2% and enabled emasculation to be done very easily by non-skilled workers. The cost of hybrid seed by using the *ps*-2 female lines decreased by approximately 50% as a result of easiness in emasculation and pollination and by '*ms*10^{35} aa' approximately 20% as a result of elimination of emasculation of flower (Georgiev, 1991).

Use of Male Sterility with Natural Crossing

Hybrid seed yield using this method is not very encouraging (0.3-1.5 kg/acre). This method could gain popularity in practice only if suitable insect pollinators or other means are found to raise the percentage of cross pollination to the extent of at least 50 percent.

SEED PRODUCTION

Isolation: Although tomato is a highly self-pollinated crop, its seed crop should be provided an isolation distance of 50 meters from other tomato varieties or the same variety not conforming to the varietal purity requirements according to the seed certification laws. Isolation is provided to eliminate any chance of outcrossing.

Land Requirement: The tomato seed field should be free of volunteer plants. It is, therefore, advisable to plant seed crop in a field where previous crop was other than tomato. However, there is no specific condition in the seed certification standards regarding the previous crop.

Field Inspection: A tomato seed field should be inspected at least thrice during the crop season to identify and remove the off type plants. The first inspection is made before flowering and the identification of true-to-type plant is ascertained on the basis of foliage and plant type characters. Second inspection is done at flowering fruiting stage. Flower, fruit and other plant characters form the basis of this inspection. At the third inspection, the color of mature fruit and its shoulders as well as fruit shape form the basis for establishing the trueness to type of the plants.

Breeder Seed Production

1. Grow about 75-100 single plant progenies each consisting of at least 20-25 plants. After every 25 progeny rows one line should be planted with the breeder seed as check. The number of progenies can be reduced or increased depending upon the requirements of the breeder seed. The spacing should be 100 cm × 50 cm between the lines and the plants.
2. The plant progenies which do not conform to the varietal purity requirements should be rejected before flowering on the basis of plant type and at immature and mature fruit stage on the basis of fruit size, shape, colour, as well as the over all performance of the progeny. The plants showing early blight, leaf spot and mosaic (TMV) symptoms must be removed since these diseases are known to be seed borne.
3. Collect the seed separately from about 75 ideal plants to repeat the above cycle. The seed harvested from each of these plants would be sufficient to repeat the process for breeder seed production during the subsequent 2-3 years provided the seed is properly dried and stored.
4. The true-to-type progenies are bulk harvested to obtain the breeder seed for further seed multiplication.

Foundation Seed Production

1. The bulk seed harvested from the selected progenies as above is used for the production of foundation seed. The plants are spaced as indicated earlier for breeder seed production.
2. Rogue plants should be singled out and removed. The plants with off type foliage and other characters should be removed before flowering. The plants should be examined for overall performance along with shape, size, color, and interior characters of the fruit as well as the plant type during the subsequent inspections. The plants, not conforming to the varietal descriptions, should be removed.
3. The diseased plants, affected by early blight, leaf spot, and mosaic (TMV) should be rouged out. The percent diseased plants should not exceed the maximum limit prescribed under the seed certification laws.

FIELD STANDARDS

The minimum seed certification standards in India have been reported by Tunwar and Singh (1988) as follows.

Varieties

General Requirements

Seed fields should be isolated from the contaminants shown in column 1 of the table below by the distance specified in columns 2 and 3.

Contaminants	Minimum distance (meters)	
	Foundation	Certified
1	2	3
Fields of other varieties	50	25
Fields of the same variety not conforming to varietal purity requirements for certification	50	25

Specific Requirements

Factor	Maximum permitted (%)*	
	Foundation	Certified
Off types	0.10	0.20
Plants affected by seed borne diseases**	0.10	0.50

* Maximum permitted at final inspection
** Seed borne diseases shall be: Early blight (*Alternaria solani* Sorauer), Leaf spot (*Stemphylium solani* Weber), Tobacco Mosaic Virus (TMV).

Seed Standards

Factor	Standards for each class	
	Foundation	Certified
Pure seed (minimum)	98.0%	98.0%
Inert matter (maximum)	2.0%	2.0%
Other crop seeds (maximum)	5/kg	10/kg
Weed seeds (maximum)	None	None
Germination (minimum)	70%	70%
Moisture (maximum)	8.0%	8.0%
For vapor-proof containers (maximum)	6.0%	6.0%

Hybrids

General Requirements

Seed fields should be isolated from the contaminants shown in column 1 of the table below by the distances specified in columns 2 and 3.

Contaminants	Minimum distance (meters)	
	Foundation	Certified
1	2	3
Fields of other varieties including commercial hybrid of the same variety	200	100
Fields of the same hybrid (code designation) not conforming to varietal purity requirements for certification	200	100
Between blocks of the parental lines in case seed parent and pollinator are planted in separate blocks	–	5

Specific Requirements

Factor	Minimum permitted (%)*	
	Foundation	Certified
Off types in seed parent	0.010	0.050
Off types in pollinator	0.010	0.050
** Fertile segregants (pollen shedding cymes) in seed parent	0.050	0.10
*** Plants affected by seed borne diseases	0.10	0.50

* Standards for off types and fertile segregants (pollen shedding cymes) in seed parent shall be met at and after flowering and for seed borne diseases at final inspection.
** It shall be applicable in case male sterile line is used for producing hybrid seeds.
*** Seed borne diseases shall be: Early blight (*Alternaria solani* Sorauer),
                                    Leaf spot (*Stemphylium solani* Weber)
                                    Tobacco Mosaic Virus (TMV)

Seed Standards

Factor	Standards for each class	
	Foundation	Certified
1	2	3
Pure seed (minimum)	98.0%	98.0%
Inert matter (maximum)	2.0%	2.0%
Other crop seeds (maximum)	5/kg	10/kg
Weed seeds (maximum)	None	None
Germination (minimum)	70%	70%
Moisture (maximum)	8.0%	8.0%
For vapor-proof containers (maximum)	6.0%	6.0%

All certified seed lots, which have been produced by adopting emasculation and hand pollination, will be subjected to grow-out test and shall conform to the following minimum genetic purity requirements:

Class	Genetic purity (%) (minimum)*
Certified	90.0%

* During grow-out test, the off type plants (other than selfed plants) such as sergeants, out crosses and plants of other varieties should not exceed more than 1.50% out of the 10.0% plants earmarked for selfed plants.

The minumum population size of 400 plants should be maintained in two replicates of 200 each or four of 100 each throughout the test and each plant will be examined individually. The reject number will be as follows:

Class	Genetic purity (%) (minimum)	Reject number
Certified	90.00 (10 in 100)	44

HARVESTING AND SEED EXTRACTION

The tomato crop is harvested for seed when majority of the fruits are ripe and before many have rotten. Seed is physiologically mature at red ripe stage. The seed extracted from unripe green, yellow or sun scalded fruits show poor germination. Diseased and rotten fruits should also be discarded for seed harvest.

Tomato seed is surrounded by a mucilaginous material which makes seed separation difficult. The mucilage surrounding the seed can be removed by any of the following methods.

Fermentation: Tomato fruits are pulped either by hand or machine and placed for fermentation in wooden or earthen pots or a non-corrosive metal container. As suggested by Singh et al. (1978), fermentation should be allowed to continue for 2-3 days during the month of May-June in plains of North India. At lower temperature, fermentation is slow since the rate of fermentation is directly dependent on temperature. Fermentation is complete when white foam-like substance appears on the surface of the pulped tomato fruit material. During fermentation, the pulped material should be stirred every morning and evening to avoid any fungal growth.

Acid Treatment: Commercial grade hydrochloric acid at the rate of 10-15 ml per kg of the pulped tomato fruit is sufficient to disintegrate the mucilaginous material surrounding the seed (Singh and Gill, 1982). The mixture is continuously stirred and the seed is washed free of tomato pulp and skin after 30 minutes. If there is an incidence of bacterial canker, the seed should be soaked in 0.8 percent acetic acid for 24 hours to kill the pathogen (Blood, 1937).

Juice Extraction: In this method juice is a by-product. Tomato seed and pulp are separated from the juice through a juice extraction machine. Seed is then washed free of fruit pulp and skin. If there is a danger of bacterial canker, the seed should be treated with acetic acid as described above.

Mechanical Extraction: An Axial-Flow Vegetable Seed Extraction Machine designed by Verma and Singh (1986) can be used for tomato seed extraction. Tomato seed material collected from the seed outlet is put into a trough containing water. The seed being heavier settles down at the bottom of water. The floating fragments of tomato pulp and skin are removed by decantation. Clean tomato seed is obtained after two to three washings with clean water. The clean seed thus obtained is treated with HCl at the rate of 8-10 ml per kg of seed to break the mucilaginous layer surrounding the seed. The seed is again washed with water after 15-20 minutes to remove the acid. Two washings with clean water are sufficient to obtain the clean seed.

DRYING AND STORAGE

The tomato seed should be dried as rapidly as possible. After through and repeated washings with water, the seed is spread in thin layers on the drying floor or on the screen-bottom trays placed in the open where maximum exposure to sun is attained. When the summer temperature exceeds 40°C, the seed drying under partial shade is desirable to avoid any reduction in seed germination due to over heating. Under artificial conditions, the seed should be dried at low temperature (38°C) for a few hours and subsequently the temperature should be raised to 40°C without any danger to seed viability (Anonymous, 1961).

The moisture content of thoroughly dried tomato seed should not exceed 8 percent. The seed thus dried should be properly packed and placed in a cool, well-ventilated and dry place. If the seed is to be stored in moisture proof or moisture resistant container, the seed moisture should not be more than 6 percent. Excessive moisture in the sealed storage containers damages seed viability rapidly, especially if the seed is stored in ambient or high temperature conditions. Carefully and properly stored seed even under ambient conditions retains its viability for 2 to 3 years.

CONCLUSIONS

Hybrid tomato varieties will continue to predominate high input agricultural systems and may expand under some lower input systems where benefit can be demonstrated. Increased demand for hybrids can stress commercial hybrid production abilities. Therefore, advances in genetics emasculation method can prove to be very useful. Development of chemical hybridizing agents to emasculate seed parent is also one possible approach. Development in cloning technology to clone male sterile genes and inserting them into seed parents will be another way to assure availability of F_1 seeds yield and hybrid vigor of tomato can be enhanced by using the genes (QTL's) from their wild relatives. The identification of these genes and their ability to enhance yield in elite tomato backgrounds needs to be demonstrated.

REFERENCES

Anonymous (1961). The year book of agriculture USDA. Washington, DC: 295-306.

Anonymous (1999). FAO production year book. Food and Agriculture organization, Rome.

Blood. H.L. (1937). A possible acid soak for the control of bacterial canker of tomato. Science 86(2226) new series: 1999-2009.

Boswell, V.R. (1949). Our vegetable travellers. The National Geographic magazine, August, Xc VI: 145-217.

Bullard, E.T and Stevenson, E.C. (1952). Production of hybrid tomato seed. Amer. Soc. Hort. Sci., 61: 451-458.

Buscher, R.W., Sistrunk W.A., Tigchelaar E.C. and Timorthy J.N. (1976). Softening, pactolytic activity and storage life or rin and nor tomato hybrids. Hort Sci. 11: 603-604.

Dhaliwal, M.S. (1986). Heterosis breeding in tomato: Theory and practice. Seeds and Farms. 12(3): 37-39.

Georgiev, H. (1991). Heterosis in tomato breeding. In: Genetic improvement of tomato by. G. Kalloo (Ed.) Springer-Verlag, Berlin, Heidelberg, pp. 83-98.

Kanahama, K. and Saito, T. (1988). Inflorescence type of tomato. Journal of Japanese Society for Horticultural Science. 57(3), 426-432.

Kalloo, G. and Bhutani, R.D. (1993). Improvement of tomato. In Advances in Horticulture Vol. 5-Vegetable Crops: Part I Ed. K.L. Chadda and G. Kalloo Malhotra Publishing House, New Delhi

Lapushner, D. and Frankel, R. (1979). Rationale and practice of tomato F_1 hybrid breeding and seed production. In Bianchi A. (Ed.) Monografic di Genetica Agraria. Vol. IV. Roma pp. 259-273.

Laterrot, H. (1973). Resistance de la tomato an virus de la mosaique du tobac. Difficulties rencontrees pour la selection de varieties resistance. Ann. Amel Plant 23: 287-313.

McGuire, D.C. (1952). Storage of tomato pollen. Ann. Bot. 39:591-596.

Philouze, J. (1974). Genes marquehrs lies aux genes de sterilite male $ms10^{35}$ et ms 32 che 2 la tomato. Ann. Amel. Plant 24: 77-82.

Philouze, J. (1976). Les hybrids de la tomato. Pepinierist Hortc. Maraich 164: 13-18.

Rick, C.M. (1976). Tomato (Family Solanaceae). In N.W. Simmonds (Ed.) Evolution of crop plants. Longman, London.

Sidhu, A.S., Kalloo and Pandita, M.L. (1980). Studies on some important aspects of floral biology in vegetable crops. A review. Haryana J. Hort. Sci. 9: 207-217.

Singh, H., Gill, S.S. and Kanwar, J.S. (1978). Germination of tomato seed as effected by fruit maturity, fermentation and harvest dates. In: Physiology of sexual reproduction in flowering plants. Kalyani, New Delhi, pp. 473-477.

Singh, H. and Gill, S.S. (1982). Tomato seed extraction. Prog. Fmg. 18(12): 13.

Soost, R.K. (1959). Tobacco mosaic resistance. TGC Rpt. 9: 46.

Tunwar, N.S. and Singh. S.V. (1988). Indian minimum seed certification standards. The central seed certification board, Ministry of Agriculture and Cooperation, Govt. of India, New Delhi.

Verma, S.R. and Singh, H. (1986). Development and Testing of Mechanical Gadgets for Extraction of Vegetable Seeds. Final Report ICAR Adhoc Project, Ludhiana. p. 73.

Yordanov, M. Gentschev, S., and Stoyanova, Z. (1971). Morphological changes in male sterile tomato mutants treated with gibberellin. CR Accd. Sci. Agri. Bulg IV/I: 61-55.

Yordanov, M. (1983). Heterosis in tomato. In "Heterosis" by R. Frankel (Ed.) Springer Verlag Berlin Heidelberg. pp. 189-219.

An Outlook
in Hybrid Eggplant Breeding

A. S. Sidhu
S. S. Bal
T. K. Behera
Mamta Rani

SUMMARY. Eggplant or brinjal is an important crop of many countries. It is highly productive crop and consumed as a cooked vegetable in various ways. It is a hardy annual herbaceous plant. There is a lot of variability exists in the nature in color, shape, size, etc. A number of cultivars are grown across the globe depending on the market needs and consumer preference. Different worker in terms of yield, maturity, uniform harvest, etc., has showed considerable amount of heterosis. Eggplant is attacked by many insect pests and diseases like Shoot and fruit borer, Epilachna beetle, Damping off, Phomopsis blight, Leaf spot, Wilt, Little leaf, Fruit rot, Collar rot, etc. Some varieties tolerant/resistant to different biotic and abiotic stresses are developed. Further work on resistance breeding is of much importance in this regard. Hybrids are also popular in this crop. In some countries, F2 are also being marketed by different companies. The flowers of eggplant are large and showy with the corolla purple in color. Flowers are hermaphrodite. The hybrid seed production is based on hand emasculation and hand pollination. For this

---

A. S. Sidhu, S. S. Bal, and Mamta Rani are affiliated with the Department of Vegetable Crops, Punjab Agricultural University, Ludhiana 140004, India.

T. K. Behera is affiliated with the Division of Vegetable Science, Indian Agricultural Research Institute, New Delhi 110012, India.

[Haworth co-indexing entry note]: "An Outlook in Hybrid Eggplant Breeding." Sidhu, A. S. et al. Co-published simultaneously in *Journal of New Seeds* (Food Products Press, an imprint of The Haworth Press, Inc.) Vol. 6, No. 2/3, 2004, pp. 15-29; and: *Hybrid Vegetable Development* (ed: P. K. Singh, S. K. Dasgupta. and S. K. Tripathi) Food Products Press, an imprint of The Haworth Press, Inc., 2004, pp. 15-29. Single or multiple copies of this article are available for a fee from The Haworth Document Delivery Service [1-800-HAWORTH, 9:00 a.m. - 5:00 p.m. (EST). E-mail address: docdelivery@haworthpress.com].

http://www.haworthpress.com/web/JNS
© 2004 by The Haworth Press, Inc. All rights reserved.
Digital Object Identifier: 10.1300/J153v06n02_02

we need honest and skilled manpower and we have to take utmost care while undertaking the seed production program. *[Article copies available for a fee from The Haworth Document Delivery Service: 1-800-HAWORTH. E-mail address: <docdelivery@haworthpress.com> Website: <http://www. HaworthPress.com> © 2004 by The Haworth Press, Inc. All rights reserved.]*

KEYWORDS. Eggplant, genetics, breeding, heterosis, resistance breeding, hybrid seed production

INTRODUCTION

Eggplant or brinjal (*Solanum melongena* L.) is an important solanaceous crop of many countriei particularly India, Japan, Indonesia, China, Bulgaria, Italy, France, USA, and the several African countries.

In "Origin of cultivated plants" published in 1886, De Candolle, stated that the species *S. melongena* has been known in India from ancient times and regarded it as a native of Asia. According to Vavilov, the eggplant originated in Indo-Burma region. It can be grown in almost all parts of India except higher altitudes all the year round. A number of cultivars are grown throughout the country depending on the yield, consumer's preference about the color, size and shapes of the various cultivars. It is highly productive and usually finds its place as the poor man's crop. In India, it is being consumed as a cooked vegetable in various ways. It may be roasted, fried, stuffed, cooked as curry pickled, etc. The crop is extremely variable in India. The early European name eggplant suggests that the first introduction were small fruited and gradually various sizes and shapes were evolved due to selections and other breeding methods.

CYTOGENETICS

The chromosome number of eggplant is $2n = 2x = 24$. Choudhury (1973) reported the aberrations like dicentric chromosomes forming bridge formation along the acentric segment, lagging, and elimination of univalents and fragments in West African diploid *S. melongena*. The high frequency of multivalents observed in diploid indicates the presence of translocations. Choudhury (1975a) observed a haploid chromosome number of $n = 12$ in a west African *S. melongena* var. Zaria. The high frequency (29.9%) of multivalents observed suggested the occurrence of translocation interchanges. Chromatid bridges, lagging chromosome and univalents were also observed. Choudhury (1975b) also observed a chromosome number of $n = 12$ in East African diploid *S. melongena* var. Bassawa. The chiasma frequency per bivalent ranges from 1.8-2.2 at diplotene to 1.0-1.3 at diakinesis.

CROP BIOLOGY

Eggplant is an annual herbaceous plant. The plant is erect, compact, and well branched. The leaves are opposite, large, simple lobed, and the underside of the most cultivars is covered with dense wool like hairs. The flowers are large and showy with the corolla purple in color. The flowers are hermaphrodite and the stamens dehisce at the same time the stigma is receptive so that self-pollination is rule, although there is some cross-pollination by insects. The fruit is pendant and is a fleshy berry borne singly or in clusters. The color of mature fruits varies from purple, purple black, yellowish, white, green, and striped, depending on cultivars. The seeds are borne on the fleshy placenta and the placentae with the seeds completely fill the locular cavity. Its inflorescence is often solitary but sometimes it constitutes a cluster of 2-5 flowers. Solitary or clustering nature of inflorescence is a varietal character. Flower is complete, actinomorphic and hermaphrodite. Calyx is five lobed, gamosepalous and persistent. It forms a cup like structure at the base. Corolla is five lobed gamopetalous with margins of lobes incurved. There are five stamens, which are free and inserted at the throat of corolla. Anthers are cone shaped, free and with apical dehiscence. Ovary is hypogynous, bicarpellary, syncarpous and with basal placentation.

Heterostyly (the difference in position of stigma from the anther tips) is a common feature. Krishanamurthy and Subramaniam (1954) reported 4 types of flowers depending on the length of styles, viz. (i) long styled with big ovary, (ii) medium styled with medium sized ovary, (iii) pseudoshort styled with rudimentary ovary, and (iv) true short styled with very rudimentary ovary. The position of stigma in relation to stamens varies with the cultivars and can also vary in different flowers of same cultivar. Stigmas are found either above, on the same level as or below the stamens and the highest percentage of fruit set is found where the stigma is above the stamens. He also observed that at the temperature of 20-22°C and 50-55% relative humidity, pollen viability is retained for 8-10 days. Repeated pollination with pollen from different plants increases both fruit and seed set. Pal and Osvald (1967) observed that the fruit set of insect pollinated plants was much higher than that of self-pollinated ones. Flower abortion is favored by natural day light reduction and high (30°C) night temperature as reported by Saito and Ito (1973). Pal and Taller (1969) opined that the number of seeds per fruit is closely related with the type of pollination. It is highest with free pollination, lower in selfed plants and lowest in exclusively artificially crossed plants. It was observed that the flowers in the cluster are either short styled and medium or all medium styled. However, the fruiting habit in a cultivar was not directly related to the occurrence of different flower types in cluster.

Fruit setting occurs in long- and medium-styled flowers. Fruit setting in long-styled flowers normally varies from 70 to 86.7% and that in medium-styled flowers from 12.5 to 55.6%. In short-styled flowers (non-fruiting) types in which androecium is fertile but stigma is smaller with underdeveloped papillae.

HETEROSIS

Heterosis is a general biological phenomenon observed in F_1 generation, which manifests itself by greater vitality, rapid growth and development, higher productivity, resistance and uniformity. Exploitation of hybrid vigor has become a potential tool for improvement in eggplant (Pal and Singh 1949, Mishra 1961, Samandam 1962, Dhankar et al. 1980, Chadha and Sidhu 1982). Kakizaki (1931) reported that F_1 hybrids of the most of eggplant hybrid showed immediate increase in weight due to increase in size of the embryo.

The occurrence of considerable hybrid vigor was recorded as early as 1892 by Munson in U.S.A. and in 1926 in Japan by Nagai and Kida and hybrid egg-plants are now commonly used in many countries, especially in Japan. Experiments conducted mostly in Japan and India has shown distinct hybrid vigour in eggplant. Gotoh (1952) recorded marked yield increase in the F_1 generation of a series of crosses between Japanese cultivars. Odland and Noll (1948) observed that F_1 hybrids exceeded the parental mean yields by 62% and higher yield was due to more number of fruits produced rather than fruit size. In Illinois, Courter and Glover (1965) reported that Burpee hybrid and Black Magic hybrid gave the best result in eggplant. Popova et al. (1976) obtained increased yield in F_1 hybrid to a great extent. Similarly in Tennessee, Cordrey et al. (1979) observed that Black Bell hybrid was found superior to the recommended cultivar, Florida Market.

In India, several reports are available on hybrid vigor in eggplant. Pal and Singh (1949) reported that hybrid in eggplant showed 48.8-56.6% increased yields over the better parent. Mishra (1961) also observed increased yields in eggplant hybrids. Viswananthan (1973) reported that most economic combination was Muktakeshi × Banaras Giant in which the increase in yield was over 100% compared to the parental mean. Higher yield in the hybrid than both the parents due to more number of branches, increased fruit size, fruit number, and fruit weight. Pusa Anmol, a hybrid cultivar has been evolved at IARI from a cross between Pusa Purple Long and Hyderpur. It showed early and increased yields of about 80-100% more than Pusa Purple Long (Choudhury 1966). The cross combination Pusa Purple Cluster × Sel-5 exhibited 162.5% heterosis for yield over better parent (Singh and Kumar 1988). As much as 204% heterosis for fruit yield per plant over mid parent in hybrid of

Muktakeshi × Pusa Purple Long are obtained. Hybrid vigor in eggplant has also been reported by Thakur et al. (1968), Lal et al. (1974), Singh et al. (1978), Ram et al. (1981), and Salehuzzaman (1981). Some of the good general combiners for yield and yield attributing characters are APAU 1 and Sel-5 (Singh and Kumar 1988), H 4 (Chadha and Hegde 1989), Round White, Pusa Kranti, Bhubaneshwar-4, and Keonjhar-1 (Mishra and Mishra 1990), which can be used in heterosis breeding program.

The cost of hybrid seed production in eggplant is not high as compared to other vegetables and this can be further reduced by the use of male sterile lines. Fang et al. (1985) developed 2 male sterile lines by 3 backcrosses to *S. melongena* after crossing *S. geto* with *S. melongena* line 9334A and 2578A and suggested that male sterility is under cytoplasmic control. Chauhan (1986) reported naturally occurring genetic male sterile line with cultivar Pusa Purple Long. The male sterility is controlled by 2 recessive nuclear genes, ms1 and ms2. A homozygous male sterile line UGA 1-MS was derived from a male sterile line spontaneous mutant of Florida Highbush. The functional male sterility of UGA 1-MS is governed by simple recessive allele, for which the symbol *fms* is proposed (Phatak et al. 1991). Chemicals are also used to induce male sterility. Application of 2-4-D @ 10 ppm at 10 day interval induces complete pollen sterility and satisfactory number of seeds in a single cross. Good effects can also be obtained by spraying buds, two or three weeks before anthesis with 0.29% 2,3-dichloroisobutyrate. Pollen sterility may also result from 0.01% maleic hydrazide injected in the main axis of inflorescence when flowers are very young. Exploitation of hybrid vigor in eggplant is economical as each fruit contains large number of seeds as compared to other vegetables.

SYSTEM OF POLLINATION AND FRUIT SETTING

Eggplant is a self-pollinated crop but cross pollination also occurs in it. This is because of heterostyly nature of the crop. Insect pollination extends up to 60-70% while self-pollination occurs in 30-40%. Bumble bees (*Bombas* sp.) and honey bees (*Apis* sp.) are the major pollinators. The lack of pollinators favours selfing in this crop. To encourage pollination and visit of bees, *Mimosa pudica* plants should be planted in the vicinity of the eggplant.

Eggplant is a day neutral plant. Fruit setting is affected by average maximum temperature (21-27°C) and precipitation during first five days after flower opening Anthesis in eggplant flower, starts from 7-30 a.m. and continues up to 11 a.m. peak time for anthesis is 8-30 to 10-30 a.m. The pollen dehiscence starts from 9-30 to 10 a.m. It has been observed that anthesis and dehiscence are mainly influenced by the daylight, temperature and humidity

(Sidhu et al. 1980). Vijay et al. (1978) reported that anthesis in Pusa Purple Long and Supreme cultivars commenced at 5-35 a.m. which continued up to 7-35 a.m. with peak at 6-05 a.m. Deshpande et al. (1978), however, described that maximum anthesis in eggplant was between 6 and 8 a.m. and the stigma was most receptive at the time of flower opening.

HYBRID SEED PRODUCTION

According to Bal and Sidhu (1998), in eggplant, many pure-line cultivars have been developed. The presence of heterosis has been reported and exploited in many breeding programs and several eggplant hybrids have been developed. Eggplant hybrids yield 40-50% higher than the parents and have very attractive color. Punjab Agricultural University (PAU) has released two hybrids of eggplant namely Brinjal Hybrid-1 (BH-1) and Brinjal Hybrid-2 (BH-2). Three hybrids namely Pusa hybrid 5, Pusa hybrid 6, and Pusa hybrid 9 of Indian Agricultural Research Institute and one hybrid, Arka Navneet, developed by Indian Institute of Horticultural research are released for commercial cultivation. Narendra Dev University of Agriculture and Technology and Gujarat Agricultural University has released 2 hybrids each namely NDBH 1, NDBH 2 and ABH 1, ABH 2, respectively. Private seed companies have also developed a number of hybrids for cultivation in different agro-ecological zones. The seeds of these hybrids are produced through hand emasculation and pollination. Easy cross-pollination and large numbers of seeds per crossing are advantageous for commercial hybrid seed production. This accompanied by large heterosis and low seed rate for sowing have enabled the cultivation of hybrids.

To obtain highest levels of fruit setting and seed production under artificial pollination the following points are important.

- The earliest, single long styled flowers are most suitable.
- Emasculation is best done in buds that would open next day.
- Pollination performed without prior emasculation can give as much as 97% hybrid seeds.
- A large quantity of pollen increases the fruit set and number of seeds.
- Pollen can be stored for two days and pollination may be performed one day after pollination.
- Stigma receptivity and the fertilizing capacity of pollens are the highest at the time of opening of the floral buds hence pollination is to be done at this time.

AGROTECHNIQUES FOR HYBRID SEED PRODUCTION

Selection of field: Silt loam and loam soils are preferred for eggplant cultivation. Soil should be deep, fertile and well drained. The soil should not have more than 5.5 to 6.6 pH for good growth and development. Seed production plots do not have a seed crop of eggplant in previous season so that there are no volunteer plants.

Seed rate and seed treatment: The seed rate is 200 g per acre. The seed is sown in ratio of 4:1 female and male parents. Thus the seed requirement of female parent is 160 g and that of male parent is 40 g per acre. The seed of parental lines should be procured from a reliable sources. It should be treated with Thiram or Captan @ 3 g/Kg of seed before sowing.

Season and time of sowing: Eggplant can be cultivated during autumn-winter season and spring summer season. Autumn-winter crop is sown during June and seedlings are transplanted in July. Spring-summer crop is sown in March and seedlings are transplanted in April. For hybrid seed production, autumn-winter crop is more suitable because the seed setting percentage is higher in this season.

Nursery raising: The land for nursery raising should be prepared by ploughing or digging thoroughly. An area of 25 m^2 is required for raising nursery for 1 acre. Mix 5 q of well rotten FYM with the soil and prepare 1.5 m wide and 20 cm raised beds. Irrigate the beds at least 10 days before sowing. Drench the beds with 1-1.5% formalin by applying 4-5 liters of solution per m^2. Thereafter, the beds should be covered with plastic sheet/tarpaulin for 24 hours. After formalin application, the soil in the beds should be thoroughly turned once a day for 4-5 days to eliminate the adverse affects of formalin on germinating seeds. Seeds are sown 1-2 cm deep in rows, 5 cm apart. The nursery should be drenched with 4% Captan or Thiram after 5-7 days of germination. The irrigation should be withheld up to 4-5 days before transplanting to harden the seedlings but irrigation is required one day before uprooting the seedlings.

Transplanting: The seedlings become ready for transplanting in 4-6 weeks. Transplanting should be preferably done in the evening for proper establishment of seedling. Row to row spacing is kept at 75 cm and plant to plant at 60 cm.

Fertilizer application: About 10 tonnes of FYM should be incorporated in the soil thoroughly before transplanting. Apply 25 Kg of N (55 Kg urea), 25 Kg P_2O_5 (155 Kg single superphosphate), and 12 Kg K_2O (20 Kg muriate of potash) per acre at transplanting time. Apply another dose of 25 Kg N (55 Kg urea) per acre after 1 month.

Weed control: Stomp 750 ml per acre should be applied before transplanting and should be followed by one hoeing after one month.

Irrigation: First irrigation should be given two days before transplanting and the second, immediately after transplanting. The crop should be irrigated as per the need thereafter. During hot months the interval of irrigation should be 4-5 days. In light soils, frequent irrigation should be preferred. During rainy season water should not be allowed to stand in the field for more than one day.

Harvesting: Eggplant fruits are ready for seed harvest when at least one third part of the fruit from the stem-end is turned yellow. If the harvesting of the mature fruits is delayed, rotting starts at the blossom-end of the fruits after touching the ground. In addition, rodents damage the over matured fruits.

PROCEDURE OF HYBRID SEED PRODUCTION

Emasculation: Eggplant bears hermaphrodite flowers. The flowers of female parent are prepared for cross-pollination for hybrid seed production by removing anthers. The operation is called emasculation. Long- and medium-styled flowers are taken for emasculation. Selection of flower buds for emasculation is primary step. The flower buds where tip of corolla have not been separated can be selected for emasculation. Emasculation can be done at any time but it is convenient and effective to perform it in evening as pollination is done following morning. Soon after emasculation the flower is protected by a selfing bag or cotton pad.

Pollen collection: The flowers of the male parent, which are to be used for pollination, are bagged prior to opening in order to avoid contamination. Either pollen is selected from the flower of the male parent for which anther dehiscence is essential. Anther dehiscence depends on light, temperature, and humidity. There is poor anther dehiscence on cloudy days. Pollen can be collected by putting the anthers in a vial along with small iron balls for giving a beating effect.

Cross-pollination: The flowers of female parent which are to be emasculated in the evening are cross-pollinated in following morning before 11 a.m. Pollen grains are taken in the petridish or on the thumbnail and then transferred to stigma of female flower with the help of brush, needle or match stick. After pollination, selfing bag or cotton pad is placed over the flower. The pollinated flower should be left with identification mark either by chipping of calyx or by tying a tag.

Before pollination, stigmatic surface of the female flower should be checked for presence of pollen. The color of pollen grain is white whereas that of stigma is green.

Seed extraction: Seed is extracted manually or mechanically. The mature fruits are cut and crushed into small pieces. The seed along with the fruit flesh is scooped with hand. The seed is washed free of fruit material by washing with hands in a tub containing water. The seed being heavier settles at the bottom and the flesh floats over the water surface. The floating pulp is decanted off. Clean seed can be obtained by repeated washings by water. Axial flow vegetable seed extraction machine can also be used for extraction.

Seed drying and storage: Seed should be dried immediately after washing. It should be spread in a thin layer on cloth or in trays under sun for drying. Seed drying on the cemented floor under scorching heat should be avoided. Seed extraction should be done in morning so that seed is sufficiently dried during the day. By doing so, the possibility of sprouting during night is avoided. Seed can be stored for 2-3 years under cool, dry and well-ventilated conditions.

Seed Standards:

Factor	Seed standards for certified seed
Pure seed (minimum)	98%
Inert matter (maximum)	2%
Other crop seeds (maximum)	None
Weed seeds (maximum)	None
Germination (minimum)	70%
Moisture (maximum)	8%
Moisture for vapor proof containers (maximum)	6%

Efficacy of seed production: One person can emasculate 40 flowers and can pollinate 30 flowers per hour. Single fruit of eggplant yields 4-5 g of seed when Punjab Jamuni Gola or Punjab Neelum is used as a seed parent. Seed rate of eggplant hybrid is 200 g per acre. Thus one person can produce seed for 1 acre in 1 day. Hence, manual hybrid seed production in eggplant is a viable exercise.

MAINTENANCE OF INBRED LINES/PURE LINES

High quality of seeds of parents is a critical factor in hybrids seed production. Eggplant being a often cross pollinated crop, the seed of parents can be produced by isolating the seed production plots by 200 meter. The seed should be procured from an authentic source. The seed production plots should be vigorosly inspected and off type plants should be removed the agronomic practices for the seed production are the same as those of hybrids.

SEED PRODUCTION OF INBRED LINES

Cultivars of eggplant bear flowers in clusters as well as singly. The flower drop in a single flower bearing cultivar is only 1% whereas 80% is in clusters.

In Pusa Purple Long cultivar, the inflorescence may contain 4-5 flowers but there is fruit set in one or two flowers and rest of the flowers drop off.

Eggplant is a self-pollinated crop. The cone like formation of anthers favours self-pollination in the same way as in tomato. Frydrych (1964) was in the opinion that self-fertility found in eggplant was partially due to absence of pollinators.

Though self-pollinated it can out crossed to a considerable extent through insects, if more than one cultivar is grown in adjacent plots. Generally out crossing is more in case of long styled flowers. The extent of natural crossing depends on insect activity and recorded from 0-48% (Agrawal 1980).

It is, therefore, essential to isolate cultivars to produce quality and pure seeds. Agrawal (1980) suggested to give isolation distances of at least 400, 200, and 100 meters for nucleus, foundation and certified seed production, respectively. The nucleus seed plots should also be isolated from commercial seed crop. Nucleus seed is raised by massing the selfed seed of selected plants.

Rouging in many cultivars may not be possible till formation of first fruit. Leaf and stem characters can easily be detected even in nursery. The seed grower should be well acquainted with characters of the cultivar and should select true to type plants and rogue out the off types and undesirable plants at different stages of crop growth.

The agronomic practices normally followed for fruit production are generally employed for seed production also. Eguchi et al. (1958) reported that in Japan the most favourable conditions for growth and seed production occur at 260 meters above sea level, owing to the wider range of day temperatures during the growing season. In Indian condition, it can produce seed both in plains and in hills. Sowing of the seed crop should be so adjusted so that maturity does not coincide with heavy rains. The winter crop needs special protection from frost (Agrawal 1980). Singh et al. (1964) reported that in the hills of Kullu Valley in Himachal Pradesh, the eggplant seed should be sown in nursery during the months of March and April for growing seed crop. Usually after 60-65 days, the seedlings are transplanted in hills. Eggplant responded well to nitrogen fertilization at 75 Kg/ha in respect of seed yield (Seth and Dhankar 1970). There was, however, no significant response to phosphorus fertilization. They also reported that seed yield/plant and fruit quality were slightly better with a row spacing of 100 cm as compared to 75 cm.

Petrov et al. (1981) recommended that seeds should be collected from first or second tier fruits as those having higher 1000 seed weight, germination energy, and germination weight than seeds collected from fruits beyond second tier. They further suggested that fruit harvested at botanical maturity gave the highest quality seeds and seeds extracted from fruits kept for 5 days after harvest improved earliness and total yield in progeny. They also reported that for hybrid seed production ample pollen collected from many plants should be used and pollination may be repeated once or twice. Pollen should not be stored for more than 4 days and should be applied within 2-3 days of flower emasculation. Crop protection measures should be taken in time. The highest seed yield (3.6-7.1 q/ha) of eggplant was obtained when the plants were sprayed 5 times with Dimethoate at 0.72 Kg/ha followed by spraying of Carbaryl at 2.4 Kg/ha thrice.

For seed production the fruit should be harvested when fully ripe, i.e., turning normal color to yellow. The outer covering is peeled off and the flesh with seeds is cut into thin slices. In a hard skinned cultivar like Nurki (Purple Clustered Selection), both basal and upper portion of fruits, which constitutes the seeds free Zones are chopped off, and the remaining portion is used. Thus sliced materials are stored overnight so that the seeds after washings with water and sieving in the morning can be dried during the day. The washing should be always done in morning hours so that the seed is at least half dried till the evening or else there is a danger of its germination. Agrawal (1980) suggested that the seeds should be dried in the partial shade to moisture content of 8% or below. Acid extracted seeds of eggplant cv. Co.1 were conditioned to 8% moisture content and graded into sinkers and floaters by the water floatation method. The seeds that sank were of better quality than those floated, giving 83.5% germination compared with 23.1% (Selvaraj 1988).

The variation in seed yield is due to environmental factors, crop management practices and cultivars. The average seed yield is 100-120 Kg/ha (Choudhury 1976). In the region of Kullu Valley, India, a seed yield of 590-880 Kg/ha in Pusa Purple Long, Pusa Purple Round, and Pusa Clustered Selection cultivars were also reported.

RESISTANCE BREEDING

The eggplant is subjected to the attack of many diseases like damping off, phomopsis blight, leaf spot, wilt, little leaf, fruit rot, collar, rot, etc. Chakravorty and Chowdhary (1975) reported 3 eggplant lines namely, selection 212-1, 252-1-1, and selection 252-1-2 are resistant to little leaf. Kalda et al. (1976) found that *S. xanthocarpum, S. indicum, S. gilo, S. khasianum, S. nigrum,* and *S. symbrifoilum* are highly resistant to phomopsis blight and Pusa Bhairav, cultivar of *S. melongena* is found resistant. Resistance to phomopsis blight is recessive and polygenically inherited (Kalda et al. 1977). The cultivars Pusa purple cluster is also found to be resistant to bacterial wilt (Rao et al. 1976). *S. melongena* var. *insancum* carries a single dominant gene for resistance to bacterial wilt (Swamnathan and Sriniwasan 1972).

Eggplant is also susceptible to several insect pests. The most serious pest in this crop are shoot and fruit borer, jassids, epilachna beetle, white fly, and aphids. Behera and Singh (2002a) reported that *S. indicum* was immune and *S. gilo* was resistant to soot and fruit borer. In a cytological study Behera and Singh (2002b) observed the chromosomal associations and chiasma frequency of F1 hybrids between *S. melogena* and *S. gilo*. They found that the frequency of univalents were more compared to bivalents and then hybrids showed high degree of sterolitz.

In another screening study carried in Maharashtra, India, for resistance to jassids it was reported that S 188-2, Pusa Purple Long, S-34, S-258, Manjari Gota, and Dorli were resistant to jassids with 30 nymph/15 leaves. In 1987, eight varieties of eggplant were screened against epilachna beetle at Andhra Pradesh Agricultural University and moderate resistance was found in Punjab Chamlika, SM-204, and SM-195. Eggplant cultivars Chaklasi, Doli, Doli 5, and Pusa Purple Cluster were found to be resistant to jassid, shoot and fruit borer, and little leaf disease at Gujarat, India.

Resistance to abiotic factors such as extreme temperature and humidity, drought and high salinity conditions were also studied in eggplant. It was found that maximum reduction in photosynthetic rate and stomatal resistance was observed in plants stressed for 9 days during fruiting stage. The cultivar, Arka Sirish was found to be more drought tolerant. In one experiment conducted Indian Agricultural Research Institute, it was reported that JC-1, RAH-51, Pragati, and Pusa Bindu are salt tolerant.

CONCLUSIONS

The use of F_1 hybrid cultivars is likely to increase as a result of appreciation of their attributes and efforts to economize in such seed production need proper intention. Several related wild species are possibilities for getting the complete resistance sources and difficulties in such interspecific hybridization require further intensive investigations.

REFERENCES

Agrawal, R.L. (1980). Seed Technology, Oxford and IBH Publishing Co. New Delhi. pp. 198-201.

Bal, S.S. and Sidhu, A.S. (1998). Brinjal. In Dhillon B. S. and Allah Rang (eds.) Hybrid crop cultivars and their seed production. Punjab Agricultural University, Ludhiana. pp. 101-108.

Behera, T.K. and Singh, Narendra. (2002a). Interspecific crosses between eggplant (*S. melongena* L.) with related *Solanum* species. Scientia Horticulturae, 95: 165-172.

Behera, T.K. and Singh, Narendra. (2002b). Interspecific hybridization in eggplant for resistance to shoot and fruit borer. Capsicum and Eggplant Newsletter, 21: 102-105.

Chadha, M.L. and Hedge, R.K. (1989). Combining ability studies in brinjal. Indian J. Hort. 46:33-52.

Chadha, M.L. and Sidhu A.S. (1982). Studies on hybrid in brinjal. Indian J Hort. 39:233-238.

Chakravarty, A.K. and Chaudhary, B. (1975). Breeding brinjal resistance to little leaf disease. Proc. Ind. Nat. Sci. Acad. B14: 379-385.

Chauhan. S.V.S. (1984). Studies in genetic male sterile *Solanum melongena* L. Indian J. Genet. Pl. Breed. 44: 367-371.

Choudhury, B. (1966). Exploiting hybrid vigour in vegetables. Indian J. Hort 10: 56-60.

Choudhury, B. (1976). Vegetables (4th Edn.). National Book Trust, New Delhi, India: pp. 50-58.

Choudhary, H.C. (1973). Spontaneous chromosome aberrations in west African diploid *S melongena*. Genetics 74: 547.

Choudhury, H.C. (1975a). Cytological studies in west African *S melongena* L. var Zaria. Nucleus 18:33-39.

Choudhury, H.C. (1975b). Cytological studies in east African *S melongena* L. var Bassawa. Cytologia 140: 389-400.

Cordrey, T.D., Coffey, D.L. and Brown, J.F. (1979). Tennessee farm and Home Science Progress Report, 112: 41-42.

Deshpande, A. A., Bankapur, V. M. and Nalawadi, U. G. (1978). Some aspects of blossom biology in brinjal varieties (*Solanum melongena* L.). Curr. Res. 7: 174-175.

Dhankar, B.S., Mehrotra, N. and Singh, K. (1980). Heterosis in relation to yield components and shoot/fruit borer (*Leucinodes orbonalis* Gn) in brinjal (*Solanum melongena* L.). Genetica Agraria 34:215-220.

Eguchi, T., Oshika, Y., and Yamada, H. (1958). Studies on the effect of maturity on longevity in vegetable seeds. Japan Nnational Institute of Agricultural Sciences bulletin series. E-7: pp. 145-165.

Fang, M. R., Mao, R. C., and Xie, W. M. (1985). Breeding cytoplasmically male sterile lines of eggplant. Acta Hort. Sin. 12: 261-266.

Frydrych, J. (1964). Biology of flowering in eggplant. Bull Vyzkumny Ustav Zelinarbos Olomone. 8: 27-37.

Gotoh, K. (1952). Studies on combining ability on eggplant varieties. Jap J Breed 1: 196.

Kalda, T.S., Swaroop, V., and Choudhary, B. (1976). Studies on resistance to phomopsis blight in eggplant (*Solanum melongena* L.). Veg Sci. 3: 65-70.

Kalda, T.S., Swaroop, V., and Choudhary, B. (1977). Resistance to phomopsis blight in eggplant. Veg Sci. 4: 90-101.

Kakizaki, Y. (1931). Hybrid vigour in egg plant and its practical utilization. Genetics, 16:125.

Krishnamurthi, S., and Subramanian, D. (1954). Some investigations on the types of flowers in brinjal (*Solanum melongena)* based on style length and fruit set under natural conditions. Indian J. Hort. 11: 63-67.

Lal, S,, Verma, G., and Pathak, M.M. (1974). Hybrid vigour for yield and yield components in brnjal (*Solanum melongena* L.). Indian J. Hort. 36: 51-55.

Mishra, G.M. (1961). Investigation on hybrid vigour in brinjal (*Solanum melongena* L.). Indian J. Hort. 18:305-317.

Mishra, S. N., and Mishra, R. S. (1990). Diallele analysis for combining ability in brinjal. Indian J. Hort. 47: 239-243.

Odland, M. I., and Noll, C. J. (1948). Hybrid vigour and combining ability in eggplants. Proc. Amer. Soc. Hort. Sci. 51: 417-420.

Pal. B. P., and Singh, H. B. (1949). Hybrid brinjal give increased yields. Indian Farming 10: 378-380.

Pal, G., and Osvald, Z. (1967). A study of fertilization after removing different amounts of various parts of the pistil. Acta Agronomica Academiae Scientiarum Hungaricae. 16: 33-40.

Pal, G., and Taller, M. (1969). Effects of pollination methods on fertilization in eggplant (Solanum melongena). Acta Agronomica Academiae Scientiarum Hungaricae. 18: 307-315.

Pathak, S. C., Liu, J., Jaworski, C. A., and Sultanbawa, A. F. (1991). A functional male sterility in eggplant: Inheritance and linkage to the purple fruit colour gene. J. Heredity 82: 81-83.

Petrov, K. H., Doikova, M., and Popova, D. (1981). Studies on the quality of eggplant seed. Acta Horticulturae. 111: 273-280.

Popova, D., Murtazov, T., Petrov, K., and Daskalov, S. (1976). Some manifestatations of heterosis in eggplant. Biologicheskii Zhurnal Armenii. 29: 67-72.

Prasad, D. N., and Prakash, R. (1968). Floral Biology of brinjal (Solanum melongena L.). Indian J. Agric. Sci. 38: 1053-1061.

Ram, D., Singh, S. N., Chouhan, Y. S., and Singh, N. D. (1981). Heterosis in brinjal. Haryana J. Hort. Sci. 10: 201.

Rao, M.V.B., Sohi, H.S., and Vijay, O.P. (1976). Reaction of some varieties of brinjal (Solanum melongena L.) to Pseudomonas solanacearum. Veg. Sci. 3: 61-64.

Saito, T., and Ito, H. (1973). Studies on flowering and fruiting in eggplant–VIII. Effects of early environmental conditions and cultural treatments on flower development and drop. J. Japanese Soc. Hort. Sci. 42: 155-162.

Salehuzzaman. M. (1981). Investigation on hybrid vigour in Solanum melongena L. SABRAO J. 13:25.

Sambandam, C. N. (1962). Heterosis in eggplant (Solanum melongena L.): Prospects and problems in commercial production of hybrid seeds. Econ. Bot. 16:71-76.

Sambandam, C. N. (1964). Heterosis in eggplant. Econ. Bot. 18: 128-131.

Selvaraj, J. A. (1988). Effect of density grading on seed quality attributes in brinjal (Solanum melongena L.). South Indian Hort. 36:32-35.

Seth, J. N., and Dhankar, D. J. (1970). Effect of fertilizer and spacing on the seed yield and quality in brinjal variety Pusa Purple Long. Progressive Hort. 1:45-50.

Sidhu, A. S., Kallo, and Pandita. M. L. (1980) Studies on floral biology in some important vegetable crops: A review. Haryana J. Hort. Sci. 9:207.

Singh, B., Joshi, S., and Kumar, N. (1978). Hybrid vigour in brinjal (Solanum melongena L.). Haryana J. Hort. Sci. 7: 95-99.

Singh, B., and Kumar, N. (1988). Studies on hybrid vigour and combining ability in brinjal (Solanum melongena L.). Veg. Sci. 15: 72-78.

Swaminathan, M., and Srinivasan, K. (1972). Studies on brinjal hybridization II. Transference of bacterial wilt resistance from a wild brinjal variety. Agri. Res J. Kerala. 9: 11-13.

Thakur, M. R., Singh, K. and Singh, J. (1968). Hybrid vigour studies in Brinjal (*Solanum melongena* L.). Punjab Agric. Univ. J. Res., Ludhiana. 5: 490-495.

Vijay, O. P., Nath, P., and Jalikop, S. H. (1978). Correlations and path coefficient analysis of biometric characters in brinjal. Indian J. Hort. 35:370-377.

Viswananthan, T. V. (1973). Hybrid vigour in brinjal. Proc. Indian Acad. Sci. 77: 176-180.

Breeding for Hybrid Hot Pepper

J. S. Hundal

R. K. Dhall

SUMMARY. Hot pepper is an important vegetable crop of tropics and subtropics regions of the world. The pungency, red color, and oleoresin contents important characters in the crop. There are different types of hot pepper available in the market depending upon the market needs and consumer preference. The genic and cytoplasmic male sterility is available in the crop. Genetic resistance to several diseases and insects pest is also available. Hybrid seed production requires lot of care and intensive supervision. The pollination control mechanisms available are helpful in making the commercial hybrid seed production profitable. *[Article copies available for a fee from The Haworth Document Delivery Service: 1-800-HAWORTH. E-mail address: <docdelivery@haworthpress.com> Website: <http://www.HaworthPress.com> © 2004 by The Haworth Press, Inc. All rights reserved.]*

KEYWORDS. Hot pepper, crop biology, seed development, heterosis, genic male sterility, cytoplasmic male sterility, hybrid seed production

INTRODUCTION

Hot Pepper (*Capsicum annuum* L.) commonly known as chilli in India, is mostly grown in the tropical and subtropical regions of the world as an impor-

---

J. S. Hundal and R. K. Dhall are affiliated with the Department of Vegetable Crops, Punjab Agricultural University, Ludhiana 141004, India.

[Haworth co-indexing entry note]: "Breeding for Hybrid Hot Pepper." Hundal, J. S., and R. K. Dhall. Co-published simultaneously in *Journal of New Seeds* (Food Products Press, an imprint of The Haworth Press, Inc.) Vol. 6, No. 2/3, 2004, pp. 31-50; and: *Hybrid Vegetable Development* (ed: P. K. Singh, S. K. Dasgupta, and S. K. Tripathi) Food Products Press, an imprint of The Haworth Press, Inc., 2004, pp. 31-50. Single or multiple copies of this article are available for a fee from The Haworth Document Delivery Service [1-800-HAWORTH, 9:00 a.m. - 5:00 p.m. (EST). E-mail address: docdelivery@haworthpress.com].

Digital Object Identifier: 10.1300/J153v06n02_03

tant vegetable and condiment crop. It is reported to be the native of tropical America (Thompson and Kelly, 1957). In India, its introduction is believed to be through the Portuguese in the 16th century. Both sweet and hot pepper originated from *Capsicum annuum*. The pungency, red color, and oleoresin contents in chilli are specially important for processed food as well as for export purpose. The pungency principle 'Capsaicin' in chilli is not only a digestive stimulant but also prevent from heart diseases and is curative for many rheumatic troubles (Reddy and Murthy, 1988).The coloring matter 'Capsanthin' and oleoresin content are particularly important for food and spice industries. The green chilli fruits are valuable on account of their richness in vitamin A and C (Table 1).

India is one of the leading countries in the world with respect to chilli growing area (0.95 million ha) and production (0.82 million tonnes of dry chilli). During 1998-99, India exported chilli near about 55,750 tonnes with value of Rs.2101.3 millions (Peter, 1999). India is also the largest exporter of chilli followed by China, Japan, Indonesia, Mexico, Uganda, Kenya, and Nigeria (Rajput and Wagh, 1994). At present, chilli is exported in the form of dry powder or dry fruit but there is great scope to export chilli products in the form of oleoresin, chilli paste, sauces, ketchup, pickles, paprika, etc., of high quality to compete in international market. The most important chilli growing states in India are Andhra Pradesh, Maharashtra, Karnataka, Orissa, and Tamil Nadu, which together constitute nearly 75% of the total area. Chilli is also grown commer-

TABLE 1. Average nutritive value of hot pepper (100 g edible product)

Waste %	13
Dry matter (g)	34.6
Energy (Kcal)	116
Protein (g)	6.3
Fiber (g)	15.0
Calcium (mg)	86
Iron (mg)	3.6
Phosphorus (mg)	80
Potassium (mg)	217
Carotene (mg)	6.6
Thiamine (mg)	0.37
Riboflavin (mg)	0.51
Niacin (mg)	2.5
Vitamin C (mg)	96
Average nutritive value (ANV)	27.92
ANV per 100 g dry matter	80.7

cially in China, Korea, Indonesia, Pakistan, Sri-Lanka, Turkey, Japan, Mexico, Ethiopia, Nigeria, Uganda, Yugoslavia, Hungry, Italy, Spain, and Bulgaria.

CROP BIOLOGY

The hot pepper or chilli is *Capsicum annuum* var. *acuminatum* L., whereas the bell pepper or green pepper is *Capsicum annuum* var. *grossum* L. The genus *Capsicum* is a member of the Solanaceae family. All the cultivated as well as wild species are diploid ($2n = 24$). There are five major cultivated species in the genus *Capsicum* (Table 2).

TABLE 2. Cultivated species of genus *Capsicum*

Species	Synonyms	Place of origin	Characteristics
C. annuum	*C. purpureum* *C. grossum* *C. cerasiformae*	Mexico/Central America	Milky white large corolla, single flower at each node, presence of calyx teeth, yellow and smooth seeds, annual, medium to large size fruits, medium pungent
C. chinense	*C. luteum* *C. umbilicatum* *C. sinense*	Amazonia	Dull white corolla, two or more flowers at each node, devoid of calyx teeth, constriction between the base of calyx and pedicel, yellow and smooth seeds.
C. frutescens	*C. minimum*	Amazonia	Greenish white corolla, two or more flowers at each node, devoid of calyx teeth, no constriction between base of calyx and pedicel, yellow and smooth seeds, perennial, small size fruits, highly pungent.
C. baccatum	*C. pendulum* *C. microcarpum* *C. angulosum*	Peru and Bolivia	Cream to white colored corolla with yellow to green spots on each corolla lobe, one or more flowers at each node, presence of calyx teeth, yellow and smooth seeds.
C. pubescence	*C. eximium* *C. tovari* *C. cardenasii*	Peru and Bolivia	Deep purple to faintly violet corolla with white center, one or more flowers in each node, presence of calyx with small teeth, black to brown and rough seeds.

Adapted from Greenleaf, W.H. (1986)

Wild forms of all the five species except *C. pubescence* exists. Esbaugh (1980) has suggested that *C. pendulum* and the closely related species *C. microcarpum* be reclassified as botanical varieties of *C. baccatum*. In this system, *C. pendulum* becomes *C. baccatum* L. var. *pendulum* and *C. microcarpum* becomes *C. baccatum* L. var. *baccatum*.

PLANT MORPHOLOGY
(BRANCHING, LEAVES, ROOTS)

The hot pepper (*Capsicum annuum* L. var. *acuminatum*) is an annual herbaceous plant during early stage of growth but becomes woody later on. The old branches become brittle with age and are easily broken. Branching mainly depends upon cultivar, soil fertility, soil moisture, and season. The main shoot is radial, but lateral branches are cincinnate. High branching is preferred for easy picking of fruits and for effective inter-cultivation and to prevent rotting of fruits.

The leaves are simple, alternate, exstipulate, elliptic, lanceolate, and glabrous. The leaf color varies from light to dark green but in some varieties it is purple. Leaf size is also variable and differs according to cultivar as well as management practices. Leaf area per plant and leaf length varies from 1000-3000 sq. cm and 0.5-2.5 cm, respectively. The leaves are shed either due to foliar diseases especially due to powdery mildew infestation, moisture stress or due to senescence. Chilli plants infested by mites have long petiole and leaves curl downward, while thrips infested leave have cup shape appearance. Leaf thickness has bearing on pest tolerance especially thrips.

Chilli plant possess strong tap root which is usually broken or arrested during transplanting, resulting in the development of profusely branched laterals as much as 1 m long. The entire root system remains restricted to the upper soil layer of 30 cm depth but more active feeding roots are found up to 10 cm depth (both laterally and vertically) from the base of the plant. Application of organic manure and fertilizers enhances root activities. Chilli plants withstand drought better than excess soil moisture. Water stagnation or saturated condition of soil for more than 24 hours is highly detrimental to the chilli crop at any stage of its growth.

FLORAL BIOLOGY

The flowers are solitary or in clusters at the tip of the branch in the axil of the leaves. They are bracteolate, actinomorphic, pedicellate, bisexual, and hypogynous. Calyx is carpanulate, sepals usually five gamosepalous, and is shorter than fruit. Corolla is penapetalous and bell shaped. However, large fruited cultivars may have 5 to 7 corolla lobes. The stamens are normally 5 and alternate with the petals. Stigma is club shaped, subcarpitate and faintly bifid. Style is slender, terminal and linear. The ovary consists of 2 to 4 or more locules.

Most flowers open at 5 a.m. Stigma becomes receptive from a day earlier to anthesis and remains so far 2 days after anthesis. Anther dehiscence take place during morning hours up to 11 a.m. depending upon temperature. Pollen grains become fertile a day before anthesis, with maximum fertility on the day of anthesis. In chilli plant, emasculation and hybridization can be done simultaneously (Padda and Singh, 1971; Vijay et al., 1979). Chilli is basically a self-pollinated crop but cross pollination to an extent of 16% has been reported (Tikloo, 1991). Bees, ants and thrips are the principal pollinating agents. To maintain purity of chilli variety, a minimum isolation of 400 m is considered safe. Days required for flowering in chilli crop mainly depends upon the cultivar, temperature, light intensity and duration, soil moisture, fertility status, and age of the seedling at transplanting. Normally, flowering in chilli commences 40 days after transplanting with a peak flower production at 60 to 80 days after transplanting. Peak flower production in chilli is influenced by soil moisture, soil fertility, and incidence of pest especially thrips and mites. About 40-50% fruit set is observed in chilli (Shukla and Naik, 1993).

FRUIT DEVELOPMENT AND RIPENING

The fruit of chilli is 'berry' botanically. Unlike the usual barries, the seeds are not embedded in fleshy pericarp. The berry develops from bicarpellary superior ovary and has an axile placentation. The major fruit components are seed, pericarp, placenta, and pedicle. All these components vary greatly, depending on the variety and climatic conditions. The pericarp is leathery or succulent which turns from green or purple to red, orange or orange red. The placenta carries numerous seeds. The pericarp begins to dry on full ripening. The pedicle is short and thin. The fruit may be deciduous (only pedicle or calyx remains on the plant) or persistent. The position of the fruit may be pendent, intermediate and erect. Both pendent and intermediate fruits are sometimes noticed on the same plant. The fruit length may vary from 5 to 20 cm in different cultivars. The fruit may be elongate, oblate, round, conical or bell shaped. After anthesis, fruits gradually increases in size and shape. Seeds start developing 14 to 21 days after anthesis. It takes about 30-35 days from fruit set to complete development of fruit for harvest at green stage. Fruit starts ripening 45-50 days after fruit set. The first picked fruits have longer fruits than late pickings.

The most important quality characters in chilli is the pungency and the colour. The pungency is due to presence of a crystalline volatile alkaloid called 'Capsaicin.' But according to recent knowledge, capsaicin is not a single compound but a mixture of various amides. They are denoted by the common name of capsaicinoids. The maximum concentration of pungency is present in the placenta and septa (Table 3).

The capsaicin content in red dry chilli varies between 0.03 to 1.81% (Deshpande and Anand, 1988). The red color in chilli is mainly due to carotenoid pigments. Nearly 37 pigments have been isolated from capsicums (Curl, 1962), of which capsanthin is the major red pigment of chilli contributing towards 35% of the total pigments. The capsanthin content in red chilli varies between 0.08 to 0.41% (Deshpande and Anand, 1988). Color of chilli under storage is affected by moisture content, temperature, light, and fat content of chilli seeds. About 9-10% moisture content is optimum for storage. In general storage at higher temperature increases the rate of color degradation. Sunlight exhibits pronounced effect in bleaching of color and brings about maximum discoloration of the red pigments of chilli. The presence of higher amount of unsaturated fat in chilli seeds leads to quicker deterioration of color due to oxidation (Krishnamurthy and Natarajan, 1973).

SEED DEVELOPMENT AND MATURATION

Chilli seeds are compressed obicular and minutely pitted. The seeds are borne on the placental tissues. Seeds start developing 14 to 21 days after anthesis. Viability of seeds commences 35 days after anthesis and increases full maturity up to 50 days after anthesis. Seeds are generally yellow colored and smooth at maturity. Thousand seed weight varies from 3.4 to 3.6 g. Seeds remain viable for longer period in fruit than when seeds are extracted and stored (Murthy and Murthy, 1961).

TABLE 3. Capsaicin content of different fruit parts

Fruit part	Capsaicin (%)
Placenta and septa	2.5
Seed	0.7
Pericarp	0.03
Whole fruit (average)	0.6

HETEROSIS

The term 'heterosis' signifies increased or decreased vigor of the F_1 hybrids over average performance of the parents or over better parent. Positive heterosis implies hybrid vigor and negative heterosis indicates heterotic depression of the F_1 hybrid as compared to the parents. The utilization of hybrid vigor is an important tool in plant breeding for the improvement of crop plants. The magnitude of heterosis for yield, maturity, plant height, fruit weight, fruit size and, number of fruits is considered important. Heterosis in chilli was first reported by Deshpande (1933). Pal (1945) also observed that a hybrid between two Pusa types gave increased yield, matured earlier and had fruits thicker than the better parent. Heterosis in hot pepper was reported by various workers for different traits (Singh and Singh, 1978; Joshi and Singh, 1980; Singh, 1980-81; Gangadhar Rao et al., 1981; Pandey et al., 1981; Sekar and Arumugam, 1985; Joshi, 1986; Wang et al., 1986; Gopalakrishnan et al., 1987; Hundal and Khurana, 1988; Thomas and Peter, 1988; Mishra et al., 1989; Ram and Lal, 1989; Sahoo and Mishra, 1990; Bhagyalakshmi et al., 1991; Rahman et al., 1996; Zeecevic and Stevanovic, 1997; Ahmed et al., 1999, and Singh, 2001). The use of hand emasculation and hand pollination is highly uneconomical method of seed production in chilli crop. Therefore, use of male sterility for production of F_1 hybrids has been reported (Singh and Kaur, 1986), as it facilitates natural pollination and cost involved in emasculation is lowered. By using genetic male sterile line, Lakshmi et al. (1988) were able to produce 43.80% heterosis over the mid parent and Hundal and Khurana (1988) were able to produce heterosis over 'Punjab Lal' a standard variety up to 235.71% for green fruit yield and 138.00% for red ripe fruit yield. The hybrids CH-1 (MS_{12} × LLS) and CH-3 (MS_{12} × S-2530) were released for commercial cultivation in 1992 and 2000, respectively by Punjab Agricultural University, Ludhiana. Both these hybrids were tested in yield trials and observed that CH-1 and CH-3 gave an average red ripe fruit yield of 250 q/ha and 275 q/ha, respectively, which were 81.99% and 93.00% more than 'Punjab Lal'. The heterosis studies of various attributes are given in Table 4.

POLLINATION CONTROL MECHANISMS

Chilli is basically a self-pollinated crop, but cross pollination up to 30 to 40 percent has been reported, it may go up to 80 percent depending upon the prevailing conditions like pollinators population and adjoining crop. Cross pollination is an essential part of hybrid seed production. The extent of natural cross pollination is estimated by planting a variety having recessive character(s) with another one having counterpart dominant trait(s). The minimum

TABLE 4. Heterosis for various attributes in hot pepper

Cross	Attributes	References
	Total yield	Likleev (1966), Marfutina (1968), Mishra et al. (1976), Balakrishnan et al. (1983), Gill (1986) Meshram and Mukewar (1986), Wang et al. (1986), Thomas and Peter (1988), Mishra et al. (1989), Ram and Lal (1989), Sahoo and Mishra (1990)
MS_{12} × LLS	Total yield	Hundal and Khurana (1988), Singh (2001)
MS_{12} × S-2530	Total yield	Hundal and Khurana (2001), Singh (2001)
-	Early yield	Pandey et al. (1981), Sekar and Aramugam (1985), Wang et al. (1986), Zeecevic and Stevanovic (1997)
Kalinkuv × Sivreja 600 DL 603 × Sivreja 600	Early yield	Likleev (1966)
MS_{12} × S-2530	Early yield	Hundal and Khurana (2001)
-	Fruit yield/ plant	Bhagyalakshmi et al. (1991), Rahman et al. (1996), Zeecevic and Stevanovic (1997), Ahmed et al. (1999)
Jwala × Pant C-1	Fruit yield/ plant	Gopalakrishnan et al. (1987)
Pant C-1 × NP46A AC-37 × NP46A AC-37 × Kalyanpur yellow	Number of fruits per plant	Ram and Lal (1989)
5417 × 6718	Number of fruits per plant	Singh et al. (1973)
CO2 × DS-3	Number of fruits per plant	Sekar and Arumugam (1985)
Pant C-1 × NP46A	Fruit weight	Ram and Lal (1989)
Eubanella × KAU Cluster Early Calwonder × KAU Cluster	Fruit weight	Thomas and Peter (1988)
-	Fruit weight	Singh et al. (1973), Gill (1986), Gopalakrishnan et al. (1987), Mishra et al. (1989), Bhagyalakshmi et al. (1991)
-	Fruit length	Lippert (1975), Mishra et al. (1976), Singh (1980-81), Sekar and Arumugam (1985), Ram and Lal (1989), Mishra et al. (1989)
5416-4 × 6718	Fruit length	Singh et al. (1973)
Sweet Red Chilling Picking × KAU Cluster	Fruit length	Thomas and Peter (1988)
-	Plant height	Singh et al. (1973), Singh et al. (1978), Joshi and Singh (1980), Pandey et al. (1981), Joshi (1986), Mishra et al. (1989), Ram and Lal (1989), Gopalakrishnan (1987)

isolation distance for chilli is estimated to be 400 meter. A number of insects, namely, honeybee, bumble bee, housefly, ant, wasp, thrips, etc., act as pollinator especially during morning hours, honey bee being the most important. Therefore 2 to 3 bee hives are required for a seed production plot of 1 acre to ensure good pollination. Although chilli flower is less preferred by honeybee as compared to that of sunflower, onion, cucurbits, etc., but flowering season of these crops are limited. Therefore, honeybee and other pollinators ultimately visit chilli crop.

The genetic male sterility is an important pollination control mechanism which is exploited commercially for hybrid seed production in chilli. The Punjab Agricultural University (PAU), has developed MS-12, which carries genetic male sterility (GMS) controlled by recessive gene (msms). The plant having recessive gene in homozygous state (msms) are male sterile whereas those in heterozygous (Msms) and homozygous dominant (M_SM_S) state are male fertile. The progeny of male sterile plant pollinated by male fertile plant segregates in 1:1 ratio for male sterility and male fertility. The male sterile line (MS-12) is developed by transferring sterility gene from France (ms-509) into the cultivar 'Punjab Lal' through back crossing (Singh and Kaur, 1986). By using this male sterile line (MS-12), PAU has released two chilli hybrids, viz., Chili Hybrid-1 (CH-1) and Chilli Hybrid-3 (CH-3). Patel et al. (1998) also reported genic male sterile line $ACMS_2$, having mongenic recessive gene ($acms_2$ $acms_2$) but no hybrid has been developed by using this male sterility.

Cytoplasmic-Genic Male Sterility (CGMS) is also available in this crop. It was first reported by Peterson (1958) in an introduction of *C. annuum* from India (PI164835). Most authors have reported that a single nuclear gene, designated rf1, interacts with S cytoplasm to produce sterility, and the restorer allele Rf1 restores fertility. Plants with N cytoplasm are fertile regardless of whether they have the Rf1 or rf1 allele. The cytoplasmic factor S interacts with recessive nuclear gene ms and produces a male sterile line of the genotype Smsms, which is male sterile and is known as "A" line in the hybrid seed production programme. The genotype with normal cytoplasm is known as Nmsms and is male fertile. This is referred as "B" line and is used as a maintainer for the male sterile line, Smsms (A line) after repeated back crossing, the male sterile "A" line and the maintainer "B" line become almost isogenic. To produce hybrid seed, the "A" line is inter-planted with the pollinator or " C " line of the genotype NMsMs.

The main advantage of the CGMS system over the genic male sterility is that can get 100% male sterile plants for direct use as females; however, the known cytoplasm source in *Capsicum annuum* (Peterson,1958) is not exploited commercially, because of an instability under fluctuating conditions, particularly temperatures and a low rate of natural cross pollination in cultivated peppers. Novak et al. (1971), working with the Peterson's source of male sterility and broader genetic material found digenic inheritance; they also noted variable degree of sterility from complete male sterility to partial fertility. Shifriss and Frankel (1971) found S-type cytoplasm following interspecific crosses in C. *annuum*, probably identical to that found by Peterson (1958). Since these previous studies indicated lack of stability of cytoplasmic genic male sterility, the character was not utilized in breeding F1 hybrids. Shifriss and Guri (1979) developed several cytoplasmic-genic male sterile (partial fertile) cultivars in pepper and concluded that cultivar with male sterility stability, like that of "Bikura" can serve as "A" line using natural cross pollination in hybrid seed production. Woong Yu (1985) suggested to combine both genic and CMS components of sterility in order to obtain double cross hybrids. Such a programme can exploit the yielding heterosis well known among hot pepper accessions. Daskalov and Mihailov (1988) successfully tested an improved method for hybrid seed production, based on using a cytoplasmic male sterile line possessing a lethal gene with action that can be easily inhibited and a female sterile pollenizer. The lethal gene ensures 100% purity of the F1 crop. The female sterile pollenizer produces a permanent abundant flowering with excess of pollen grains that leads to increased hybrid seed production without additional labour expenses.

HYBRID SEED PRODUCTION

Today, hybrids are gaining more popularity due to their high productivity, improved quality, built in resistance, environment adaptation and earliness; which result into better monetary returns to vegetable growers. Inspite of several constraints, there are some success stories of chilli hybrid seed production in certain specific regions of this country. Punjab Agricultural University (PAU), Ludhiana has taken lead by developing chilli hybrids, viz., CH-1 (Hundal and Khurana, 1993) and CH-3 (Hundal and Khurana, 2001), which have out yielded all the recommended chilli varieties by a margin of 80-100%. Both these hybrids (CH-1 and CH-3) use genetic male sterility (GMS) for hybrid seed production but there is segregation for male sterile and fertile plants in GMS system; and male fertile plants are to be removed before pollination. The male sterile line (female), MS-12, is common in both these hybrids. The male parents are Ludhiana Local Selection (LLS) and Selection-2530 (S-2530) in CH-1 and CH-3, respectively. The private seed companies also released a number of chilli hybrids (Table 5) but CH-1 and CH-3 have significantly higher yield over all these hybrids due to possession of resistance to diseases and insect pests. On the basis of their high yield potential, multiple disease resistance and quality attributes, their acceptance has been very fast and, consequently, the acreage under chilli almost tripled in the last 4-5 years in Punjab

TABLE 5. Private sector hybrids in chilli

Name of hybrid	Seed companies
Tejaswini, MPH-55, MPH-58, MPH-59	Mahyco
HOE-808, HOE-888, HOE-818	Hoechest
NS-1101, NS-1420, NS-1701	Namdhari
Agni, SHPH-54, SHPH-35, SHPH-47, Picador	Novartis
ARCH-006, ARCH-001, ARCH-226, ARCH-228, ARCH-236	Ankur Seed
BSS-141, BSS-131, BSS-273, Gayatri	Beejo Sheetal
Champion	Seoul
Delhi hot, Hot green, Skyline	Hung Nong

State and likely to further increase in the near future. There is great acceptance of CH-1 in other states especially Haryana and Rajasthan.

PUBLIC SECTOR CHILLI HYBRIDS

CH-1: This hybrid is developed from a cross of MS-12 × LLS. Its plants are medium tall (92.6 cm), vigorous with light green foliage. Fruits are medium sized (6.62 cm length and 1.6 cm width) and weigh 2.7 gm each. Fruits are of light green when immature but turn deep red with shining surface at maturity. Due to its high pungency (0.83%) and red color content (88.75 ASTA), its fruits get premium price over other hybrids. Its dry matter content is 23.8%. Fruit span is longer (June to October) due to its tolerance to diseases (fruit and wet rot) and viruses (mosaic and leaf curl). Low incidence of fruit borer and sun scald were recorded in this hybrid. Its fruits are highly suitable for drying and used as salad. Yield of red chilli is 250 quintals per hectare but in case fruits are picked at green stage the yield is almost double.

CH-3: This hybrid is developed from a cross of MS12 × S-2530. Its plants are 69.2 cm, medium tall, vigorous with dark green foliage. Fruit length and fruit weight is higher than CH-1, i.e., 8.20 cm and 5.9 g, respectively. Immature fruits are dark green but turn dark red at maturity. It is about 7-12 days early in maturity than CH-1 and coloring matter also higher (145 ASTA) than CH-1 but capsaicin content is less (0.52%). It is also tolerant to diseases (fruit rot, wet rot, die back, wilt) and viruses (mosaic and leaf curl). Yield of red chilli is 275 quintals per hectare.

AGRONOMIC PRACTICES

Selection of field: The field should be selected where chilli or bell pepper was not grown during last year. Soil should be sandy loam and field should be well drained and preferably at higher level so that water may be drained off during rainy season. Low lying, water logged fields should be avoided.

Land preparation: The land should be prepared by 4 to 5 ploughings and plankings. The FYM may be added at the time of land preparation. All weeds and stubbles should be carefully removed from the soil otherwise these may lead to termite infestation and serious damage to chilli crop.

Seed rate and seed treatment: Good quality seed is required for successful hybrid seed production. The seed must be true to type and should have high germination, vigor, and good health. It should be bold and have uniform color. The seed which is black, shrivelled, light in weight and has black spot in the center, will not germinate and should not be used. Female parent (MS-12) 300 g and male parent (LLS or S-2530) 100 g is enough for one acre. The seeds should be treated with Thiram or Captan at the rate of 3 g per kg of seed before sowing, to check surface borne pathogens causing damping off during germination. The seed of parental lines should be purchased from a reliable source like PAU.

Nursery raising: An area of 12 m^2 is required to raise nursery for 1 acre. Mix 2 quintals of well rotton Farm Yard Manure (FYM) with the soil and prepare 1.0 m wide and 20 cm raised beds. Before sowing the seeds in nursery beds, it should be drenched with 1.0 to 1.5% formalin by applying 4 to 5 litres of solution per m^2. Thereafter, the beds should be covered with plastic sheet for 72 hours. After formalin application, the soil in the beds should be thoroughly turned once a day for 2 to 3 days to eliminate its adverse effect on germinating seeds. Seeds are sown 1 to 2 cm deep in row, 5 cm spaced apart. The nursery should be drenched with 0.4% Captan or 0.1% Bavistin after 5 to 7 days of germination. The irrigation should be with held 4 to 5 days before transplanting to harden the seedlings.

HYBRID SEED PRODUCTION PROCEDURE

The hybrid seed is produced in the open in an isolated field, called ' hybrid seed production block.' The hybrid seed production block must be at least 400 m isolation distance from other different or similar chilli or sweet pepper cultivars/hybrids. The female and male lines are planted alternatively in the ratio of 2 rows of female line to one row of male line. The planting distance in row should be 60 cm in both female and male lines and plant to plant distance should be 20 cm and 40 cm in female and male lines, respectively. The female line (MS-12) produces both the male fertile and male sterile plants in the ratio of 1:1. In large fields, it is useful to keep about 2 or 3 bee hives per acre to ensure large population of honey-bees to do the pollination. The female line (MS-12) is first prepared for cross pollination for hybrid seed production by removing the male fertile plants before pollen shedding. Daily rouging of pollen bearing plants and other off types in the female line (MS-12) in the morning, can be continued up to evening. This operation may take 10 days or more. All the sterile plants should be tagged to avoid rechecking. Three or four days after this operation, remove all fruit set (small or large) from male and female parents.

A hybrid seed crop is inspected at different stages of plant growth and development to ensure the genetic purity of seed crop. The first inspection is made before flowering. The off type plant for foliage (leaf size, leaf shape, leaf color) and plant type characters should be removed. The diseased plants and extra early flowering plants should also be rouged out. The second inspection is conducted at flowering stage. The plants which do not confirm to the purity requirements regarding the flower orientation, flower color, spread of the plant, and leaf characters such as size, shape, color, etc., should be removed. Third inspection is conducted at the fruiting stage. The plants showing variation for fruit shape, color, size and position of fruit (erect, pendent or semi-pendent) should be rouged out. Removal of the off type plants at this stage will help to avoid mechanical admixtures and further chances of out-crossing of true to type plants with the off types.

Identification of Male Sterile and Male Fertile Plants in Female Parent

Identification of male sterile and male fertile plants in female parent is done when plants are in blooming stage. Generally anthesis occurs between 5 to 7 a.m. and anthesis is followed by dehiscence. After dehiscence the plants can be checked for the presence or absence of pollen in the flower which can be further verified on a black paper or cloth. The characteristics of male fertile and sterile plants in female parent (MS-12) are presented in Table 6.

Harvesting and Seed Extraction

Chilli fruits are ready for seed harvesting when fruits are red ripe. The seed harvested from male sterile (MS-12) plants will be hybrid seed. The seed harvested from male parent can be used for the next year. Common practice of seed extraction involves a thoroughly drying of red ripe fruits in the sun followed by crushing of fruits and winnowing to separate the fragments of the dried fruits. At some places where labor is very cheap the seed is removed by squeezing the individual fruits between hand fingers either from the freshly

TABLE 6. Characteristics of male fertile and sterile plants in female parent (MS-12)

Character	MS-12	
	Male fertile plants	Male sterile plants
Pollen (White powdery substance)	Present	Absent
Anther color	Light grey	Purple or yellow
Anther size	Normal	Reduced to less than half
Anther bursting	Anther burst to shed pollen	No bursting
Fruit set	Heavy	Low

harvested or from the half dried fruits. For large scale hybrid seed production, Axial Flow Vegetable Seed Extracting Machine can be used to extract seed from freshly harvested red ripe fruits. The seeds are dried in sunlight after extraction. The viability of seed is one year under ordinary storage conditions and can be increased by storing the seed under low temperature and low relative humidity in sealed polythene bags or glass jars.

MAINTENANCE OF FEMALE (MS-12) AND MALE LINES

Genetic constitution of male sterile plant is 'msms' and that of male fertile plant is MsMs (homozygous) and Msms (heterozygous). The male sterile plant can be maintained by crossing it to the heterozygous male fertile plant because progeny of such crossing produce the population of male sterile and male fertile plants in 1:1 ratio (Figure 1A). The female parent/male sterile (MS-12) is planted in a separate block called 'Male Sterile Maintainer Block' by keeping an isolator of 400 m from other chilli or bell pepper varieties or hybrids and also from hybrid seed production block. The seedlings are planted on ridges, 60 cm apart with a 45 cm distance between plants. Rouging is done at three different stages as described under hybrid seed production. In this male sterile maintainer block, the female parent (male sterile) bears both the male fertile and male sterile plants in the ratio of 1:1. At the time of flowering, the male sterile plants are identified and tagged. The male fertile plants in this block will serve as a pollinator for the male sterile plant. There is no need to rouge out the male fertile plants in this block. The fruits from male sterile plants are harvested when red ripe, dried, and their seed is extracted. These seeds from male sterile plants will serve as female parent (MS-12) for subsequent seed production programmes. To ensure pollination 2 to 3 bee hives per acre may be kept near the seed production field. The average yield of male

FIGURE 1. Procedure for utilization of genetic male sterility for chilli hybrid seed production

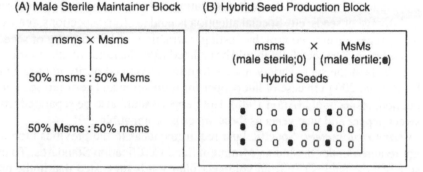

(A) Male Sterile Maintainer Block   (B) Hybrid Seed Production Block

MsMs and Msms = male fertile; msms = male sterile

sterile (MS-12) is about 40 kg per acre. The fruits of male fertile plants are used for non seed purpose.

The male parent is maintained as such in the hybrid seed production block because seed collected from male parent (LLS or S-2530) serve as pure male seeds and can be stored for subsequent seed production programmes. The different stages of rouging are described earlier. The agronomic practices for seed multiplication of parents are the same as described under hybrid seed production. The average yield of male parents is about 1 quintal/acre.

SEED PRODUCTION

Due to increasing pressure on land through urbanization and industrialization, it is not feasible to increase the area under vegetables for increasing the vegetable production commensurate to our requirements, hence the preciousness of high quality seeds becomes much more significant. The seed production of open pollinated varieties is done by growers mostly without any technical knowledge. It causes decline in a variety with the time. Since the seeds of open pollinated varieties can be produced even without any knowledge of their parents, it is difficult to check its piracy and the originating institution of particular variety may not enjoy all the benefits of its popular variety. Large areas being under different open pollinated varieties, chances of contamination are always there which affects the purity of a variety, in general. The market price and the productivity of open pollinated varieties are low in comparison to hybrids.

SEED PRODUCTION TECHNIQUE

Raising of chilli seed crop is highly specialized job and demands intimate knowledge of crop management, preventing outcrossing, curing, processing and storage of seeds, etc. Special attention is paid to crop inspections, removal of off-types, disease control, harvesting, extraction, and processing of seed.

Isolation: The seed field should be isolated from the fields of other varieties and fields of the same variety not conforming to varietal purity requirements by 400 and 200 m in case of hot pepper for foundation and certified seed production, respectively (Anon,1988). Hot pepper should also be separated from sweet pepper and vice versa as these are cross compatible.

Land requirement: Specific land requirements as to the previous crop are not required under the Indian Minimum Seed Certification Standards. To ensure that the land is free from volunteer plants, it is suggested that in the preceding season a chilli crop should have not been raised on the same piece of land.

Field inspection and rouging: A seed crop is inspected at different stages of plant growth and development to ensure the genetic purity of seed crop. The first inspection is made before flowering. The off-type plants for foliage and plant type characters should be removed. The extra early flowering plants should also be rouged out. The second inspection is conducted at full bloom and fruiting stage. The plants which do not conform to the varietal purity requirements regarding the fruit shape, color, size, position of the fruit (erect, pendent or semi-pendent), flower color, spread of the plant and leaf characters such as size, shape, color, etc., should be removed. Third inspection is conducted just before first fruit picking. The plants showing true to type color of the fruits and various other fruit characters are retained for seed harvest and the rest are uprooted. Removal of the off type plants at this stage will help to avoid mechanical admixtures and further chances of outcrossing of the true to type plants with the off-types. The field standards for chilli varieties are given in Table 7.

Harvesting and seed extraction: Chilli fruits are ready for seed harvest when fruits are red ripe. Common practice of seed extraction involves a thorough drying of the red ripe fruits in the sun followed by crushing the fruits and winnowing to separate the fragments of the dried fruits. At some places where labor is very cheap, the seed is removed by squeezing the individual fruits between hand fingers either from the freshly harvested or from the half dried fruits. Seed can also be extracted from red ripe fruits immediately after picking with the use of axial flow vegetable seed extraction machine (Verma and Singh, 1988) developed by PAU. It has been observed that quality of seeds, particularly seed color, is affected by sun drying but extraction through machine retains natural color of seed (Karivarathararaju and Palanisamy, 1984). The average seed yield of hot pepper varieties varies from 160-300 kg/acre.

Seed drying and storage: When the seed is extracted from freshly fully ripe fruits, its moisture content is very high. It is, therefore, essential to clean and dry the seed as quickly as possible. Seed must be thoroughly dried to a moisture content below 8 percent for proper storage. Seed can be dried either by spreading it on the drying floor in the sun or on screen bottom trays directly exposed to sun. The viability of seed is one year under ordinary storage conditions and can be increased by storing the seed under low temperature and low relative humidity. The seed germination should not be less than 60 percent. The seeds having 8 percent moisture content can be packed in cloth bags or the seeds with 6 percent moisture content can be packed in sealed containers (Table 8). Proper labelling of the seed is very essential to maintain the identity of the seed lot.

TABLE 7. Field standards for chilli varieties

Factor	Max. permitted (%) in certified seed
Off-types	0.2
Plants affected by seed borne diseases like:	0.5
Leaf blight (*Alternaria solani* Sorauer),	
Anthracnose (fruit rot and die-back)	
(*Colletotrichum capsici* (Syd.) Butter and Bisby)	

TABLE 8. Seed certification standards for chilli varieties

Factor	Seed standards for certified seed
Pure seed (minimum)	98%
Inert matter (maximum)	02%
Other crop seeds (maximum)	10/kg
Weed seeds (maximum)	10/kg
Germination (minimum)	60%
Moisture content (minimum)	08%
Moisture content for vapor proof	06%
Container (maximum)	-

REFERENCES

Ahmed, N., Tanki, M.I., Jabeen-Nayeema and Nayeema, J. (1999). Heterosis and combining ability studies in hot pepper. *Appl. Biol. Res.* 1: 11-14.

Anonymous. (1998). Indian Minimum Seed Certification Standards. The Central Seed Certification Board, New Delhi.

Ashantkar, C.M. and Jaipurkar, M.A. (1988). Heterosis and inbreeding depression in chilli (*Capsicum annuum* L.). *Ann. Pl. Physiol.* 2: 193-203.

Bhagyalakshmi, P.V., Ravishankar, C., Subramanyam, B. and Ganesh Babu, V. (1991). Heterosis and combining ability studies in chillies. *Indian J. Genet.* 51:420-423.

Curl, A.L (1962). The carotenoids of red bell peppers. *J. Agric. Food Chem.* 10: 504.

Daskalov, S. and Mihailov, L. (1988). A new method of hybrid seed production based on cytoplasmic male sterility combined with a lethal gene and a female sterility pollenizer in *Capsicum annuum* L. *Theor. Appl. Genet.* 76:530-532.

Deshpande, A.A. and Anand, N. (1988). Chilli breeding work at Indian Institute of Horticultural Research Bangalore. *Proc. Nat. Seminar on Chillies, Ginger & Turmeric*, held on 11-12 Jan, 1988 at Hyderabad, pp. 17-24.

Deshpande, R.B. (1933). Studies in chilli III. The Inheritance of some characters in *C. annuum* L. *Indian J. Agric.* Sci. 3: 219-300.

Esbaugh, W.H. (1980). The taxonomy of the genus capsicum. *Phytologia* 47:153-166.

Gangadhar Rao, S.V.S, Murthy, N.S. and Reddy, N.S. (1981). Heterosis and inbreeding depression in chilli. *Andhra Agric. J.* 28: 191-199.

Gill, N.S. (1986). Comparison of inbreeds vs. F_1 tester for combining ability studies in *Capsicum annuum* L. MSc Thesis, Punjab Agricultural University, Ludhiana, India.

Gopalakrishnan, T.R., Gopalakrishnan, P.K. and Peter, K.V. (1987). Heterosis and combining ability analysis in chilli. *Indian J. Genet* 47: 205-209.

Greenleaf, W.H. (1986). Pepper breeding. In M.J. Bessett (ed.). *Breeding Vegetable Crops*, pp. 67-134, AVI Publ. Co.

Grubben, G.J.H. (1977). Tropical vegetables and their genetic resources. *IBPGR*, Rome, pp. 197.

Hundal, J.S. and Khurana, D.S. (1988). Heterosis potential in chilli (*Capsicum annuum* L.). *Proc. Nat. Seminar on Chillies, Ginger and Turmeric*, held on 11-12 Jan., 1988 at Hyderabad, pp. 33-37.

Hundal, J.S. and Khurana, D.S. (1993). 'CH-1'–A new hybrid of chillies. *Prog. Fmg.* 29:11-13.

Hundal, J.S. and Khurana, D.S. (2001). A new hybrid of chilli 'CH-3'–Suitable for processing. *J. Res. P.A.U.* 39(2):326.

Joshi, S. (1986). Results of heterosis breeding on sweet pepper (*Capsicum annuum* L.). *Capsicum Newsl.* 5:33-34.

Joshi, S. and Singh, B. (1980). A note on hybrid vigour in sweet pepper (*Capsicum annuum* L.). *Haryana J. Hort. Sci.* 9: 90-92.

Karivarathararaju, I.V. and Palanisamy, T.U. (1984). Effect of seed extraction methods on seed quality in chillies. *South Indian Hort.* 32: 243-244.

Krishnamurthy, M.N. and Natarajan, C.P. (1973). Colour & its changes in chillies. *Indian Food Packer* 27:39-44.

Lakshmi, N., Parkash, N.S. and Harini, L. (1988). Genetic and breeding behaviour of male sterile mutant in capsicum. *Proc. Nat. Seminar on Chillies, Ginger & Turmeric*, held on 11-12 Jan., 1988, at Hyderabad, pp. 28-32.

Likleev, G. (1966). A comparative study of large fruited varieties of red pepper in the Sandanski-Petric area Gradian. *Lozar Nauk Hort. & Viticult*, Sofia 3: 227-232.

Lippert, L.F. (1975). Heterosis and combining ability in chilli pepper in diallel analysis. *Crop. Sci.* 15:323-325.

Marfutina, V.P. (1968). Heterosis in red pepper. *Sci. Mem. Voronez Seet. All-UN-Bot. Soc.* pp. 105-113.

Meshram, L.D. and Mukewar, A.M. (1986). Heterosis studies in chilies (*Capsicum annuum* L.). *Scientia Hort.* 28: 219-225.

Mishra, R.S., Lotha, R.E. and Mishra, S.N. (1989). Heterosis in chilli by diallel analysis. *South Indian Hort.* 37:179-180.

Mishra, S.P., Singh, H.N. and Singh, A. (1976). A note on heterosis in chilli (*Capsicum annuum* L.). *Prog. Hort.* 8: 61-64.

Murthy, N.S.R. and Murthy, B.S. (1961). Chilli seed viability in relation to period of storage. *Andhra Agric. J.* 8: 246-247.

Novak, F., Betlach, J. and Dubovsky, J. (1971). Cytoplasmic male sterility in sweet pepper (*Capsicum annuum* L.). Phenotype and inheritance of male sterile character. *J. Plant Breed.* 65:129-140.

Padda, D.S. and Singh, J. (1971). Studies on some important aspects of floral biology in chillies. *Indian J. Agric. Res.* 5: 217-218.

Pal, B.P. (1945). Studies in hybrid vigour. II. Notes on the manifestation on hybrid vigour in gram, sesamum and chillies. *Indian J. Genet. Pl. Breed.* 5: 106-121.

Pandey, S.C., Pandita, M.L. and Dixit, J. (1981). Studies on heterosis in chilli (*Capsicum annuum* L.). *Haryana J. Hort. Sci.* 10:116-121.

Patel, J.A., Shukla, M.R., Doshi, K.M., Patel, S.A., Patel, B.R., Patel, S.B. and Patel, A.D. (1998). Identification and development of male sterile line in chilli (*Capsicum annuum* L.). *Veg. Sci.* 25: 145-148.

Peterson, P.A. (1958). Cytoplasmically inherited male sterility in *Capsicum. Amer. Naturalist* 92:111-119.

Peter, K.V. (1999). Spices research and development: An updated overview. *Agro-India.* August, pp. 16-18.

Rahman, H., Hazarika, G.N., Alam, S. and Thakur, A.C. (1996). Genetic variability and heterosis in few crosses of chilli. *Proc. of Seminar on Problems & Prospects of Research and Development in North-East India*, Assam Agricultural University, Jorhat. India. November 27-28, 1995, pp. 78-82.

Rajput, J.C. and Wagh, R.G. (1994). *Maharashtratil Bhajipala lagwad* (book in Marathi)

Ram, B. and Lal, G. (1989). Heterosis and inbreeding depression in chilli (*Capsicum annuum* L.). *Prog. Hort.* 21: 368-372.

Reddy, K. Govinda and Murthy, N. Srirama Chandra (1988). *Proc. Nat. Seminar on Chillies, Ginger & Turmeric*, held at Hyderabad from Jan.11-12, 1988, pp. 37-38.

Sahoo, S.C. and Mishra, S.N. (1990). Mean performance and residual heterosis of F_2 diallel crosses in chilli (*Capsicum annuum* L.). *Orrisa J. Agric. Res.* 3: 87-92.

Sekar, K. and Arumugam, R. (1985). Heterosis in chilli (*Capsicum annuum* L.). *South Indian Hort.* 33:91-94.

Shifriss, C. and Frankel, R. (1971). New sources of cytoplasmic male sterility in cultivated peppers. *J. Hered.* 62:254-256.

Shifriss, C. and Guri, A. (1979).Variation in stability of cytoplasmic genic male sterility in *Capsicum annuum* L. *J. Amer. Soc. Hort. Sci.* 104:94-96.

Shukla, V. and Naik, L.B. (1993). Chilli. In: Advances in Horticulture Vol. 5 (K.L.
Chadha and G. Kalloo, eds.), Malhotra Pub. House, New Dellhi, P 384.

Singh, A. and Singh, H.N. (1978). Genetic divergence & heterosis in chilli (*Capsicum
annuum* L.). *Plant Sci.* 10:17.

Singh, A., Singh, H.N. and Mittal, R.K. (1973). Heterosis in chilli. *Indian J. Genet. Pl.
Breed.* 33: 398-400.

Singh, J. (1980-81). A note on hybrid vigour in chilli (*Capsicum* spp.). *Punjab Veg.
Grower,* 15-16: 31-33.

Singh, J. (1992). Improvement of chillies. In: Advances in Horticulture, Vol. 5-Vege-
table Crops Part-I (1993). Ed. K.L. Chadha and G. Kalloo, Malhotra Pub. House,
New Delhi, pp. 69-86.

Singh, J. and Kaur, S. (1986). Present status of hot pepper breeding for multiple disease
resistance in Punjab. *Proc. VI EUCARPIA Meeting on Genetic and Breeding on
Capsicum & Eggplant,* held at Zaragoza (Spain), Oct., 21-24 (1986): 111-114.

Singh, R. (2001). Heterosis and combining ability studies in chilli (*Capsicum annuum*
L.) for oleoresin and related traits. PhD Dissertation, Punjab Agricultural University, Ludhiana, India.

Thomas, Pious and Peter, K.V. (1988). Heterosis in inter-varietal crosses of bell pepper
(*Capsicum annuum* var. *grossum*) and hot chillies (*Capsicum annuum* var.
fasciculatum). *Indian J. Agric. Sci.* 58: 747-750.

Thompson, C.H. and Kelly, C.W. (1957). Vegetable Crops. McGraw-Hill Book Co.,
Inc., USA.

Tikloo, S.K. (1991). Crop management of Solanaceous vegetables: Tomato, brinjal
and chillies. In: Recent Advances in Tropical Vegetable Production. Kerala Agricultural University, Vellanikara, India.

Verma, S.R. and Singh. H. (1988). Development of an axial-flow vegetable seed extraction machine. *J. Agric. Engg.* XXV(1):98-104.

Vijay. O.P., Singh, D.P. and Yadav, I.S. (1979). Studies on flower development, fruit
maturity and floral biology in sweet pepper (*Capsicum annuum* L.). *Veg Sci.* 6:
5-10.

Wang, M., He, X.M. and Ma, D.H. (1986). Heterosis and correlation between F_1 hybrids and their parents in pepper (*Capsicum annuum*). *Servicio De Investigacion,*
Agraria, pp. 61-65.

Woong Yu, I. (1985). Inheritance of cytoplasmic male sterility in pepper (*Capsicum
annuum* L.). MSc Thesis, Kyung Hee University, South Korea.

Zeecevic, B. and Stevanovic, D. (1997). Evaluation of heterosis for yield & yield components in intervarietal crosses of pepper. *Selekoija-L-Semenarstovo.* 4: 177-183.

Perspectives of Bell Pepper Breeding

Subodh Joshi

Terry Berke

SUMMARY. The Solanaceae family is the most important, representing a diverse group of different vegetables that are indispensable in every kitchen. Among the fruit vegetables, bell pepper has a specific identity. Bell pepper fruits are generally blocky, square, thick fleshed, three to four lobed and nonpungent. It is a versatile crop plant and its consumption is on the increase all over the world. With its increasing demand, world wide emphasis is being given to the development of hybrids. Flowers are bisexual borne in ones, twos or rarely in clusters at the apex of the main shoot as well as on the axils of the lateral branches. Many of the *Capsicum* varieties and hybrids are in the market. Considerable amount of heterosis is reported in this crop for many economical characters. Genetic and cytoplasmic male sterile systems are available in the crop. Chemical gametocides may compliment the use of male sterility in bell pepper for better hybridization. Micropropagation could be used for maintenance and multiplication of parental lines. Double haploids could be useful in fixing homozygosity of the population. In bell pepper, single cross hybrids are the only type of commercial hybrids. The hybrid seed yield is low but becomes economical as the seed price is relatively high. The hybrid seed is commonly produced by hand emasculation and hand pollination. Genetic and cytoplasmic male sterility is

Subodh Joshi is affiliated with the Division of Vegetable Crops, Indian Agricultural Research Institute, Pusa Campus, New Delhi 110012, India.

Terry Berke is affiliated with the Seminis Vegetable Seeds, Inc., 37437 State Highway 16, Woodland, CA 95695 USA.

[Haworth co-indexing entry note]: "Perspectives of Bell Pepper Breeding." Joshi, Subodh. and Terry Berke. Co-published simultaneously in *Journal of New Seeds* (Food Products Press, an imprint of The Haworth Press, Inc.) Vol. 6, No. 2/3, 2004, pp. 51-74; and: *Hybrid Vegetable Development* (ed: P. K. Singh, S. K. Dasgupta, and S. K. Tripathi) Food Products Press, an imprint of The Haworth Press, Inc., 2004, pp. 51-74. Single or multiple copies of this article are available for a fee from The Haworth Document Delivery Service [1-800-HAWORTH. 9:00 a.m. - 5:00 p.m. (EST). E-mail address: docdelivery@haworthpress.com].

also being used but on a limited scale. Seed production required lot of care of the seed plot, timely inspection, proper crossing/hybridization, harvesting, seed extraction, drying, and storage. *[Article copies available for a fee from The Haworth Document Delivery Service: 1-800-HAWORTH. E-mail address: <docdelivery@haworthpress.com> Website: <http://www. HaworthPress.com> © 2004 by The Haworth Press, Inc. All rights reserved.]*

KEYWORDS. Bell pepper, crop biology, pollination control mechanisms, heterosis, inbred lines, male sterility, seed production, hybrid seed production

INTRODUCTION

Vegetables play an important role in the human diet by supplying plentiful nutrients and provide good returns to the farming community. The genus *Capsicum* is used as a food condiment and medicine and also aesthetically as an ornamental crop in the garden among flowers. Growing vegetables commercially or in the home garden provides enjoyment, satisfaction, and fulfils many of our needs. Their vitamins and minerals provide nutritional security besides generating self-employment to some extent. Solanaceous vegetables contribute greatly to total vegetable production (Kalloo, 1996). Worldwide, chilli is grown in the largest area followed by tomato, eggplant, and bell pepper.

The Solanaceae family is the most important, representing a diverse group of different vegetables that are indispensable in every kitchen. The potato, tomato, and chilli are important constituents of many foods; they can be used singly or blended with different foods like eggplant and bell pepper. Among the fruit vegetables, bell pepper has a specific identity. It is known by various names such as *Capsicum*, sweet pepper, bell pepper, Simla Mirch, vegetable paprika, etc. Bell pepper fruits are generally blocky, square, thick fleshed, three to four lobed, and nonpungent. They are used for salads, stuffing, cooked as a vegetable, pickled, or processed and are appreciated world wide for their flavor, aroma, color and as an important source of provitamin A (at the red stage) and vitamin C. Its fruits are important constituents of many recipes. Its consumption is increasing all over the world with the increase in the fast food industries and many hidden uses.

Bell pepper is a versatile crop, its many uses make it a major commodity even though it is listed as a minor crop in many countries. The consumption of bell pepper is on the increase all over the world. It has become a multibillion dollar industry, as well as a part time hobby for home gardeners. Fruits (actually inflated berries) are available in the market year round with different shapes, color, and sizes. With its increasing demand, world wide emphasis is being given to the development of hybrids. To achieve the required goals of increasing productivity, utilizing heterosis breeding for bell pepper was advocated by Joshi and Singh (1980). During the recent past, considerable improvement has been made in the development of new pepper varieties by utilizing indigenous and exotic germplasm.

Bell pepper requires ample heat to grow and set fruit, preferring temperatures between 18-28°C. It is widely cultivated throughout temperate and subtropical Europe, USA, Africa, India, East Asia, and China (Govindarajan, 1985). Its cultivation for the last 400 years in different terrains, climatic conditions, and cultivation techniques has resulted in natural hybridization and selection, providing us with a range of varieties. Many important quality traits are present in bell pepper, including taste, aroma, flavor, color, vitamins, and carotenoid pigments. Its research and development is focused on varieties that may also be suitable for non-traditional areas where bell pepper is yet to be introduced. The development of hybrids and economic hybrid seed production technology is rapidly being adopted by seed companies.

CROP BIOLOGY

The *Capsicum* genus represents a diverse plant group, with fruits ranging from the sweet green bell pepper to extremely pungent habaneros (Bosland, 1996). Capsicums are historically associated with the voyage of Columbus's first expedition (1492-93), when he observed that natives of the new world used a colorful red fruit called aji or axi with most of their foods (Govindarajan, 1985). He introduced it to Europe, and from there the Portuguese introduced it to Africa and Asia. He thought that he had reached India and called it red pepper to differentiate it from the black pepper already known in Europe. Tropical South America, especially Brazil, is thought to be the center of diversity of peppers (Shoemaker and Teskey, 1955).

The phylogenetic taxonomy of bell pepper as described by Somos, 1984 is

- Spermatophyta
- Angiospermae
- Dicotyledones
- Solanales
- Solanaceae
- *Capsicum*
- *annuum*

In India, the term *Capsicum* is often used for the fruits of bell peppers. In actual sense, *Capsicum* is a genus with five domesticated species including *Capsicum annuum* L. (origin Mexico), having a toothed calyx, *C. frutescens*

L. (origin Amazonia), having a greenish flower and non-toothed calyx, *C. baccatum* L. (origin Amazonia), having a corolla with yellow/brown spots, *C. chinense* Jacquin (origin Amazonia), having a constricted toothed calyx and *C. pubescens* Ruiz and Pavon (origin Peru and Bolivia), having black colored seeds and purple flowers. It has also 22 wild species. Bell pepper may have evolved from an ancestral form in the Bolivia/Peru area. All bell pepper that are commercially grown, belongs to *C. annuum*. Its classification is based on fruit characteristics such as pungency, color, shape, flavor, size, and end use.

The flower structural formula of the family is calyx, K (5-3); corolla, C (5-8); androecium, A (5-8); gynoecium, G (2-4). Its plants are herbaceous, tender, woody shrubs. Leaves are spirally inserted with whole or slightly sinuate blades. The shape of the corolla is short tubiferous, and the carpels are located slantwise to the medium plant, because the peduncle frequently gets twisted. The stamens are aduate to the corolla base. The fruit is an inflated berry. The fruits are divided by septa and have 3-5 locules. The ovules are borne on the central receptacle and partly on the septa (Somos, 1984). Its leaves are long petiolated, leaf blades ovate to broad lanceolate, green to dark green in color. The corolla is mostly white, but may be purple. Calyx dentation is acuminate. The anthers are cylindrical with a slight incision on the lower part, greyish violet in color. Flowers are mostly borne singly. The seed is straw colored, flat and slightly curved with a narrow hilum, and a pointed obtuse rostrum.

Pickersgill (1977) stated that the basic chromosome number of bell pepper is 2n = 24 which is characteristic of the family. She also states that in *C. annuum* one of the karyotypes having two pairs of acrocentric chromosomes is confined to Mexico. It is thus assumed that domesticated *C. annuum* is from Mexico.

Following Watts (1980), Somos (1984), Bosland and Votava (2000), and others a brief description of plant characteristics is given as follows.

Seedlings: It is a dicotyledonous plant with epigeal germination. Cotyledonary leaves are narrow at both ends, may vary in size, and are generally green. The main taproot and its lateral branches anchor it in the soil at the base of the cotyledonary leaves.

Roots: The main root axis consists of well-developed spindle shaped roots. The lateral roots are uniform and evenly distributed. When uprooted for planting the main root frequently gets damaged and new adventitious roots develop. Later the diarchous root system develops into a dense tassel. The full-grown plant has a root system like a web 40-60 cm vertically and 25-50 cm horizontally. Its root mass is approximately 10% of the total biomass of the plant.

Shoots: The main axis develops dichotomous branching which is racemose at the bottom and cymose on the top. Side shoots are either absent or poorly developed. The length of the internodes reduces as it develops towards the growing tip, and on the top-most region growth is restricted when several fruits are present. The stem is soft at early stages, later on it becomes hard (lignified) and green in color. In some varieties violet color patches develop on the nodes due to anthocyanin. If the fruits are harvested regularly the height of the plant increases.

Foliage: Leaves of bell pepper varieties may vary in their size and shape but are always simple, entire, and petiolate. Leaves are spirally arranged in the main axis, either in clusters or one of two. Leaf blade is ovate, elliptic and lanceolate. Old leaves are darker green and larger in size than newly grown leaves. Varieties that have large leaf size and more leaves help in photosynthesis and protect the growing fruit from exposure to direct sun, which causes sunscald.

Flower: Flowers are borne in ones, twos, or rarely in clusters at the apex of the main shoot as well as on the axils of the lateral branches. The flowers are bisexual attached to the shoot by the peduncle. The peduncle is green with five to seven ribs, fused with sepals to form a calyx tube that ends in calyx teeth. The corolla consists of five to six white petals fused to form a corolla tube. Five to six stamens are attached to the corolla tube on white filaments. Anthers are short, elliptic in shape, and yellow or violet blue in color. At flowering, the anthers dehisce lengthwise and pollen falls out immediately. At the base of the filaments, nectar glands are located that excrete nectar for insects. The pollen grain is elliptic, three segmented and light yellow in color. Hirose (1957) reported that 11,000 to 18,000 pollen grains develop in one anther and are almost equal in size. The pistil consists of the ovary, style, and stigma. The ovary is round or slightly elliptic and funnel shaped stigma is attached to the ovary by the style, which varies in length according to variety. The stigma may be on an equal level to the anthers, may be exerted above the anthers, or may be inserted below the anthers. Most bell pepper varieties have a stigma equal to the anther height. Receptivity of the stigma to pollen is up to seven days (Popova, 1963). Up to 100 flowers develop in the plant during the growing season (Kormos, 1954). The fruit set of new flowers is negatively correlated to the number of fruits developing on the plants. A fruit generally sets in the axil of the main shoot.

Fruit: Botanically bell pepper fruit is an inflated berry (Somos, 1984). The peduncle helps in holding the fruit, and the stem continues growing inside the fruit in the central placenta. Fruits may be borne upright, drooping or semierect, but drooping is most common. Seeds are located on the surface of the placenta. The fruit develops from 3-5 carpels of the ovary after completing the process of fertilization. Fruit apex may be lobed, pointed, obtuse, blunt or enfolded depending upon variety. Pericarp thickness is an important quality trait of bell pepper. A thick, fleshy, firm pericarp is desirable. The fruit cavity inside is large and the septa protrude into it. Generally bell pepper has 3-4 septa depending upon number of lobes. Fruit color may be waxy cream, purple, or green at the immature stage. On ripening, it may be red, dark red, orange, yellow or chocolate colored. Productivity of the plants depends upon the variety, regular harvesting, cultural practices, the growing environment, and the presence of insect and diseases.

Seed: A campylotropic ovule develops into a seed. Fully matured seeds are kidney-shaped, 3-4 mm in diameter and 0.5 to 1 mm thick, with a rim-like thickening on the edge (Somos, 1984). Seeds are located inside the fruit on the central placenta. One thousand seed weight ranges from 5-10 g. The seed is straw colored, covered by a parchment-like seed coat. Its ovule is in a bent state, with one radicle turned towards the hilum, while the remainder of the seed is filled with endosperm.

Varieties: Varieties should be chosen considering the growing climate and economic condition of the crop producing area. It should be highly productive, early maturity, with good fruit size and shape and tolerance to important biotic and abiotic stresses.

The bell pepper fruits range from 8-12 cm long and 7-10 cm wide with smooth, thick flesh. Shape is blocky, blunt, three to four lobed, and square to rectangular.

Germplasm: Many of the *Capsicum* varieties as reported by Somos (1984) were known in the 16th century. Wafer (1699) referred to bell peppers in his diary. Among older varieties Campana (1774), California Wonder (1828), and Ruby King (1884) were listed by Bassett (1986). Bose et al. (1993), and Joshi and Munshi (2000) have mentioned common cultivars as: Nonpungent, green, turning red at maturity–California Wonder (various strains), Yolo Wonder (various strains), Bull Nose, Ruby King, World Beater, North Star, Bell Boy, Keystone Resistant Giant, HC-201, Arka Mohini, Arka Gaurav, Pusa Deepti, Florida (various strains), Chinese Giant, Early Giant and Jingle Bells. Nonpungent, green turning orange at maturity–Arianne, Golden Bell, Romanian, and Golden California Wonder. Pungent bell peppers include Bull Nose Hot, Romanian Hot, and Mexibell. Beside the above listed cultivars, there are many open pollinated and hybrid varieties developed and recommended for cultivation in different countries. Poncavage (1997) has given a brief description of different types within bell pepper:

- Hottest bell pepper–Mexibell–Large bell is normal in every way, takes 70 days from transplant to green maturity, ripening to red, pungent in taste.
- Sweetest bell pepper–Arianne–Thick walled, orange when ripe. Yellow and orange bells of all varieties are sweeter than bells of other colors

(Steve Czeplewski, pers. comm.). Arianne reaches mature green stage in 68 days after transplant and takes 2 weeks more to turn orange.

- Earliest ripening sweet bell–Ace, an old Japanese hybrid that matures to full green size in 50 days after transplant, to red 2 to 3 week later. These small tapered bells are about 8 cm long and have thin crispy walls.
- Biggest stuffing pepper–Big Bertha–Largest elongated bell pepper, 18 cm × 10 cm thick walled, blocky fruit reach full size green about 73 days after transplanting.
- Tiniest sweet bell pepper–Jingle Bells–Exceptionally little, about 4 cm × 4 cm, early fruits are ready after 60 days of transplanting.
- Oldest bell pepper–California Wonder–The first smooth and blocky thick walled pepper to be known as a bell pepper was released in 1828 by C.C. Marse. Takes 70 days to green and 90 days to red maturity stage after transplanting. Still a popular bell pepper variety.
- Most productive bell pepper–Large number of fruits plant $^{-1}$–Jingle Bells–> 60 fruits plant $^{-1}$, Jupiter–20 thick walled fruits plant $^{-1}$, and North Star–17 fruits plant $^{-1}$.

The first step is introduction of genetically diverse germplasm to enrich the base as a source of variability. Plant introduction involves exploration, collection, purification, conservation, evaluation, storage and utilization. Wild forms of cultivated species, wild species, land races, old and new cultivars, and other genetic materials are important. In bell pepper, International Board of Plant Genetic Resources (IBPGR), Rome, and Asian Vegetable Research and Development Centre (AVRDC), Taiwan, are important international sources for enriching the germplasm base beside personal collection from institutes and research stations working for the improvement of this crop. Varieties introduced from different countries are commonly used as a source of new variability, especially open pollinated cultivars like California Wonder, Yolo Wonder, Bull Nose, Chinese Giant, World Beater, etc. In India, heterozygous material obtained from different sources has been utilized to develop new varieties by following pure line breeding methods. New germplasm should be observed under isolation for plant quarantine to avoid the possibilities of introducing new pathogens. The isolated germplasm found free from pathogens can be utilized in the improvement of bell pepper. If variability is not available in the germplasm as in the case of bell pepper it can be generated by hybridization.

Heterosis

The goal of increasing productivity in the quickest possible time can be achieved by utilizing heterosis breeding. Hybrid vigor in plant hybrids was first studied by Kolreter in 1763. After reporting the phenomenon of heterosis by Shull (1914) many-fold improvements have been made in various aspects of exploiting heterosis in vegetable crops. An F_1 hybrid variety is the result of a cross between two homozygous (but genetically distinct) pure lines. The prerequisite is that all the F_1 plants should resemble each other phenotypically. *Capsicum* has a tendency towards autogamy and even though it is hermaphroditic in nature, self-fertilization cannot be avoided to 100 percent. Heterosis in sweet pepper has been reported for early and total yield by Diku (1976), and Novak and Chemalo (1975). Expression in the F_1 depends upon the selected parents' combining ability. It also depends upon the genetic divergence; generally the wider the divergence, the higher the heterosis. The economic feasibility of hybrid cultivars depends upon the cost of hybrid seeds in relation to the value of increased yield. Heterosis breeding is feasible in bell pepper due to its low seed rate hectare $^{-1}$ and only 400-600 grams of seed are sufficient to raise the seedlings for planting one hectare.

Capsicum tolerates inbreeding, although there is some degree of heterosis for reproductive traits including seed yield component (Mishra et al., 1991). Heterosis averaging between 2 and 55 percent has been reported for various traits including early maturity, plant height and certain yield components (Gopalkrishnan et al., 1987; Khalil et al., 1987; Lee and Shin, 1989; Lippert, 1975; Joshi et al., 1995; Meshram and Mukewar, 1986). Combining ability studies for yield components have indicated that additive effects in general were significant for yield components such as fruit length, fruit circumference, fruit weight, primary branches per plant and fruit weight per plant; few components showed significant dominance effects (Pandey et al., 1981; Joshi and Singh, 1983). Leskovar et al. (1989) reported that hybrid bell pepper had more vigorous shoot and root growth than open pollinated varieties, however, gain in productivity were realized only under favorable environmental conditions.

The breeding objectives for fresh market bell peppers are color, lack of pungency, uniform shape, size, lobe number, flavor, firmness, exocarp thickness and resistance/tolerance to important biotic and abiotic stresses. Yield under open field conditions is desirable from eight to nine fruits plant $^{-1}$ with an exception of 15 fruits plant $^{-1}$. In green house cultivation, it may be forty to fifty fruits plant $^{-1}$. The major drawbacks of pepper hybrid adoption are high seed cost, high cost of production inputs, and fresh seed purchase every time.

Open pollinated varieties have traditionally dominated home garden and commercial bell pepper cultivation. Nevertheless, in the last 30 years hybrid pepper has attained commercial status (Poulos, 1994). Many successful commercial hybrids are grown, primarily in temperate zones, but most pepper production in the tropics relies on land races or commercial open pollinated varieties, particularly in Asia, Africa, and Latin America. The seed sector has developed in these regions now, so there is a tendency towards hybrid development. Farmer-adoption of hybrids must be combined with careful seedling and transplant management to justify the cost of hybrid seed. Many hybrids are now available for commercial utilization in Solanaceous crops. Bell peppers grown from hybrid seeds are more uniform and usually higher yielding than open-pollinated varieties. Capsicum breeding has evolved through pure line selection of local land races to hybrid development (Poulos, 1994) for high productivity. Observing the feasibility of heterosis in bell pepper (Joshi and Singh, 1980), use of genetically diverse genotypes to find superior hybrid combinations has been taken up by the private sector. Today, many hybrids are commercially viable in the world market. Besides high fruit yield, hybrids may provide disease resistance and improved quality attributes through favourable dominant gene combinations. Uniformity of produce due to genetic homogeneity is an important requirement for a sustainable bell pepper trade. Therefore, to achieve the desired goals it is necessary to conduct genetic analysis in bell pepper to evaluate different desired traits for their suitability in developing hybrids.

It is often desirable to introduce exotic germplasm in bell peppers and study their performance to know the nature and magnitude of variation of different desired horticultural characters. Another technique is to study the combining ability and nature of gene action, using diallel crosses with an ultimate aim to identify superior parents for a hybridization programme. A third area of study is the inheritance pattern and gene action of economic traits. The best inbred lines can be used for developing hybrid varieties. To have the proper understanding in the above aspects, recent reviews and research papers in *Capsicum* can be of great help, including Verma and Joshi (2000), Bosland (1992a, 1996), Joshi et al. (1995), Poulos (1994), Bosland and Votova (2000), Pickersgili (1991), Greenleaf (1986), Morrison et al. (1986a,b), Muthukrishan et al. (1986), and Somos (1984). Areas of particular interest include genetic resources, mating systems, breeding methods, breeding objectives (yield, disease and pest resistance, abiotic stress tolerance, horticultural types, quality traits, etc.), gene symbols, and genetic mapping.

POLLINATION CONTROL MECHANISM

In bell pepper, hybrid varieties have been successfully developed. However, the major constraint of using these varieties for commercial cultivation is the high seed cost, because there is a lack of efficient and economic hybrid seed production technology. Male sterility can be used to minimize the hybrid seed cost, which is high because of research involved in selecting parents, testing progeny, assessing performance, maintaining different lines, etc., before recommending an F_1 hybrid for commercial cultivation. Techniques for controlled pollination, such as male sterile lines to economize hybrid seed production. Male sterile plants are occasionally observed as spontaneous mutants in commercial fields (Bosland, 1996). Cytoplasmic male sterility (CMS) was first discovered by Peterson (1958) in *C. annuum* P.I. 164835 from India. But this line was unstable in fluctuating environments and produced pollen at low temperatures. Some commercial seed companies use the CMS system to produce F_1 hybrids in hot pepper but not in sweet peppers. The most successful and viable technique currently practised is hand emasculation and hand pollination.

All breeding work starts with a flower and results in the formation of seed. Seed is produced by the female cells of the plant (the ovules inside the ovary). The main requirement for hybrid development in bell pepper is homozygous parental lines that ideally would give a large number of seed per fruit from a single pollination. The main task of improvement is to assemble into a cultivar superior genetic elements necessary for increasing yield, quality and protection against abiotic and biotic stresses. To achieve this, one should be familiar with pepper genetics and its reproductive behavior. Bell pepper flowers are complete, that is they have calyx, corolla, anthers, and stigma. It can cross-pollinate because of its protogynous condition (Bosland and Votava, 2000) and exerted position of stigma in some cases. Pickersgill (1977) has reported 2 to 90% cross-pollination in open pollinated crops. Therefore, breeders make controlled crosses to transfer genes from one plant to another through pollen. The genetic variability in bell pepper is very narrow, thus introducing new variability, via crosses, is a goal of many breeding programs.

Crosses in bell pepper can be made any time during the daytime, but the best time is early morning or late afternoon (Greenleaf, 1986). Before preparing the female flower for emasculation, it should be ensured that all the plants of the breeding line are pure (true to type). From segregating populations desired plants are selected following one or more methods, such as pedigree, single seed descent, recurrent, or back cross. Selection of pure lines showing resistance against biotic (diseases, insects, nematodes, etc.) and abiotic (environment and soil conditions) stresses is routinely done during the inbreeding process.

INBREEDING

The parental lines should be homozygous in order to obtain a uniform hybrid. Inbreeding may be practiced on single plants for five to six generations so that the amount of heterozygosity is reduced by half in each selfed generation. Homozygosity is adequate by S_6 generation for using a parent for development of commercial hybrids. Watts (1980) states that homogenous cultivars/pure lines can also be used as parental lines if their combining ability is good.

Back crossing has been an essential method for introgressing genes from hot pepper into sweet pepper and from interspecific crosses. Several successful synthetic varieties were developed from a Hungarian breeding programme where up to 12 parental lines were crossed in a diallel pattern before developing the synthetic variety (Zatyko et al., 1983). Poulos (1994) advocated single seed descent as a useful method for rapid generation advance to be used for bell pepper improvement. Recurrent selection may be useful for pepper breeding to improve quality traits (Lippert, 1975; Somos, 1984; Venkata Rao and Chhonkar, 1984), but it is labor intensive for emasculation and pollination. The incorporation of male sterility could reduce the labour requirement and facilitate the adoption of recurrent selection in bell pepper breeding. At AVRDC, Taiwan, mass selection with progeny tests was being used to develop hot and sweet pepper populations for heat tolerance and rainy season adaptation (Poulos, 1994). The source populations were bulks of F_3 selections derived from two way or three-way crosses targeted for disease resistance and heat tolerance.

Bell peppers can be screened for chemical composition by high performance liquid chromatography (HPLC). Near infrared reflectance spectroscopy (NIRS) was used for screening germplasm and breeding populations with reduced cost and time compared to HPLC at the AVRDC (Poulos, 1994). Capsaicin, color, crude fibre, fat, sugars, vitamin C, and dry matter are some of the traits evaluated. Early generation selection for most of the quality traits has been explored in connection with genotype × environmental interaction and quantitative inheritance (Ahmed et al., 1984; Ribeiro et al., 1990; and Tewari, 1990). Seeds in bell pepper are usually removed before cooking or processing. Therefore, there is great potential for developing seedless (parthenocarpic) fruits with proper shape and size (Shifriss and Eidelman, 1986).

Depending upon the production system, heat tolerance, and horticultural type, there is great variation in yield of bell pepper, particularly in the tropical zone. Yield depends on number, and weight of fruits, cultural practices, and climatic conditions. Fruit weight and number of fruits plant^{-1} are the primary yield components in bell peppers. Reports of high heritability for yield components suggest good feasibility for early generation selection (Lee and Shin, 1989; Singh and Singh, 1977). Studies on genotypic and phenotypic correlations among traits, path coefficient analysis and regression analysis have helped to formulate indirect selection indices (Joshi and Singh, 1983; Kim et al., 1985). Several studies indicate fruit yield is highly correlated with fruit number plant^{-1} rather than individual fruit weight (Legg and Lippert, 1966; Gill et al., 1977).

A pure line is a selection from a single, selfed, homozygous individual plant developed by in breeding over several generations from heterozygous breeding stocks or germplasm. Pedigree selection method in bell pepper is most widely used for developing varieties with high yield, earliness, quality, resistance to diseases and to abiotic stresses. This method can be employed following Johansen (1960). For fruit quality, disease resistance, or any other specific attribute a pure line can be improved by backcrossing to incorporate novel genes. In bell pepper generally 4-5 backcrosses are required to produce near-isogenic lines for the novel gene in the pure line. Backcrossing is used whenever disease resistance, improvement of male sterile line or any other simply inherited morphological trait is to be transferred because its application is convenient in bell pepper. In a convergent method, desirable characters of two or more varieties are incorporated in a new genotype which can be further utilized for the production of hybrids. Single seed descent method is used occasionally in bell pepper for the traits having additive genetic inheritance. In bell pepper, fruit quality is a highly recessive trait therefore in a hybridization program both parents must have desirable horticultural traits.

The consumption of bell peppers as salad or stuffed as a fresh fruit makes it desirable to have disease and insect resistant varieties to reduce pesticide application. Therefore choice of parents should be made keeping this aspect in the program. Parents are generally selected on the basis of phenotypic (morphological) characters and their respective combining ability. Parents with high GCA for the desired traits are ideal.

MALE STERILITY

In bell pepper, genetic male sterile (gms) lines are being used to a limited extent for hybrid seed production. GMS lines are maintained by hand pollination using heterozygous fertile plants as males. Poulos (1994) reported that genic male sterility was the first controlled pollination strategy adopted by seed industry in the production of F_1 hybrid seed after a long history of hand pollination and emasculation. Although much hybrid seed is still produced by hand emasculation and pollination, sterile lines are being used to reduce the labour cost and to improve hybrid purity. A number of gms genes have been reported and spontaneous mutants occur frequently (Shifriss, 1973; Csillery, 1989; Yu, 1990). The discovery of molecular and physical seedling markers that are linked to gms alleles is impending. Cytoplasmic male sterility (cms) has been tested by Daskalov and Mihailov (1988) and Yu (1990) for commercial hybrid seed production in bell pepper. They have reported that there are some problems in the stability of sterility and its restoration. Daskalov and Mihailov (1988) proposed a scheme using a cms seed parent containing a lethal gene and a female sterile pollen parent to improve the efficiency of hybrid seed production without hand pollination. Chemical gametocides may complement the use of male sterility of bell pepper hybridization but limited success has been reported (Salgare, 1989). Novel sterility systems are likely to be adopted in an effort to curtail the labor cost of producing hand pollinated hybrids and to assure hybrid purity.

GENETIC MALE STERILITY (MS)

Shifriss and Frankel (1969) found spontaneous genetic male sterile plants in the cultivar All Big. Male sterility was determined by a single recessive gene. Due to this sterility, parthenocarpic fruits developed in unfertilized flowers throughout the season. Shifriss and Rylsky (1972) discovered a second gene encoding genetic male sterility in the cultivar 'California Wonder'. It too, was found to be inherited as a single recessive gene. The authors suggested that the sterility discovered in 1969 may be described as *MS1* and second as *MS2*. The use of genetic male sterility is limited in hybrid seed production due to the inefficiency in rogueing out male fertile plants, since 50 percent of the plants are sterile and 50 percent are fertile. Therefore, 50 percent fertile plants must be rouged out from the field. Ideally you would use a molecular marker to identify male sterility in the early growth stages so that fertile plants can be rouged at the seedling stage. The mechanics of using genetic male sterility for F_1 hybrid production require (i) normal fully fertile inbred line as male and (ii) female inbred line which is maintained by crossing together, known heterozygous (*Msms*) and male sterile (*msms*) plants. Seed must always be harvested from *msms* plants and should give equal proportions of plants of the same genetic constitution as the parents. Msms plants should not be allowed to intercross with one another since some homozygous male fertile (*MsMs*) will be produced and will upset the system (Watts, 1980). In practice, it would be advisable to cross groups of up to five *msms* plants with a single *Msms* plant, to ensure that other gene differences do not build up.

Shifriss and Pilovsky (1993) increased the ratio of *ms* plants in a line. They crossed two isogenic lines that have different male sterility genes. Their intention of this digenic cross was to produce male sterile plants containing both *ms1* and *ms2* genes. This plant was then crossed to a fertile plant that was heterozygous for both genes, i.e., *ms1ms1 ms2ms2* × *Ms1ms1 Ms2ms2*. The progeny of this cross segregated in a ratio of three male sterile and one male fertile plant. Only one-quarter of the plants have to be removed from the hybrid seed production field. The problem was that both parents have to be maintained asexually with proper protection from viruses (Shifriss, 1997).

CYTOPLASMIC MALE STERILITY (CMS)

CMS is another system of sterility by which hybrids are produced. In this system a maintainer male can regularly produce 100 percent male sterile plants in the female. These plants are used in F_1 production with a male fertile line, which produces fully fertile F_1 progenies. Sterility results from an interaction of nuclear and cytoplasmic factors. Peterson (1958) described a CMS system in peppers but it was found unstable, and fertile pollen developed under cool conditions. Several studies (Novak et al., 1971; Shifriss and Frankel, 1971; Shifriss and Gurl, 1979) indicated from the study of Peterson's CMS material that additive factors affect pollen sterility and stability. In general, seed set on pollinated male sterile plants was only half that of normally pollinated plants and hand pollination has to be performed although emasculation is avoided. Some red pepper hybrids were developed by using CMS and exhibited heterosis for early and total yield (Diku and Anakecko, 1975).

MICROPROPAGATION

Multiplication of F_1 hybrids through micropropagation if permitted is potentially cheaper for a crop. It is also potentially useful in maintenance of parental lines and production of artificial seeds through somatic embryogenesis. Regeneration of plant tissue explants from pepper has been achieved (Agarawal et al., 1989; Fari, 1986, Harini and Sita, 1993; Morrison et al., 1986a; Sripichitt et al., 1987). Success with anther culture for tissue regeneration and differentiation is genotype-specific and dependent on environmental conditions (Ochoa-Alejo and Ireta-Moreno, 1990).

Despite the varying degrees of out-crossing in bell pepper, populations are managed most commonly as for self-pollinated species, with the final goal of developing homozygous lines. Using doubled haploids as a tool for reducing generation time and fixing homozygosity has been a long time interest of many workers. Haploid-derived diploids may occur spontaneously or be induced by colchicine treatment in peppers. Haploids may arise from poly embryony, chemical mutagenesis, inter specific hybridization, or anther culture. Anther culture seems to be the most applicable for rapidly fixing new genotypes for the development of elite breeding lines (Li and Jiang, 1990; Morrison et al., 1986a). Depending on the genotype, haploid regeneration from anther culture in pepper is generally between 1 to 20 percent, although success up to 50 percent has been reported (Dumas de Vaulx et al., 1981; Morrison et al., 1986b; Munyon et al., 1989).

HYBRID SEED PRODUCTION

In bell pepper, hand emasculation and hand pollination are mostly used for production of F_1 hybrid seed on a commercial scale. Mature buds (a day before flower opening) are selected for emasculation and anthers are removed by forceps. Pollen is collected from the male parent and transferred to the stigma with the help of forceps, brush, thumb nail or any other suitable tool. The stigma is fully mature in the bud stage and anthers mature either in the bud or after two to three hours of flower opening. Popova (1963) studied the condition of flowering in bell pepper. She reported that the period immediately preceding the opening of petals is the most suitable for emasculation. Pollination with more pollen grains increases the number of fruit and the quantity of seed in the fruit. For pollination, morning hours (0700 to 1100 hrs) are best. However, the number of seeds developing in the fruit depends on the females. For pollen collection, flowers that have not opened are picked in the afternoon and pollen can be collected by vibrating the flowers or by removing the petals preferably in the morning hours. Pollen can also be stored for 3 to 4 days under cool dry conditions depending upon temperature and humidity. Pollination is done in the emasculated flowers after 15 to 20 hours. Pollination can also be attempted just after emasculation if matured buds are emasculated but the percentage of fruit setting may be less. Pollinated flowers are marked with a clip or a string. The ratio of female and male parent plants is normally kept 4:1. In bell pepper, maximum fruit setting is observed in the temperature range of 20-27°C, with the night temperature being the critical temperature. In bell pepper, single cross hybrids in which two inbred lines are involved are the only type of commercial hybrids (Kalloo, 1986). The hybrid seed yield is low but it becomes economical as the price of the seed is relatively high. Bell pepper fulfils the basic requirement for F_1 hybrid production, i.e., homozygous parental lines that produce a large number of seeds from a single hand pollination. The lines with the greatest seed quality potential should always be used as female parent if there is no difference in the reciprocal hybrids.

Companies in Japan, Korea, Denmark, Holland, Italy, and USA sell hybrid seed. In India, it is produced in the states of Karnataka, Maharashtra, Himachal Pradesh, and Gujarat. F_1 hybrids of sweet pepper are very common in USA, Europe, and other advanced regions.

HAND EMASCULATION AND HAND POLLINATION

Sharma (1995) suggested it is the most reliable method in this crop. Emasculation can be carried out on flower buds one day before flower opening, preferably in the evening. Emasculated buds may be protected by non-absorbent cotton plugs, perforated butter paper bags, or cheesecloth, but it is not necessary. The buds are tagged with the description of male parent, date of pollination, etc., or use different color threads or twist ties in the pedicels such as red, yellow or blue for hybrids and white for selfs. For pollination, pollen is collected from full grown buds at the time of anthesis and placed on the stigma by brush or needle. The stigma is generally receptive before anthesis. After pollination, no protection is necessary.

The set fruit should be protected carefully, using optimum plant protection measures to avoid its rotting or insect damage. Allow the set fruit to mature on the plant till they turn red/golden depending upon the variety. The seeds are extracted by cutting the fresh mature fruit. Dry the seeds to the moisture content <7 percent before packing. A plant in the open field condition with 8 to 9 cross-pollinated would give good yield of hybrid seed. For this, crossing should be attempted on 10 to 12 flowers plant^{-1}. However, under controlled conditions hybrid seed yield can be enhanced by pollinating more flowers and leaving 15-18 cross-pollinate fruits plant^{-1} in the extended period of growth.

HAND POLLINATION WITHOUT EMASCULATION

Use of male sterility for economizing hybrid seed production was first reported by Shifriss and Rylski (1973). Breuits and Pochard (1975) developed the hybrid Lamuyo by using male sterile gene ms-509. Use of androcides has also been suggested by Kohli et al. (1981) for sterility induction. Spraying GA_3 @ 1000 mg l^{-1} at 10 days interval from the onset of flowering for three times has caused complete male sterility which lasted throughout the cropping season. Chauhan (1980) used malic hydrazide and dalapon to develop ms plants in *Capsicum*. These gametocides caused complete male sterility. In some cases hybrid seed yield can be doubled if the female line has conical shaped fruits, 15-20 plant^{-1}, crossed with round/large blocky male parents. Some desirable combinations such as 19 × yw and 19 × 02, yw × 70, yw × 02 and 201 × 02 have been reported by Joshi et al. (1995) for commercial exploitation.

MAINTENANCE OF INBRED/PURE LINES

In bell peppers, pure lines are maintained and used as inbreds as in other self-pollinated crops. The uniformity of hybrids solely depends upon the uniformity of parents, therefore 2 to 3 years of selfing to remove all heterogeneity from the parents should be followed. Only homozygous pure lines should be used for crossing. For inbreds seed production, self-pollination should be done in isolation. The multiplication of elite pure lines starts with the positive selection of mother plants. The progenies of the plants are kept separately as A lines. The seeds of individual A lines are usually sufficient to plant B lines in a block experiment where original stock of the variety is planted as a purity check. Further generations of selfing, designated as C and D lines in increasing amounts are needed to provide sufficient quantity of the bulk seed required for hybrid seed production in isolation. Treating the seeds with trisodium phosphate can minimize contamination with infectious diseases. Selection of true type plants is the first step. The number of plants to be selected depends upon the requirement of breeder seed stock. Selection should be done rigorously and plants appearing off-types as inferior in any regard or differing in any character should be rouged discarded. Critical examination of selected plants for stem, leaf, and fruit characteristics (bearing habit, color, shape, size, etc.) is absolutely necessary for the production of foundation seed of inbreds. The selected individual plants can be covered by a thin three-ring muslin cloth bag (preferable 40 to 45 cm diameter and 60-80 cm length) to prevent cross-pollination. At full maturity stage, true to type selfed fruit can be harvested from each of the individual plants. Seeds of the above individual plant progenies should be grown in isolation where the same crop was not grown in the preceding year. An isolation distance of 400 m is minimum. Individual plant progeny rows should be observed very carefully. A minimum of 3 to 5 field inspections is required for bell pepper. In early stages of plant growth, inspection may be based on vegetative morphological characters of leaf, stem, plant type, etc. Second and third inspections may be based on fruiting habit, fruit size, shape, color, etc. The fourth and fifth inspections should be made at full maturity stage looking for true to type fruits. If there is any off-type plant, it should be removed. Only selfed fruits are taken for extracting seeds. Fruits of individual plants may be harvested separately after attaining physiological maturity (Munshi and Sharma, 1997). Fruits of individual plants are critically examined and if there is no heterogeneity, they can be bulked for extracting seeds. Seeds of all inbreds are maintained separately. After proper cleaning and drying, the seed is stored at 15°C and 40% relative humidity for use as parental lines. Bosland and Votava (2000) suggested use of tissue culture to regenerate plantlets from pollen microspore tissue. Plantlets produced from pollen microspores are haploid. By treating with colchicines, the chromosome number is doubled. These double haploid plants are homozygous at every locus and save many generations of self-pollination.

SEED PRODUCTION TECHNOLOGY

Bell pepper requires a frost-free, moderately cool climate for 4-5 months. The cultural practices recommended are for the hills of Himachal Pradesh, India, but can be suitably modified according to regional conditions. In the plains of East, West, and South India, the crop can be planted in *kharif* and *rabi*, i.e., during May-June and September-November but in the hills there is only one planting season, during summer (April-May). In the plains of North India, the crop is sown in early October and transplanted in November. Optimum temperature for seed germination is 30°C and for plant growth 24-27°C with a maximum of 32°C. Sowing/planting time is more or less the same as chilli pepper.

Bell pepper can be grown in any fertile, well drained soil with pH 6.5-7.0. Seed rate for nursery planting of hybrid seed is 250-350 grams ha^{-1} where germination is around 80-90 percent. *Capsicum* is preferably transplanted at a spacing of 45-45 cm plant to plant and row to row. Land preparation is done by three to four ploughing to obtain fine tilth. Farmyard manure (FYM) @ 20-25 tonnes ha^{-1} is mixed with the soil at the time of land preparation. If FYM is not available, well decomposed compost could also be used. The following general fertilizer recommendations may be followed under normal field conditions. Adjustment is necessary based on soil test. Basal dose of inorganic fertilizer @ NPK 60:100:60 kg ha^{-1}, followed by top dressing 3 weeks after transplanting @ NPK 20:0:20 kg ha^{-1} and a final top dressing 6 weeks after transplanting @ NPK 40:0:40 kg ha^{-1}. However, Sharma (1995) has reported that cultivar Pusa Deepti yielded maximum when a fertilizer dose @ 240:60:80 NPK kg ha^{-1} was applied. Seedling raising in the nursery should be attended carefully. For the best germination, well drained soil is selected and 10-12 beds of 7 m long, 1.2 m wide and 15 cm deep are prepared. Apply 15-20 kg well decomposed compost and 500 g of 15:15:15 NPK fertilizer to each bed 15 to 20 days before sowing. Nursery beds may be treated with formalin (40 liter water and 1 liter formaldehyde) by sprinkling on the surface of the seed bed and then the beds are covered air tight with polythene sheets or thick tarpaulin for 48 hrs. Later the soil is kept open till there are no fumes of formalin left. If formalin is not available then drenching can also be done first by water followed by Captaf or Foltaf (3 grams liter^{-1} of water @ 5 liter m^{-2}). After sowing the seed, beds are covered with dry paddy straw or with dry leaves to give warmth, retain moisture, avoid dislocation of seeds at the time of watering and cracking of soil layer on drying. Keep the seed beds moist but not too wet. Remove the mulch when the seedlings start to emerge. The seedlings are ready for transplanting at 4-6 true leaf stage. Seedlings are first hardened by gradually withholding water 6-9 days before transplanting and exposing them to direct sun light. Thoroughly water the seedlings 12 to 14 hrs before uprooting. Healthy seedlings may be transplanted in the evening hours, at a spacing of 45 cm apart. Apply 3 grams Furadon 1G in each furrow by mixing well with the soil to avoid cut worm damage. Irrigate the field after transplanting. Fertilizer application should be as per schedule however, if the growth of the plants is poor give urea spray @ 2 grams litre^{-1} of water after 30 days of planting.

Plant protection measures include spraying Endosulfan at the rate of 1-2 ml liter^{-1} of water if insects like thrips, mites, aphids, fruit borers, or hairy caterpillars are noticed. Crop may be sprayed with fungicides like Dithane M-45 (2 grams liter^{-1} of water), copper oxychloride (2.5 grams liter^{-1} of water) and wettable sulphur (Sulfex 3%) if diseases like anthracnose or powdery mildew are observed. Crop should be kept weed free. Weeding, hoeing, earthing up, and providing proper drainage should be attended at the proper time.

SEED PRODUCTION

The cultural practices for seed production plots are almost the same as those adopted for the commercial crop. One important point for the seed crop is it should be transplanted (following proper crop rotation) as direct seeding delays ripening of the fruit. Second important point for the seed crop is proper sowing time and top dressing with nitrogenous fertilizers. It should be avoided in the second half of the vegetative period as it delays ripening and causes excess growth of the plants. Proper isolation is the third requisite. Although a self-pollinated crop, it readily crosses with chilli and paprika. An isolation distance of 400 m for certified and foundation seed is reported to be minimum to prevent contamination (Tanwar and Singh, 1988). For breeder seed, an isolation distance of 800 m is desirable. It can be reduced wherever two varieties are grown with natural barriers of tall crops such as maize, sorghum, or okra. Seed crop needs at least three to four field inspections during the vegetative growth period to rogue out off types, diseased plants, and those showing poor growth. In the early period of growth, inspection may be based on leaf and stem characters. Second inspection may be based on fruiting habit, fruit size, shape, and color. Rogue out the plants showing different fruit types and pinch off the undesirable small fruits that set after the main setting of 5-6 fruits plant^{-1}. True to type fruits (depending upon shape, size, color, and free from insect and disease) only should be kept for seed extraction.

Verma et al. (1994) recommended removal of the flower bud from the first node followed by pinching buds from the 2nd and 3rd nodes to help in getting higher seed yield. Pinching the buds stimulates the development of bigger, stronger plants. If the conditions are dry with intense sunshine and wind, transplanting two plants to a hole should be tried. Plants should be set 15 cm apart so that they will shade each other and protect the fruits. Harvesting is done at 5-10 day intervals for a period of 1-2 months. Fruits should be after-ripened for 3-4 days in the shade, and then seed is extracted from the ripe fruits. During seed extraction one more selection can be made to remove deformed, diseased or very small fruits and the ones in which the placenta has turned brown, black or mouldy. Harvesting and seed extraction should be avoided during the rainy season as viviparous germination can occur under high humidity and may result in substantial loss of germination. Seeds are removed from the placenta by water or by hand. The wet method is better as empty seeds float on the surface and can be removed easily. The seeds obtained are either dried in a drier at 35-40°C (hot air blower) or spread thinly in the sun. Mechanical as well as sun drying gives straw colored seed, brown colored seed should be discarded. The moisture content of the seed should be less than 7 percent in case of normal packing. For vapor proof packaging the seed moisture content should be less than 6 percent. On an average 110-150 seeds are obtained from a single fruit with an average yield of 600-700 gram of seeds from 100 kg of fruits. If the plant population is high, seed yield up to 125-140 kg ha^{-1} can be obtained from California Wonder or Yolo Wonder.

CONCLUSION

Bell pepper varieties have evolved from local land races to advanced hybrid development. Use of biotechnology will likely aid in the future development of improved bell pepper varieties. Besides conventional breeding methods, molecular genetics, tissue culture, and genetic engineering are becoming increasingly important. Hybrid seed production in many vegetable crops has produced a substantial increase in vegetable production. Exploitation of heterosis in vegetable crops has attracted attention of vegetable breeders to the vegetable requirements of the increasing population. There are alternative systems to produce hybrid seed, but the production of bell pepper hybrid seed commercially still relies on making crosses between two parents by conventional hand pollination, a very labor intensive and expensive process. The future outlook for bell pepper is good.

REFERENCES

Agrawal, S., Chandra, N. and Kothari, S.L. (1989). Plant regeneration in tissue culture of pepper (*Capsicum annuum* L. cv. Mathania). *Plant Cell Tissue and Organ Culture*, 16, 47-55.

Ahmed, N., Singh, J. and Bajaj, K.L. (1984). Genetics of capsaicin content in chilli (*Capsicum annuum* L.). *Journal of Research, Punjab Agricultural University*, 21(3), 465-466.

Bassett, M.J. (1986). Breeding Vegetable Crops. AVI Publications. The AVI Publishing Company, Inc. West Fort, Connecticut.

Bose, T.K., Som, M.G. and Kabir, J. (1993). Vegetable Crops of India. Naya Prakash, Calcutta, India.

Bosland, P.W. (1992a). Chillies a diverse crop. *Hort. Tech.*, 2(1), 7-10.

Bosland, P.W. (1996). *Capsicums*, innovative uses of an ancient crop. In: J. Janik (ed.), Progress in New Crops, ASHS Press Arlington, VA, pp. 479-487.

Bosland, P.W. and Votava, E.J. (2000). Peppers, vegetable and spice *Capsicums*. CABI Publishing, New York, USA.

Breutis, G. and Pochard, E. (1975). Essin de fabrication de l'hybride 'Lamuyo-INRA' avec utilization d'une sterile male genigue (ms). *Ann. Amehor. Plantes*, 25, 399-409.

Chauhan S.V.S. (1980). Effect of malic hydrazide, FW-450 and Dalapon on anther development in *Capsicum annum. J. Indian. Bot. Soc.* 59: 133-136.

Csillery, G. (1989). More efficient pepper hybrid seed production by double male sterile mother lines. In: EUCARPIA VIIth Meeting on Genetics and Breeding on *Capsicum* and Eggplant, 27-30 June, 1989, Kragujevac, Yugostavia, pp. 129-133.

Daskalov, S. and Mihailov, L. (1988). A new method for hybrid seed production based on cytoplasmic male sterility combined with a lethal gene and a female sterile pollenizer in *Capsicum annuum* L. *Theoretical and Applied Genetics*, 76: 530-532.

Diku, S.P. (1976). Hybrids of sweet pepper bred using male sterility. *Bynll. Vses, Nar. Rasteinnieved*, 64: 18-19.

Diku, S.P. and Analkecko, V.S. (1975). Heterotic hybrids of red pepper bred using male sterility. In: Testy Doks knuf Selektsiyaigenet Ovosch. Kultur, Vol. 3, Suinea, Kishiner, Moldavian, SSR, 71.

Dumas de Vaulx, R., Chambonnet, D. and Pochard, E. (1981). *In vitro* culture of pepper (*Capsicum annuum* L.) anthers, high rate plant production from different genotypes by +30°C treatments. *In French Agronomie*, 1(10): 859-864.

Fari, M. (1986). Pepper (*Capsicum annuum* L.). In: Biotechnology in Agriculture and Forestry, Vol. 2, Crop I, Berlin, Germany, Springer Verlag, pp. 345-362.

Gill, H.S., Asawa, B.M., Thakur, P.C. and Thakur, T.C. (1977). Correlation, path coefficient and multiple regression analysis in sweet pepper. *Indian Journal of Agricultural Sciences*, 47(8): 408-410.

Gopalkrishnan, T.R., Gopalkrishnan, P.K. and Peter, K.V. (1987). Heterosis and combining ability analysis in chilli. *Indian J. Genet. Plant Breed.*, 47(2): 205-209.

Govindarajan, V.S. (1985). *Capsicum* Production Technology, Chemistry and Quality. Part I, History, Botany, Cultivation and Primary Processing, CRC, *Critical Reviews in Food Science and Nutrition*, 22(2): 109-125.

Greenleaf, W.H. (1986). Pepper Breeding. In: Breeding Vegetable Crops, Westport, USA, AVI Publishing Co., pp. 67-134.

Harini, L. and Sita, G.L. (1993). Direct somatic embryogenesis and plant regeneration from immature embryos of chilli (*Capsicum annuum* L.). *Plant Science*, 89(1): 107-112.

Hirose, T. (1957). Studies on the pollination of red pepper. Flowering and the germinability of the pollen. *Plant Breeding Abst.*, 585/1959.

Johansen, W. (1960). In: Classic Paper in Genetics, Peters, J.A. Ed., Prentice Hall, Engevead, Cliffs, NJ, pp. 960.

Joshi, S. and Munshi, A.D. (2000). *Capsicum* seed production. Summer School on Advances in Hybrid Seed Production Technology, 25 April to 25 May, 2000 in the Division of Seed Science and Technology, IARI, New Delhi, India, pp. 193-196.

Joshi, S. and Singh, B. (1980). A note on hybrid vigour in sweet pepper (*Capsicum annuum* L.). *Haryana J. Hort. Sci.*, 9: 90-92.

Joshi, S. and Singh, B. (1983). Genotypic and phenotypic paths to fruit yield in sweet pepper (*Capsicum annuum* L.). *Progressive Horticulture*, 15(3): 222-225.

Joshi, S., Thakur, P.C. and Verma, T.S. (1995). Hybrid vigour in bell shaped paprika (*Capsicum annuum* L.). *Vegetable Science*, 22(2): 105-108.

Kalloo, G. (1986). In: Vegetable Breeding, Vol 1, CRC Press Inc. Boca Raton, Florida.

Kalloo, G. (1996). Solanaceous Vegetables. In: 50 years of Crop Science Research in India, ICAR Publication, pp. 574-592.

Khalil, R.K., Malash, N.M. and El-Sayed, M.M. (1987). Manifestation of heterosis in first generation hybrids in pepper (*Capsicum annuum* L.). In: Scientific International Technical Development Symposium on Hungarian Paprika (red pepper), 17-19 September, Kalocsa Szeged, 1987, pp. 100-113.

Kim, Y.C., Park, G.H. and Choi, S.H. (1985). Interrelationships and path coefficients of pericarp characters in red pepper (*Capsicum annuum* L.). In: Korea Research Bulletin of the Institute of Agril Sci. and Technol., Kyungpook National University, 2, 17-22.

Kohli, V.K., Dua, I.S. and Saini, S.S. (1981). Gibberellic acid as an androcide for bell pepper. *Scientia Hort.*, 15: 17-22.

Kormos, J. (1954). Investigations on sterility in paprika. *Ann. Inst. Biol.*, 22: 235-252.

Lee, Y.M. and Shin, D.Y. (1989). Genetic analysis of quantitative characters in diallel crosses of pepper (*Capsicum annuum* L.). In: *Korean J. Breed.*, 21(2): 138-142.

Legg, P.D. and Lippert, L.F. (1966). Estimates of genetic and environmental variability in a cross between two strains of pepper (*Capsicum annuum* L.). *Proc. Am. Soc. Hort. Sci.*, 89: 443-448.

Leskovar, D.L., Cantliffe, D.J. and Stoffella, P.J. (1989). Pepper (*Capsicum annuum* L.) root growth and its relation to shoot growth in response to nitrogen. *Hort. Sci.*, 64(6): 711-716.

Li, C. and Jiang, Z. (1990). The breed successful of 'Hai-hua-no 3' sweet pepper new variety by anther culture. *Chinese Acta Hort. Sinica*, 17(1): 39-45.

Lippert, L.F. (1975). Heterosis and combining ability in chilli peppers by diallel analysis. *Crop Sci.*, 15: 323-325.

Meshram, L.D. and Mukewar, A.M. (1986). Heterosis studies in chilli (*Capsicum annuum* L.). *Scientia Hort.*, 28: 219-225.

Mishra, S.N., Sahoo, S.C., Lotha, R.E. and Mishra, R.S. (1991). Heterosis and combining ability for seed characters in chilli (*Capsicum annuum* L.). *Indian J. Agril. Sci.*, 61(2): 123-125.

Morrison, R.A., Koning, R.E. and Evans, D.A. (1986a). Pepper. In: Hand Book of Plant Cell Culture, Vol. 4, Techniques and Application, New York, USA, Macmillan Publishing Co., pp. 552-573.

Morrison, R.A., Koning, R.E. and Evans, D.A. (1986b). Anther culture of an interspecific hybrid of *Capsicum*. *J. Plant Physiol.*, 126: 1-9.

Munshi, A.D. and Sharma, R. (1997). Nucleus seed production technique of chillies (*Capsicum annuum* L.). *Spice India*, November, 1997, p. 2.

Munyon, I.P., Hubstenberger, J.F. and Phillips, G.C. (1989). Origin of plantlets and callus obtained from chilli pepper anther cultures. *In vitro Cell Dev. Biol. J. Tissue Cult. Assoc.*, 25(3): 293-296.

Muthukrishnan, C.R., Thangaraj, T. and Chatterjee, R. (1986). Chilli and *Capsicum*. In: Vegetable Crops in India Solanaceous Crops. Calcutta, India, Naya Prakash, pp. 343-384.

Novak, F., Bettack, J. and Dubovsky, J. (1971). Cytoplasm male sterility in sweet pepper (*Capsicum annuum* L.). *Zeitschrift Pflanzenzuchtg*, 65: 129-140.

Novak, F.J. and Chemalo, V. (1975). The inheritance of features of the system of water balance in peppers (*Capsicum* spp.) II. Genetic analysis in a diallel cropping system and composition of 2X and 4X forms. *Bull. Vyjk. Ustav. Zeitudfsky*, 19: 95.

Ochoa-Alejo, N. and Ireta-Moreno, L. (1990). Cultivar differences in short forming capacity of hypocotyl tissue of chilli pepper (*Capsicum annuum* L.) cultured *in vitro*. *Scientia Hort.*, 42: 21-28.

Pandey, S.C., Pandiata, M.L. and Dixit, J. (1981). Line x tester analysis for the study of combining ability in chilli (*Capsicum annuum* L.). *Haryana Agril. Univ. J. Res.*, 11(2): 205-212.

Peterson, P.A. (1958). Cytoplasmically inherited male sterility in *Capsicum*. *Amer. Nat.*, 92: 111-119.

Pickersgill, B. (1977). Chromosomes and evolution in *Capsicums*. In: Pochard, E. (ed.), Capsicum 77. Comptes Rendus du., Montfavet Avignon, pp. 27-37.

Pickersgill, B. (1991). Cytogenetics and evolution of *Capsicum*. In: Chromosome Engineering in Plants, Genetic, Breeding Evolution. Pt. B. Amsterdam, Netherlands, Elseiver Science Publishers, pp. 139-160.

Poncavage, J. (1997). The OG Book of Peppers. *Organic Gardening*, 44(2): 30-36.

Popova, D. (1963). Studies on the effect of the age of pollen and egg cell in pollination of blossoms on the heterosis in green pepper. *Comptes Rendus Acad. Bulg. Sci.*, 16(3): 317-320.

Poulos, M.J. (1994). Pepper breeding (*Capsicum* spp.): Achievements, Challenges and Possibilities. *Plant Breeding Abst.*, 64(2): 143-154.

Ribeiro, A., Costa, C. and De, P. (1990). Inheritance of pungency in *Capsicum chinense* Jaeq. (Solanaceae). *Ravista Brasileira de Genetica,I* 13(4): 815-823.

Salgare, S.A. (1989). Induction of male sterility by dalapon in *Capsicum frutescens* L. *Adv. in Plant Sci.*, 22: 527-529.

Sharma, P.P. (1995). Production of hybrids and hybrid seed of *Capsicum*. Hybrid Vegetables and Their Seed Production Training Course. FAO, Division of Vegetable Crops, IARI, New Delhi, pp. 34-37.

Shifriss, C. (1973). Additional spontaneous male sterile mutant in *Capsicum frutescens* L. *Euphytica*, 22: 527-529.

Shifriss, C. (1997). Male sterility in pepper (*Capsicum frutescens* L.). *Euphytica*, 93: 83-88.

Shifriss, C. and Eidelman, E. (1986). An approach to parthenocarpy in peppers. *Hort. Sci.*, 21(6): 1458-1459.

Shifriss, C. and Frankel, R. (1969). A new male sterility gene in *Capsicum frutescens* L. *J. Am. Soc. Hort. Sci.*, 94: 385-387.

Shifriss, C. and Frankel, R. (1971). New source of cytoplasmic male sterility in cultivated peppers. *J. Heredity*, 62: 254-256.

Shifriss, C. and Gurl, A. (1979). Variation in stability of cytoplasmic genic male sterility in *Capsicum frutescens* L. *J. Am. Soc. Hort. Sci.*, 1044: 94-96.

Shifriss, C. and Pilovsky, M. (1993). Digenic nature of male sterility in pepper (*Capsicum annuum* L.). *Euphytica*, 67: 111-112.

Shifriss, C. and Rylsky, I. (1972). A male sterile (ms-2) gene in California Wonder pepper (*Capsicum annuum* L.). *Hort. Sci.*, 7: 36.

Shoemaker, J. and Teskey, B.J.E. (1955). Practical Horticulture. John Wiley and Sons Inc. New York.

Shull, G.H. (1914). The genotype of maize. *Am. Nat.*, 45: 234.

Singh, A. and Singh, H.N. (1977). Note on heritability, genetic advance and minimum number of genes in chilli. *Indian J. Agril. Sci.*, 47(5): 260-262.

Somos, A. (1984). The Paprika. *Akademiai Kiado*, Budapest.

Sripichitt, P., Nawatu, E. and Shigenaga, S. (1987). *In vitro* shoot forming capacity of cotyledon explants in red pepper (*Capsicum frutescens* L. cv. Yatsufua). *Jap. J. Breed.*, 37(2): 133-142.

Tanwar N.S. and Singh S.V. (1988). Indian minimum seed certification standards. The central seed certification board, Dept. of Agriculture and cooperation, Ministry of Agriculture, Govt of India. New Delhi.

Tewari, V.P. (1990). Development of high capsaicin chillies (*Capsicum frutescens* L.) and their implications of export. *J. Plantation Crops*, 18(1): 1-13.

Venkata Rao, P. and Chonkar. V.S. (1984). Genetic analysis of fruit weight and dry matter content in chilli. *South Indian Hort.*, 32(1): 26-32.

Verma, T.S. and Joshi, S. (2000). Spice Crops of India. Kalyani Publishers, Ludhiana, India.

Verma, T.S., Joshi, S., Ramesh Chand and Lakhanpal, K.D. (1994). Effect of debudding and defruiting on the seed yield of sweet pepper var. California Wonder. *Vegtable Sci.*, 21(1): 45-49.

Wafer, I. (1699). A new voyage and description of the Isthmus of America, reprinted from the original edition. Burt. Franklin, New York, p. 107.

Watts, L. (1980). Flower and Vegetable Plant Breeding. Grover Book, London.

Woong Yu, Ii (1990). The inheritance of male sterility and its utilization for breeding in pepper (*Capsicum* spp) PhD dissertation Kyung Hee University Korea p. 69.

Zatyko, L. Csillery, G. and Tobias, I. (1983). The role of resistance in the integrated control of diseases and pest in pepper. *In* Proceedings of an International Conference on Integrated Plant Protection, 4-9 January 1983, Budapest, Hungary. pp. 42-45.

Current Trends in Cabbage Breeding

Zhiyuan Fang
Yumei Liu
Ping Lou
Guangshu Liu

SUMMARY. Cabbage is an important vegetable crop of the family Brassicaceae and is grown in many countries of the globe. There are different types of cabbage depending upon shape, size, and color. It is a highly cross-pollinated crop where heterosis in F_1 hybrid progeny has been exploited so largely in developing countries too. Due to high yield, strong disease resistance, wide adaptability, good quality, and uniform economic characters. The rate of conversion from open pollinated varieties to hybrid varieties and seed replacement ratio is very high in this crop. There are different temperature requirements for fresh vegetable and seed production purposes. The self-incompatibility and male sterility systems are present in the crop, which facilitates easy and cheaper hybrid seed production. For the successful seed production program we need suitable production area, congenial environment, and stable parents. The present review deals with the vegetative, reproductive biology, breeding methods, heterosis breeding, maintenance and multiplication of self-incompatible, and male sterile lines bio-technology as well as the commercial approaches for hybrid seed production of the crop. *[Article copies available for a fee from The Haworth Document Delivery Service: 1-800-HAWORTH. E-mail address: <docdelivery@haworthpress.com> Website: <http://www.HaworthPress.com> © 2004 by The Haworth Press, Inc. All rights reserved.]*

Zhiyuan Fang, Yumei Liu, Ping Lou and Guangshu Liu are affiliated with the Institute of Vegetables and Flowers, Chinese Academy of Agricultural Sciences, Baishiqiao Road 30, Beijing 100081, Peoples Republic of China.

[Haworth co-indexing entry note]: "Current Trends in Cabbage Breeding." Fang, Zhiyuan et al. Co-published simultaneously in *Journal of New Seeds* (Food Products Press, an imprint of The Haworth Press, Inc.) Vol. 6, No. 2/3, 2004, pp. 75-107; and: *Hybrid Vegetable Development* (ed: P. K. Singh, S. K. Dasgupta, and S. K. Tripathi) Food Products Press, an imprint of The Haworth Press, Inc., 2004, pp. 75-107. Single or multiple copies of this article are available for a fee from The Haworth Document Delivery Service [1-800-HAWORTH, 9:00 a.m. - 5:00 p.m. (EST). E-mail address: docdelivery@haworthpress.com].

http://www.haworthpress.com/web/JNS
© 2004 by The Haworth Press, Inc. All rights reserved.
Digital Object Identifier: 10.1300/J153v06n02_05

KEYWORDS. Cabbage, breeding methods, hybrid seed production, hybrid variety

INTRODUCTION

Head cabbage (*Brassica oleracea* L. var *Capitata*), commonly called cabbage is a member of *Brassica oleracea*, belonging to Cruciferae family. The chromosome pattern is $2n = 2x = 18$.

Cabbage cultivars descended from wild non-heading *Brassica* and originated in the eastern Mediterranean and Asia Minor. The Greeks and Romans have cultivated several wild varieties since at least 2000 B.C., and later Romans disseminated cabbage throughout Europe. After a long time of artificial cultivation and selection, variation within and between subspecies of *B. oleracea*, such as cabbage, collard, Brussels sprouts, broccoli, cauliflower, and kohlrabi, came into use (Figure 1). The heading types were created in Europe in the 13th century, and was introduced to Canada and China in the 16th century, to U.S. in the 17th century, and introduced to Japan in the 18th century.

Cabbage, known as a highly nutritious vegetable, is used in slaw, sauerkraut, and cooked dishes and its outer leaves are also used as animal fodder.

Cabbage has wide adaptability, high disease and stress resistance, high yield potential and strong transporting tolerance, which makes it the most widely growing vegetable in the world. Cabbage is cultivated in more than eighty countries in according to the FAO (1999).

The classification of cabbage is mostly based on the botanical characteristics and the shape of head.

According to its botanical characteristics, cabbage can be classified as:

- *Common cabbage*, smooth leaves, green to dark green with protruding central rib, no apparent crinkle. It is the most common cultivated variation of *Brassica oleracea.*
- *Purple cabbage*, the leaf blade is the same as common cabbage, but its outer leaves and the inner leaves are purple. Usually for fresh use and far less cultivated than common cabbage.
- *Savoy*, with crinkled or curling outer leaves that have deep veins, but the color of leaves looks the same as common cabbage, good flavor and can be used as a cooked dish, also less cultivated.

The common cabbages are of three types based on head shape:

- *Drum head*, most of the drum headed varieties are medium or late maturing, resistance to immature bolting, diseases, and heat. When they are cultivated in spring, they hardly show any immature bolting.

- *Round head*, round head type has round and nearly round head with characteristics of early or mid-early maturing, hard and crisp head of good quality. But this type is prone to immature bolting than pointed type if it is sown early or improperly managed in early spring.
- *Pointed head*, pointed head cultivars are mostly early maturing, have strong immature bolting tolerance and chilling resistance, but are susceptible to disease and heat stress

BIOLOGY OF CABBAGE

Botanical Characteristics

Root: The cabbage root is characterized as subulate taps root system, with a hypertrophy taproot on the base and root tip forward. A lateral or secondary root is grow at the base of taproot and numerous fibrous roots grow on the taproot or lateral root. The major roots penetrate the soil to a depth of 60 cm and concentrate within a radius of 30 cm. The root system is spread to a semidiameter within 80 cm. Cabbage root has a high regeneration system so it is easy to transplant.

Stem: There are two types of stem, short stem in the vegetative stage and flower stalk in the sexual reproduction stage. The stem remains short in the vegetative stage although prolong in the rosette stage or head forming stage weakly. The short stem with outer head leaves is called the outer short stem, and inner short stem with inner head leaves, is called the core. The core length is one of the standards to test quality; usually the cabbage with short core also has a packed head with good quality. After the vegetative stage the plant begin to bolt and form flowering stalk with branches and leaves, then the inflorescence is formed.

Leaf: In the different development stage the plant has several types of leaves, such as cotyledon, basal leaves, seedling leaves, skin-forming leaves, head leaves and stalk leaves, varying in the shape. The cotyledon is oppositely arranged with nephric shape. The first true leaves (basal leaves), is paired and vertical with cotyledon and long leaf petiole; The seedling leaves ordinarily have ovate or oblong shape, reticulate venation, and the apparent petiole grows on the short stem. As the plant grow the rosette leaves, appears. The early varieties usually have 12-16 outer leaves and the medium or late varieties have 18-30 leaves. From the frame stage to the form of head, the outer leaves become wider and wider and leaf margins reach the base of leaf petiole, sessile leaves come into being. All this can be used as the index for planting management. Leaf color varies from light green to dark green; some also with red or purple leaves. Most of leaf surface is glabrous with no leaf hairs and

FIGURE 1. Variation of wild cabbage under artificial selection and breeding. 1. Wild Cabbage, 2. Kale (2A branching types, 2B non-branching types, 2C marrow like types, 2D high-stem types), 3. Cauliflower (3A biennial types 3B annual types), 4. Cabbage, 5. Savor, 6. Kohlrabi, and 7. Brussel sprouts

some types have crinkle leaves. The surface of some leaves is covered with a powdery substance, which has a positive correlation with the drought and heat resistance. The leaf midrib grows inwards and covers the axil, which forms a solid head as they develop. The leaves, which form the head, have yellow white color and are with no petiole. Stem leaves (growing on the flowering stalk) are arranged in an alternate fashion and its small leaves have short petiole or no petiole.

Flower. The first flower branches are generated in the leaf axil of the main flower stalk and the second in the leaf axil of the first ones. Plant growing under sufficient nutrient and good management practices can generate even the third and fourth branches.

Due to their differences in branching habit, the stock plants of different ecotypes vary greatly in plant shape. Round-headed varieties have strong main stalk and a few weak branches. In the beginning of bolting, there is only a main stalk and the first branches generate slowly. Comparing to the round-headed varieties, the pointed or drum-headed varieties have relatively weaker main stalk, but a larger number of strong branches that include even the third ones. In the same variety, the habit of bolting and branching varies among individuals.

A healthy cabbage stock plant can generate 800-2000 flowers depending on the varieties and agronomic management.

The flowering occurs on the main flower stalk first, followed by the first order of branches from top to the bottom and the second, third and fourth branches in that order. On one anthotaxy, flowers open orderly from the bottom to the top.

Flower time varies greatly in different varieties. Generally, varieties with point or drum head, flower 7-15 days earlier than those with round head under similar agronomic management conditions, flowering lasts for 30-50 days in a given variety.

Cabbage has a complete flower with calyx, corolla, androecium, and gynoecium. Each calyx has four sepals, and the corolla is with 4 petals are arranged in a cross. Two short and four long stamens are found inside the petals with each having an anther on the top. Maturing pollen grains are released by anther dehiscence.

Cabbage is a typical, cross-pollinated plant with strong self-incompatibility. Mainly insects mediate natural pollination. The percentage of hybrid seeds can be as high as 70% when two varieties are planted together and pollinated naturally.

Stigma receptivity and pollen viability are highest on the first day of flower opening. A stigma can be fertilized 6 days before or 2-3 days after flower opens. Pollen are viable 2 days before or 1 day after the flower opens, but the viability can be maintained for more than 7 days if pollen grains are stored in a desiccated state at room temperature and even longer at temperatures below 0°C.

It takes about 36-48 h from pollination to fertilization at optimum temperature of 15-20°C. When the pollen grains land on the stigma, it takes 2-4 h for the pollen tube to begin elongation; 6-8 h later the pollen tube penetrates the stigma tissue and fertilization is accomplished within 36-48 h thereafter. Pollen germination slows down below 10 and the normal fertilization is affected at temperature higher than 30°C

Fruit and seed. The fruit of cabbage is glabrous silique like beads, with two rows of seeds lying along the edges of replum.

There are 600-1500 pods on each plant, depending on management practices. The available pods are concentrated mainly on the first order of branches and then on the second order and main branches.

There are approximately 20 seeds in each pod. The seed number is less in pods on the top, but more in pods in the middle and bottom. The color for mature seeds is reddish brown or dark brown e seeds and the 100 seed weight is 3.3-4.5 g. A healthy plant can yield about 40-50 g seeds.

Seed development of cabbage varies among the different heading types and varieties. The maturing time for round-headed type is longer than the time for pointed or drum-headed types, and is much faster at high temperatures compared to low temperature.

Dry and cool environment is favorable for seed storage. Fully matured seeds can be stored for 2-3 years under favorable room condition, but only 1-2 years in humid conditions. If the seeds are stored dry, high germination rate can be ensured for 8-10 years.

GROWTH AND DEVELOPMENT

Cabbage is a cool season biennial crops. The plant starts from seeds and produce vegetative structures and food storage organs in the first season. After low temperature in the winter to complete vernalization and next spring's long photoperiod, flowers, fruit and seeds develop to complete the cycle of growth from seed germination to seed production. The whole life cycle can be divided into vegetative growth stage and sexual reproductive development stage (Figure 2).

VEGETATIVE STAGE

Germination Stage

From sowing to the unfolding of first pair of leaves is called germination stage, commonly referred to as the frame. The time to complete this stage depends on the season, usually 8-20 days in summer or autumn and 15-20 days in winter or spring. From germination stage to the first few days after emergence, the new plant is nourished by stored reserves present in the cotyledons; so fully matured seeds and fine seedbed are required to ensure good germination.

Seedling Stage

The stage lasts from the first true leaves unfolding to the first leaf whorl forming and till the head shell or outer skin come into being. Ordinarily 25-30 days are required to complete this stage depend on the planting date. The fertilization and irrigation are needed to acquire good seedlings.

Rosette Stage

This stage starts from the appearance of the second leaf whorl and lasts till head formation requiring about 25-40 days and varies in different varieties. The leaf and root grow fast at this stage, so it is important to control the fertilization and irrigation in order to force the root development downwards and hinder the outer leaf growth too faster. This is the base of forming big and packed heads.

Head Formation

Firm heads begin to form filled with sessile fleshy leaves, usually in about 25-30 days depending upon the varieties. Irrigation and band fertilization can help the head leaves from expanding and forming the packed head.

SEXUAL REPRODUCTIVE DEVELOPMENT STAGE

Bolting Stage

From the transplantation of seed stock to the flower stalk bolting, bolting usually lasts for 25-30 days.

Flowering Stage

Flowering begins with the opening of the first flower and ends with the last flower. Flower time varies greatly in different varieties. Flowering usually lasts for 25-40 days.

Ripening Stage

The ripening stage begins with the petals falling from the last formed flower and ends when lower pods had been ripened and fully changed yellow, in about 40-50 days.

Cabbage can only be vernalized at the seedling stage after the plant has attained a certain size. Usually stem diameter, number of leaves or leaf area is used for describing the stage to complete vernalization. The vernalization re-

FIGURE 2. Life cycle of cabbage. 1. Over wintering with tiny seedling, 2. Over wintering with normal seedling (flower differentiation under low temperature), 3. Pre-maturity bolting (high temperature and long photoperiod), 4. Bolting inside the head, 5. Heads formation from axillary buds, and 6. Flowering (high temperature and long photoperiod)

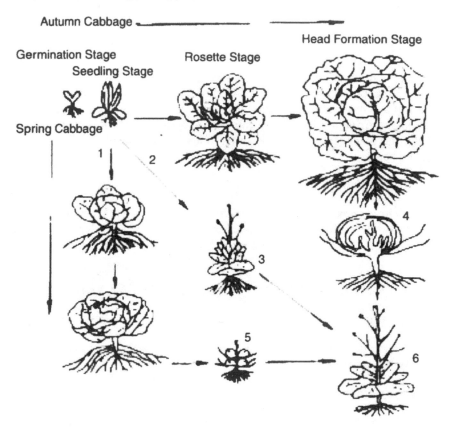

sponse varies among different cabbage cultivars and types. To vernalize different varieties need not only be given different duration of low temperature exposure but different plant size is also an important factor. For early maturing varieties proper vernalization needs only a stem diameter more than 0.6 cm, leaf number more than seven and 45~50 days of low temperature exposure. But medium or late maturing varieties need only a stem diameter more than 1 cm, leaf number 10~15 and 60~90 days duration of low temperature exposure. Cabbage easily bolt early after complete vernalization in spring.

As to the effect of day length on the bolting of cabbage, long days promote bolting, but the photoperiodic response varies greatly in different ecotypes.

Generally, point and drum-headed varieties are less sensitive to day length than round-headed ones. Round-headed varieties have a strict need for day length to bolt and flower in spring.

Environmental Conditions

The Requirement of Temperature, Irrigation, and Light

Temperature: Cabbage is a cool season crop, with proper resistance to chilling and high temperature. The optimum temperature range for cabbage production is 15 to 25°C, but optimal temperature varies with growth stage, being 7-25°C during outer leaves growth, 15-20°C during head formation. Optimum temperature is 18-25°C, at which germination will occur in 2-3 days after sowing.

The tolerance to high temperature also varies in different developmental stage. In the seedling and rosette stage, cabbage can bear temperatures up to 25-35°C. In the head formation stage, it requires cool temperature and high temperature block head formation. In drier atmospheres the leaves tend to be more distinctly petiolate, the quality of head is impaired and much of its delicate flavor is lost and even it makes the head split. Extremely high temperature also hinders pollen growth and fertilization, which lead to reduced seed number per flower or fruit.

Cold tolerance becomes stronger depending upon the grown stage. The healthy plants with 6-8 leaves can withstand temperatures as low as −1°C to −2°C for a long time and −3°C to −4°C for a while. The seedling after low temperature exposure even can bear freezing as low as −8°C to −10°C temporarily. Although the head is less low temperature tolerant than the seedling, it also can grow slowly at the temperature of about 10°C without causing damage to the curd. The head of early varieties can bear −3°C to −5 °C while the medium and late varieties can bear up to −5°C to −8°C. In the bolting and flowering stage, the plant has poor low temperature tolerance, and when the temperature is below 10°C, the plants bear abnormal seed. The flowering stalk can be destroyed if exposed to temperatures of −1°C to −3°C.

Irrigation: The plant is shallow rooted and with big outer leaves, so it does not tolerate water stress in light to medium-heavy soil. The atmospheric moisture content ranging from 80-90% and 70-80% field capacity is optimum plant growth. Water stress can be significantly hinder the growth and head formation, so irrigation in time is needed when the weather and soil become drier. Cabbage is also very sensitive to waterlogged soil, if the roots are under water for a long time, plants will die.

Light: Light intensity and duration are important for cabbage growth and development in the vegetative stage. Summer cabbage can have a good yield when intercropped with a high crop such as maize in the high temperatures. Long photoperiod weakly promotes the bolting and flowering of stalk plant.

Requirement for Soil and Fertilizer

Soil: Cabbages grow well on a wide range of soils from light sand to heavier clay. The soil pH should be in the range 5.5-6.5 for ideal growth but it is not tolerant of acidic soils. If the pH is below 5.5, the disease club root can appear. So application of lime and micronutrients is necessary for the acidic soils. Cabbage is also salt tolerant and can form head normally in saline soil.

Fertilizer: Cabbage is a heavy feeder on fertilizer and nutrients. Water, nitrogen, phosphorus, and potassium in the soil are absorbed and transported to the leaves, and these elements are needed for efficient light harvesting, to complete the transformation of carbon dioxide to sugars, and for transport, storage, and utilization of sugars to form root, stem, leaf, and head.

Cabbage is a heavy user of nitrogen, especially in the seedling stage and rosette stage. The nitrogen is absorbed in the form of nitrate and ammonia in the soil. Some of them are used for synthesis of amino acid and protein. Less nitrogen is usually applied in the fertile soils and more in lean soil.

Phosphorus is essential for vigorous and healthy growth at all stages of development. The demand for P is greater during head formation. In poor soils, application of low phosphorus results into stunting.

Cabbage requires little K at the beginning of growth but needs more and more after the head has been formed and the requirement reaches its peak at the harvest time.

In the whole life cycle the requirement of NPK is in the ratio of 5:6:4.

Some minerals, although required by plants in amounts much less than for NPK, can be critical for productivity, for example, calcium (Ca). Calcium can make the process of metabolism go smoothly by neutralizing potassium, buffering acidity, and detoxifying organic acids. Calcium deficiency affects the margins of fast growing leaves, leading to tip burn. The deficiencies of magnesium, boron, manganese, molybdenum, and iron also can lead abnormal patters of growth although the plant needs them in minor quantities.

BREEDING AND UTILIZATION OF HETEROSIS IN CABBAGE

Breeding new varieties is an important way to improve yield, quality, and wider adaptability of cabbage. Many countries, such as The Netherlands, Japan, USA, China, and Russia, pay more attention to breeding and utilization of heterosis in cabbage. Usually the Institutes and Universities focus on the study of the genetic resources and materials for breeding, and the seed companies use these materials to breed new varieties.

Breeding Objectives of Cabbage

High, reliable yields: Yield, the traditional first priority for breeders, is still a major goal, although now most breeders now pay more attention to quality and other parameters.

Quality: A new variety should have properties like crisp compact heads, core length within half of the head height, good flavor and high nutrient value.

Disease resistance: Clubroot (*Plasmodiophora brassicae*), black rot (*Xanthomonas campestris pv. campestris*), fusarium yellows (*Fusarium oxysporum* f. *conglutinans*), downy mildew (*Peronospora parasitica*), turnip mosaic virus (TuMV), powdery mildew (*Erysiphe polygoni* D.C.), bottom rot (*Rhizoctonia solani* Kuhn) and tip burn are the main diseases of cabbage, so breeding multiple-disease resistant cabbage is also a very important breeding objective.

Resistance to early bolting: Cabbage is a major spring consumed vegetable throughout the world. Early bolting can cause heavy economic losses to both growers and breeders, especially in the year with cold current in spring season, so it is important to breed early bolting resistant varieties suitable for spring growing.

Cold and heat tolerance: In order to breed cabbage with heavy yield and early maturity characteristics cold and heat tolerance is required. Usually sowing, and transplanting is done earlier and the cabbage is left in the field during winter, this requires good cold resistance; Meanwhile for the rainy and hot weather in summer and autumn it requires heat tolerance

Storage ability: Cabbage is usually stored in the winter, which is also needed to be carried from the field to the market; so compact head and storage ability are also one of the important breeding goals.

Continuity: In order to have a highly diversified all year round cabbage, breeding of early, mid, and late type completes the set of varieties in cabbage besides sowing at different time.

Crop uniformity: A uniform brassica field makes grading much easier and reduces harvest time. The final objective is to have a single-harvested field of uniform quality.

Appearance: Color and shape is another important trait. The rapid transformation of the vegetable market with improved packing and display facilities, the large offer of commodities all the year round, and increased presence of colorful and appealing fruit and salad vegetables has forced brassica producers to raise their presentation and quality standards.

Breeding Methods for Cabbage

The main breeding methods include pedigree selection, combination breeding, utilization of hybrid vigor, and utilization of biotechnology for breeding.

Pedigree Selection

The purpose of selection is to point out the most promising genotypes in the genetic resources using single and mass selective methods to breed new variety. Besides breeding new variety, the materials with special characteristics can also be selected by pedigree selection. For example, single or multiple disease resistance, stable self-incompatible lines, stable male sterile lines and inbred lines with good quality.

Combination Breeding

Crossing two genotypes of the same species, then testing in different generations, helps to acquire the new varieties at last. Combination breeding is also applied to breed new parental lines and original materials.

Utilization of Hybrid Vigor

A cabbage hybrid possesses the characters of uniformity, high yield and disease resistance, and being able to fully display the parent's valuable characteristics. Utilization of hybrid vigor is a main method to improve cabbage production by many countries in the world. Early in 50s and 60s Japan and US began to use cabbage hybrid varieties widely, and West Europe, North America, and Asia also have applied hybrid varieties since 1970s. There are two breeding systems available in *Brassica*, self-incompatibility and male sterility

Utilization of Biotechnology in Cabbage Breeding

In recent years, biotechnology, integrated with classical breeding, has improved the breeding level of cabbage greatly. Biotechnology includes cell engineering, chromosome engineering, genetic engineering, and molecular markers. For example, using protoplast fusion the US researchers successfully worked out the problem of pale-yellow leaves accentuated at low temperature when using cabbage CMS line; Chinese researchers have shortened the years for cabbage breeding through microspore embryogenesis, and the transgenic pest resistant cabbage has also been reported in China.

Hybrid Vigor Behavior in Cabbage

Growing Vigor

Comparing to its parent lines, the F1 hybrid has fleshy cotyledon, thick stem, and the leaves grow faster than parent lines. All this characters benefit for heavy yield.

Yield Vigor

Many previous tests for heterosis in cabbage showed that there was significant yield heterosis in F1 hybrids. The authors tested for heterosis in cabbage with 345 F1 hybrids in 1982 and the results showed that 90.43% F1 hybrid yielded higher than the parental lines, and 8.41% between the parental lines and 1.16% lower than the parental lines.

It was also observed that hybrids between genetically distant groups or with different head types showed greater heterosis than within the group combinations, for example the round head type crossed with the drum head type or the pointed head type. However, these different types do not flower simultaneously, so for the process of seed production we need to regulate the flowering time. The crossing with same head type but with difference in head color or other characters, such as dark green leaves crossed with yellow green leaves, green with dark green, also can give good yield vigor. The combination of this types is easy to use in seed production as their flowering time meet well and the F1 hybrid have more uniformity.

Disease Resistance Vigor

Through inbreeding we can separate some material having high disease resistance. By using this high resistant materials as the parent lines to produce hybrid, we can acquire the variety which have higher disease resistance.

Early Maturity

Among the early mature varieties, we select the inbred strains, which have good early maturity characteristic. Then using these inbred lines as parent lines, we can develop new early maturing F1 hybrid.

Quality

The F1 hybrids usually have good commercial quality such as desirable head shape, uniform head, good head density and short core length. The nutritive value of F1 hybrid lies between the parents line and can be improved through proper selection.

Approaches to Hybrid Seed Production

It is very important to produce F1 hybrid seed in cabbage by hand emasculation and pollination because of its small floral organs. Therefore, the application of scientific, rational and operative methods to produce hybrid seed is the key issue for utilization of heterosis in cabbage. Theoretically, hybrid seed production in cabbage can be achieved through the following approaches.

Intervarietal Crossing or Crossing Between Inbred Lines

Two suitable varieties or normal inbred lines are selected as parents to produce F1 hybrids. Then the parental lines are planted together at a ratio of 1:1 in the same isolated plot for open pollination. The seeds harvested are F1 hybrid seeds. However the hybrid purity can reach only 70%, which has restricted the use of this method in hybrid seed production.

Application of Marker Characteristics at the Seedling Stage

Hybrid seedlings may be distinguished from their parents by differences on morphological characters of the parents and F1 hybrids, but typical and identical marker characters have not been found in cabbage so far. Even if some characters are identified they cannot be used as markers because of their genetic complexity.

Using the Self-Incompatible Line for Hybrid Production

As self-incompatibility widely exists in cabbage stable self-incompatible lines are used as the parents to produce hybrid. In the conditions of simultaneity in flowering time the seeds from parents are all high purity hybrid F1 seeds, so it costs less and easy to use. Until now, almost all cabbage hybrid seeds have been produced using self-incompatible lines all over the world.

Using Male Sterility for Hybrid Seed Production

The instability and complex inheritance of the self-incompatibility mechanism makes its use difficult. Pollination by hand in bud stage to propagate parent lines may cost mass labor force and the parent lines after repeated self-pollination may lose vigor, but use of male sterile lines can remove all these disadvantages. There are four types of male sterile lines in cabbage, two types of male sterility–cytoplasmic male sterility and dominant genetic male sterility–have recently become economically significant, and are used by plant breeders in some countries. The main advantage to use male sterile line lies in two aspects: the low labor cost and high purity hybrid.

For the reason of their importance and wide application in cabbage hybrid seed production now, use of self-incompatible and male sterile lines are discussed in details.

Breeding of Self-Incompatible Lines

Self-incompatibility widely exists in cabbage and it is quite easy to breed a self-incompatible line in cabbage. An incompatible line with very low seed-set or even no seed-set when pollinated between plants within the same line can be developed by continuous self-pollination and oriental selection. Furthermore, when two self-incompatible lines are used as parents to produce hybrids, the reciprocal crossed seeds can be harvested as hybrid. Nagaoka No.1, which was the first hybrid in cabbage in the world, was developed through self-incompatible lines in Japan in 1995. From 1960s to 1970s, self-incompatible lines were used extensively in the production of hybrid seed in various countries in Europe, America, and Asia.

The superior self-incompatible lines for hybrid seed production should possess the following characters.

Stable self-incompatibility: A self-incompatible line requires that each flower set not more than one seed by flower pollination between plants within the same inbred line. Only two self-incompatible lines with stable self-incompatibility are used as parents and for obtaining pure hybrid seed.

High seed-set of self-pollination at bud stage: Self-incompatible lines can not set seeds by self-pollination at flowering stage, but can do so at bud stage. Therefore, self-incompatible lines can be propagated with self-pollination at bud stage. The better the seed-set at bud pollination is for propagation of self-incompatible lines, the lower the cost of seed production of parental line. At least 5-10 seeds per flower should be set by selfing at bud pollination.

Favorable and uniform economic characters: Less growing vigor degeneration from selfing can ensure that F1 hybrids possess superior and uniform economic characters.

Desirable combination ability: This can ensure that F1 hybrid have stronger heterosis. So we should choose the perfect self-incompatible line from the following aspects.

Choice of Self-Incompatibility

Firstly, the individuals with commercial characters or good combination ability are selected, and the plant has to be covered with paraffin bag before the bud open. After pollination the flower must be covered with paraffin bags immediately. When the fruit are ripening the seed set is investigated and the individual with low SI degree are selected. The average number of seeds per developed pod after self-pollination in flowers (SF) and buds (SB) were determined. The relative number of seeds $Rs = SF/SB \times 100$ or only the SF number are chosen as a measure of SI degree. A plant is considered SI if the Rs ranges from 0.0 to 20.0 or SF from 0.0 to 3.0, as partial SI with Rs from 20.1 to 80.0 or

SF from 3.1 to 5.0 and the completely self-compatible (SC) with Rs above 80 or SF above 5.

In order to acquire the offspring of the self-incompatible individuals, self-pollination is needed at the bud stage with film bag; meanwhile flower pollination is also done in other branches at the same time, because the self-incompatible line can set seeds at the bud stage. The self-incompatible line also can separate at the next generation so we need continuous flower pollination and test the self-incompatibility, usually the line with SI index below 1 and high SF plant is left until they acquire the stable self-incompatibility.

A self-incompatible line requires that each flower set no seeds by flower pollination with the same line and also need that no seeds are set by pollination between plants within the same line. This is important to reduce the false seeds in seed production. There are two ways to test the self-incompatibility between the plants within the line. One is through mixed-pollination, mixed pollen is collected from the same line but from different individuals (4-5 plants) for sister pollination at flowering stage, if they all show self-incompatibility they are also self-incompatible between plants within the same line. The other way is by paired pollination; paired pollination is done between the sister plant within the line at the normal flowering stage. If there are still some combinations showing self-compatibility, we should continue to select until the line acquires stability.

Tips for the Selection

1. Self-incompatibility should be accurate, so tweezers and hands have to be sterilized with alcohol when pollination of one line is finished and changed to another. The flower used must be fresh and it is preferred to use the flowers opening in the day or previous day, usually about 30 flowers are tested. We should pay attention to the unsuitable conditions, it affects the self-incompatibility. For example, it is not suitable to pollinate at high temperature or under humid conditions.
2. For the inbred lines with good commercial characters and combination ability, but self-compatible, they shouldn't be eliminated through selection early because they can be separated by continuous self-pollination.
3. Some plants show self-incompatibility at the normal flowering stage but at the flowers produced late and the flowers on the top of the flowering branches can also set seeds. These plants can affect the production of the F1 hybrid seeds, so usually should be discarded.

The Choice of Commercial Characters and Self-Pollination Decline Degree

The lines with good characteristic should be selected first and then individuals within the line. Besides of the integrate characteristic, the plant with special characters should also be selected. The selected line needs to have a good combining ability. The numbers for selection depend on the purity and degree. Better individual within the line is preferred to be selected first. Big populations are needed when the material is not pure and should be selected more strictly for an individual. More individuals and less line populations are needed when the material is pure.

When selecting the decline degree the material with the slow and less decline degree are selected. It is important to overcome the self-pollination decline.

Choice of Seed Set Characteristic at the Bud Stage

Self-pollination is often made at the bud stage for the seed production of self-incompatible line. So a good self-incompatible line also needs better seed set characteristic at the bud stage. It shows that the characteristic of seed set at the bud stage are quite different between the inbred lines and have the property of inheritance. We can acquire up to ten seeds per pod of the plant and its offspring with high bud seed set ratio. Some plant with poor bud seed set ratio can only set only few seeds per pod, sometime no more than one seed per pod can be obtained. So the plants with high bud seed set ratio are usually selected and the lines with poor seed set are discarded early.

Breeding and Utilization of Male Sterile Lines

Hybrid seeds production can reach 100% purity at low cost by using male sterile line as female parents. The first recessive nuclear male sterility was reported by Cole (1959), Nicuwhof (1961), and Sampson (1966). However, no marker genes linked to male sterility were found, so it was difficult to use this kind of male sterility. The second type of male sterility controlled by nigra nuclear-cytoplasm interaction was reported by Pearson (1972). The defect of this type of material is that the petals cannot open fully and the nectar gland is less developed, and doesn't attract insect pollinators. The third type, Ogura male sterility, was transferred from radish cytoplasm. The main problem of such type is chlorosis at low temperatures and suppressed nectar gland development. In recent years, scientists in France and US have used the protoplast fusion to improve male sterile lines with radish cytoplasm. While the chlorosis problem at seedling stage has been solved, the improved materials showed reduced nectar production, inability to attract insects and low seed setting by open pollination. The fourth type is the dominant male sterile gene in cabbage. The result from trials of several years verified that a pair of dominant nuclear genes controls this type of male sterility gene. This male sterile material has been used to develop homozygous dominant male sterile lines. Several superior hybrids from male sterile lines have been developed and used for commercial seed production.

By now, the most widely used male sterile lines in the hybrid seed production are improved radish CMS and dominant GMS.

Breeding and Utilization of Improved Radish CMS

Male sterility conferred by ogura cytoplasm of *Raphanus stativus* has been transferred to many crops through repeated back crossing and selection. A limitation to this system is much-reduced nectarines and the frequent occurrence of malformed flowers. So it is still not widely used in breeding. Now scientists have acquired new improved CMS without the above problem by protoplast fusion, and have also transferred it to cabbage by backcrossing. As the new improved CMS line also has the problem of dead buds at the flowering stage, which is caused by negative effect of cytoplasm, so it is not widely used in the production of cabbage seeds.

More attention should be paid to the following aspects when a radish CMS is selected.

Firstly, the male sterile line with good commercial character and high nectar production should be selected as the female parent for backcross.

Secondly, the inbred line with good commercial character should be act as male parent, and after repeated backcross for 5-6 generations, choose the normal sterile plant as female parent in each generation.

Thirdly, during the repeated backcross, the plant with good combination ability should be chosen at the same time.

The parents are planted at the ratio of 1:3 for hybrid seed production. When flower period ends the male parents are removed so that we get the seeds only from female parents. Other management is the same as the self-incompatible lines.

Utilization and Breeding Dominant Male Sterile Line in Cabbage

Fang et al. (1997) first found dominant male sterility in cabbage from a natural population. The sterility was controlled by one dominant gene, and are environmental sensitive. The male sterile lines could become partial fertile at relative lower or higher temperature situations. By now DGMS lines have been used to produce new cabbage varieties, and are also being used for hybrid seeds production in China (Figure 3).

The key to produce hybrid seeds using DGMS line as following:

Firstly, acquire the pure dominant male sterile line by repeated backcross, selfing and side across; then cross the male sterile lines with inbred lines with economical characteristic, and acquire the hybrid combination. After combining ability and field-testing, the best combination is selected.

FIGURE 3. The cabbage flower of a dominant genetic male sterile line (DGMS)

The technique for Fl seed production and field management using DGMS line are the same as CMS lines.

SEED PRODUCTION OF CABBAGE

Seed Production of Cabbage Variety

The Choice of Seed Production Area

Since cabbage is a typical biennial plant that vernalize in winter and flowers in the next spring, two main climatic factors should be taken into consideration while selecting the area for cabbage seed production. The most important one is the temperature during winter, which is critical for vernalization. The other one is the precipitation in the season for flowering, seed maturing, and harvesting.

Vernalization should occur for a period of more than 60 days at a low temperature. On the other hand, since the temperatures below − 10 °C are generally injurious to the plant, the safe temperature of stock plant for over wintering should be higher than this limit.

In tropical or subtropical areas, the temperature in winter is too high for cabbage to pass through the vernalization stage, and can cause irreversible injury to the over wintering stock plants. So these regions are not suitable for cabbage seed production. Most areas of commercial cabbage seed production are in the temperate zone where it is easy to find favorable temperature conditions for vernalization and safe over wintering of cabbage stock plants.

Precipitation mainly affects seed setting. Rain at flowering stage limits the activity of pollinating insects, causes dropping of flowers, and consequently reduces seed yield. Excessive rainfall during seed maturity and harvesting may cause many kinds of diseases like seed rot, and pre-harvest sprouting in pods, causing deterioration of the seed quality.

The Methods of Cabbage Seed Production

Three major methods, based on their over wintering, is applied for cabbage seed production.

Maturity stock plant for seed production in autumn: The seeds of cabbage variety are sown in time and the head formation is completed before winter; the stock plants are selected for next year's seed production. The stock plants for storage are either cut apart at the head or not cut. For heads, which had not been cut the whole stock plants were kept, to over wintering, and at the next spring a crosscut was made on the head with knife to help easy bolting and flowering. There are two ways to cut the head, one is to remove the outer head leaves and cut the outer part of the head, leaving the stock plant with the core of the head. The other is to cut the whole head and just leave the root for over wintering, and the branches arising from the root were used for seed production in next spring.

The advantage of this method is that we can select the plant strictly according to their characteristic, so we can get high amount of pure seeds. We usually use this way to produce the parent lines for autumn cabbage. As we cannot select the wintering and head characteristic in spring; it is not suitable for the seed production of spring cabbage.

Pre-maturity stock plant for seed production in autumn: Sow the seeds of the cabbage in late autumn and let the plant over wintering before they reach full head formation stage, and next spring for seed production. The advantage of this way is that it costs less and high seed yield is obtained, but the stock plant cannot be selected the strictly. We usually use this way to produce the seed of varieties.

Using root for seed production in spring: The plants are selected from a cabbage field in spring, the heads are cut and the root and the rosette leaves are kept, when the axillary buds grow 4-6 leaves, a portion of stem is removed with leaves attached, and then the lower leaves are removed and planted in a rooting medium. After the stem tip cutting is done, it should be shielded to avoid sunlight and rain, and it is watered everyday to keep the soils moist. After head formation in autumn, the plant are stocked for over wintering to produce seeds in next spring. These methods can keep the good characteristics of spring cabbage, but it costs high and needs more labor. The maturity stock plant and root methods are used alternately in seed production of cabbage.

Tips for Seed Production

Cultivation for stock plant: The sowing time varies according to different varieties and area; the most suitable sowing time is to keep the plant at semi-head stage before over wintering. Before sowing, the plots for seedbed should have fertile soil with high position. Before field planting, transplanting is needed when the seedlings have 2-3 leaves, and to the field when the seedlings have 7-8 leaves. Control of cabbage worms and aphids is very essential besides water and fertilizer management.

Stock selection: The stock plants can be selected at seedling stage, the head formation stage and bolting and flowering stage. The plants with the characteristic of the varieties are selected. To keep the characteristic of the variety, the selected plants should be over fifties and pollinated with mixed pollen.

Over wintering management: The plant can over winter in the open fields in warm regions while in the cold regions protection facilities are required to help cabbage stock over winter.

Transplanting and Field Management

To acquire pure seeds, the plot for cabbage seed production should be isolated at least 1000 meters where no cauliflower, kohlrabi, broccoli, kale or other cabbage types are present. The stock plants are planted in field in spring in the cold area with a density of 60,000 plants/ha. After planting, the field should be covered with film, which can help the plants to recover and get high seed production. After watering two inter-tilling in time is necessary to increase soil temperature and promote the development of root. Trimming the head with cutter can help the stock plant bolting. After bolting, the old and infected leaves are removed. Proper watering is needed in the flowering stage, usually once in 5-7 days in the blooming stage. In the pod set stage watering should be decreased and watering should be stopped when the pod turns yellow, but in the drier area with high temperature, proper watering is still needed. Aphids and cabbage worm control are important during the whole growth cycle.

Harvest of Seeds

Harvest the seed as the pods turn pale or become dry. The pod should be harvested in the mid-morning to avoid the pod splitting. Plant are dried in time to avoid rain, the seeds should also be dried in proper time and excess exposure to sunlight should be avoided.

HYBRID SEED PRODUCTION IN CABBAGE

Propagation of Basic Seeds for Parents

Nursery of Stock Plants Before Winter

Stock plants can only bolt and flower after vernalization at low temperature, therefore it has to be grown in the autumn of previous year. The sowing time is dependent on maturity of parents. Normally late sowing is recommended because parents have weak stress resistance after continuous selfing. Maintenance of small head has the advantages of reducing disease damage at seedling stage by late sowing. Loose heads is formed with resistance to diseases and suitable for over winter storage. The stock plants grow in the coming year and produce high yield of hybrid seeds. However, the sowing date cannot be too late otherwise the head is so small that proper vernalization cannot be achieved. Flowering and seed setting are also affected. Generally, seeds of self-incompatible lines with medium and medium-late maturity can be sown for raising seedlings in the mid-July or directly sown in the early August. Seeds of self-incompatible lines with early and medium-early maturity are sown for raising seedlings in early August. Seedbeds ought to be selected in fertile soil at higher terrain. Before emergence, seedbed should be covered with shading net to protect the seedlings against storm and exposure to sun. Seedlings with 3-4 real leaves are transplanted at a plant spacing of 10 cm and finally transplanted into the open fields when 7-8 true leaves have emerged. If complete solid head is used to produce seeds, the plant and row spacing for mid-late and early maturing lines are 43×50 cm^2 and 33×40 cm^2, respectively. In case of small head, the plant and row spacing for mid-late and early-maturing lines are 36×43 cm^2 and 27×33 cm^2, respectively. Spraying chemicals in addition to water and fertilizer management during whole growing season should control cabbage worms and aphids.

Stock Plant Selection for Seed Production

Generally, stock plant selection is conducted at the seedling stage, head formation and bolting.

Selection at seedling stage: The seedlings are selected as stock plants with no disease infection, vigorous growth and characteristics of inbred lines such as, leaf shape, leaf color, leaf margin, leaf surface wax, and leaf stalk, etc.

Selection of complete solid head and small head before winter: The healthy plants with uniform outer leaves and head are selected as stock plants.

Selection in the bolting and flowering stage: The stock plants are selected according to the branching habit, color of flower stalk and leaflets on the stem, bud shape and bud color at the flowering stage. Off types are eliminated.

Stock Plant Storage in Winter

In temperate zone, stock plants can be over wintered in the open field. For exceptional cold seasons, it is necessary to cover the stock plants with soil or other materials. In cold temperate zone, the selected stock plants are required to be moved for over winter storage such as, cold bed, small tunnels, temporary transplanting, pit storage, etc.

Heeling in the cold bed or small tunnels: The selected stock plants are dug up together with root. The dry and infected leaves are removed. The stock plants are temporarily planted in the cold bed or small tunnels and irrigated. The plants should be covered with straw mat or reed mat when the temperature is lower than 0°C at night. The mats should be rolled up during the day time for ventilation in order to keep the temperature at 0°C at night and 15°C in the day time in the cold bed and small tunnels.

Trench storage: The trenches with 80-90 cm in depth and 100 cm in width are dug up in the higher terrain for easy drainage. The stock plants are dried in the sun for 3-4 days after harvest and stored in the trench before the ground freezes. The stock plants are covered with soil or other materials gradually as the air temperature decreases. The thickness of covering material depends on the prevailing temperature. The principle of storage is to protect the stock plants against both cold damage and decay because of very thin or very thick covers.

Storage in vegetable cellar: The harvested stock plants are stripped off their dry and infected leaves. The stock plants are dried in the sunlight for 3-4 days after harvest and stored in the vegetable cellar before the ground freezes. For efficient use of the space in the cellar, a kind of shelf made of wood stick or bamboo can be made in the cellar. The stock plants are put on each layer of the shelf for storage. The temperature is kept at about 0°C and the humidity at 80-90% within the vegetable cellar.

Management of Stock Plant Transplanting

Basic stock plants of self-incompatible lines are mostly propagated under protected conditions such as cold bed and small tunnels. This method has the following advantages: it is easier to isolate from other cabbage seed production plots; stock plants can flower earlier; it is easier to find labor to do the hand pollination before a busy season. Under protected conditions, stock plants are transplanted in the period of February. A corridor should be left for hand pollination when stock plants are transplanted at a plant space of 30 cm and row space of 30-40 cm. The temperature should be kept at 5-10°C at night and about 15°C in the daytime from transplanting to bolting and budding. During this period, controlled irrigation, and inter-tilling should be adopted in order to increase soil temperature and stimulate the growth of root system. In the bolting, flowering and pollination stages, fertilizer and irrigation must be adequately supplied. Ventilation in greenhouse and small tunnels is required to keep the temperature at about 10°C at night and 25°C in the daytime. At the same time, some measures must be taken to control cabbage worm and aphids. Greenhouse or mini-tunnels have to be covered with insect-preventing net to avoid contamination.

At the bolting and flowering stage, the selection of stock plants has to be carried out according to the flowering characters in order to ensure seed purity.

Bud Pollination

Care must be taken not to damage stigmas during stripping. It is difficult to obtain seeds of self-incompatible lines by self-pollination and sister pollination at the flowering stage. Their basic seeds can be propagated by hand pollination at bud stage (Pearson, 1931). The procedure is as follows: the bud top is removed by tweezers and strippers to expose stigma, which is then pollinated with pollen collected from the same line. It requires specially trained persons to do the pollination carefully to avoid contamination. Tweezers and hands have to be sterilized with alcohol when pollination at one line has been finished. The seed production plot must be covered with net to avoid contamination by bees or other insects. Pollen must be freshly collected from the opened flowers on the same day or the day before. If the pollen is collected two days before pollination the seed setting will be enormously reduced. The mixed pollen collected from the same line should be used for pollination to avoid viability depression from continuous selfing. If the bagging isolation is applied to propagate the basic seeds, the flowering branch has to be covered with paraffin bag before the bud opens. When the flowers of the lower inflorescence bloom, bud pollination should be started by using the mixed pollen collected from the same plant under a covered bag. After pollination, the flowering branch has to be covered with paraffin bags immediately. The flower buds need to be stripped tenderly and carefully without twisting the flower stalk. The bud size neither too small nor too big should be selected for pollination. Bud pollination conducted 2-4 days prior to flowering can give the highest seed set. According to the bud position on the flowering branch, the bud pollination carried out between 5th-20th flower buds above the blooming flower on the inflorescence can be the best for seed setting. But for less vigorous plants, the bud pollination made between 5th-15th flower buds above the blooming flower can set more seeds.

Propagation of basic seeds of incompatible line by bud pollination is labor-intensive and costly. In consideration this disadvantage, the electricity-aided pollination (Roggen and Van Dijk, 1973), wire brush pollination, thermal-aided pollination (Roggen and Van Dijk, 1976), CO_2 enrichment (Nikanishi, Esahi and Hinata, 1975), etc., have been suggested. However, each one of them has its limitations and has not yet been used on commercial scale. In recent years, spraying a solution of 5% common salt has been used to overcome the self-incompatibility and increase the seed set by scientists in China. This method has succeeded in the propagation of basic seeds.

Harvesting the Basic Seed

Harvesting will start 70-80 days after pollination when the seed pods start getting yellow. If the seeds are not mature and are harvested too late, the seeds will be lost due to pod split. The seeds of some self-incompatible lines have the tendency to germinate within the pod; hence they should be harvested at an earlier stage. After drying, the basic seeds should be maintained in desiccators or in fully sealed container to ensure high germination rate. Specially trained persons should manage seed harvesting drying and conservation. During the whole process, mechanical mixture should be avoided.

Production of F1 Hybrid Seed

The F1 hybrid seed production using self-incompatible approach involves seedling cultivation at fall season, selection of stock, over-wintering of stock plants, flowering in the following spring, cross-pollination, seed set, and harvesting of hybrid seed. The whole process takes about 10 months. Each step should be managed properly to obtain high yield and quality of seeds.

Seedling Cultivation for Stock Plants

Sowing: As cabbage requires vernalization, the seedlings of a certain size are needed for vernalization. As the parental lines for F1 hybrid seed production are selected from continuous selfing, their tolerance to stress is relatively poor. Relatively late sowing helps to reduce disease infection at seeding stage, makes stock plants produce less tight heads favoring storage during winter and vigorous growth in the coming year. However, if it is sown too late and the seedlings grow too small, it affects elimination of off types and bad types. The worst leads to a large number of plants not getting vernalized causing abnormal flowering and seed-setting. The most suitable sowing time is to keep stock plants at semi-head stage before over wintering. Parental lines with middle and late maturity would be sown from late July to early August. For early maturity parent lines, it is better to do sowing from middle to late August. Before sowing, the plots for seed beds should have fertile soil with facilities easy irrigation and drainage. The previous plots used for cabbage seed production should not be chosen as seed beds in order to avoid mixture of remaining seeds from previous crop with parental lines. A level seedbed that is 1.0-1.3 m wide and 7-10 m long is formed. Between beds, a ditch measuring 30-40 cm wide and 10-15 cm deep is dug and linked with field drainage canal. To prevent seedbed against strong sunshine and heavy shower, shading booth should be established over seedbed. The shading booth is covered with reed mat or reed curtain or shading plastic film. If there are showers, it is necessary to add plastic film on the top of booth and protect seedlings from rain damage. The lower edge of shading booth should be at least 30 cm from seedbed in order to have fresh air flow and prevent high temperature during days with strong sunshine. The reed mat or curtain should be covered or removed, in proper time. On clear days, it should be covered at 9 a.m. and removed at 5 p.m. During cloudy days, it is not necessary to cover. When all seedlings have emerged, the booth should be removed. If the soil is rich in nutrients, it is not necessary to apply fertilizer on seedbed. If fertilizer is needed, it should be mixed completely. Before sowing seeds, the seedbed should be fully irrigated. A thin layer of soil should be added on seedbed after irrigation water seeps into the soil. Sowing seeds in a volume of 2-3 g/m^2 is recommended and a layer of 0.5-1 cm depth is covered with sieved soil.

Transplanting: Before field-planting transplanting is needed to get uniform seedlings. The transplanting bed should be prepared similar to that of the seedbed. When seedlings grow up 2-3 leaves, transplanting can be done with spacing of 10 × 10 cm^2 and preferably on a cloudy day or late afternoon. Watering should be done immediately after transplanting. Bed should be shaded for 3-5 days and then irrigated once for seedling establishment. Inter-tilling and hardening are required, and spraying is important to control aphids and cabbage worm at the seedling stage.

Field planting: The field plots for fixed planting should have rich soil, easy irrigation and drainage. If the previous crop was not a crucifer crop, it could reduce disease damage. Base fertilizer should be applied in the field plots in advance and the surface leveled after the till. Field planting should be undertaken when seedlings reach 6-7 leaves. Seedlings with roots in a lump of soil will lead to quick recovery after planting. The density for parental lines with middle or late maturity is about 35 × 45 cm^2 and for early parental lines 27 × 33 cm^2. It is better to plant on a cloudy day or late afternoon to protect them from strong radiation. Planting and watering should be done at the same time. After seedling establishment, inter-tilling and hardening are done at proper time. The field management for hybrid seed production will not be the same as that of cabbage production in autumn that requires much more watering and fertilizer to produce big cabbage heads. Control of viruses, black rot, downy mildew, etc., is very essential. More attention should be paid to control cabbage caterpillars, aphids, flea beetle, and cabbage moth. Before winter, a less tight head should be formed, otherwise the stock plants are too small and will affect the bolting, flowering, and seed-setting in the coming spring. When autumn temperature is low, more careful management should be taken to obtain less tight heads.

Stock Selection

It involves eradication of off-types and undesirable plants in the field. The off-type plants, which have no characteristics of parents, may come from biological mixture during reproduction of basic seeds or physical mixture during harvesting and drying in the sun. The undesirable-type plants refer to diseased, weak and damaged plants. Elimination of off-types and undesirable-types can be done when seedlings are transplanted. When seedlings grow up and the canopy of seedlings has not yet completely covered plots, extraordinary big or small plants and diseased plants are removed. When the plants reach a lotus flower shape, inspection should be done. All plants that have no characteristics of parents should be eliminated firstly based on leaf shape and distribution of parents should be eliminated firstly based on leaf shape and distribution of leaf vein, and secondly on leaf color and waxy appearance on leaf surface. The proper time to discard the off-types and undesirable-types is when the stock plants have formed small head before over-wintering because stock plants have showed characteristics of parental lines under normal cultivation management. All eradication should be done before frost occurs. Otherwise frost-damaged plants show abnormal leaf shape and color, which influence stock selection. When plants start to flower, final rejection should be done based on performance of leaf color, leaf shape, plant vigor, branching, bud shape, bud color, petal color, etc., in comparison with the parental characters.

Over Wintering Management of Stock Plants

The process of managing the stock plants of cabbage for over wintering depends on climate of different regions. Plants can over winter in the open field of warm temperate regions where the lowest temperature is not less than $-10°C$ while in the cold temperate regions the protection facilities are required to help cabbage stock over winter.

Over winter in the open field: Normally cabbage stock plants have strong tolerance to cold and can even survive -6 to $-8°C$ for a short time after low temperature hardening. However, the plants get damaged if low temperature persists for a longer time. The following measure can be taken to prevent the stock plants from cold damage during over wintering.

Ridge culture: The stock plants are grown on the slope of ridge from east to west so that ridge can be used as a wind proof. By this way plants are placed in warm and enough sunshine microenvironments, which help stock in safe over wintering.

Proper watering: During over wintering cabbage stocks still grow slowly. Before severe winter, watering not only meets the water demands of plants but also avoid sharp decreases of soil temperature.

Proper hilling-up: After watering in the winter it is important to intertill in order to keep soil moisture and hill-up the stem base of stock plants. The proper hill-up height can be as high as half or one-third of the plant height.

Over wintering in storage: There are several methods for cabbage over wintering in the protected areas such as small tunnel, cold bed, storage hidden in the earth, pit storage, etc.

Mini-tunnel: The parental line with round head type should be grown in the small plastic tunnel for over wintering. Because this type of parents needs strong sunshine during over wintering, they can bolt and flower well in the coming year. The tunnel is covered with plastic film and is further covered with straw at night, which is removed during the daytime. Stock plants can receive enough sunshine and grow at the proper temperature. The intensive management should follow according to weather change during winter. In the tunnel, high temperature and spindling should be prevented. Tight head formation is not good for bolting and flowering in the coming year. The stock plants are planted in small plastic tunnels when freezing appears in the open field. After planting watering is needed. Several days later, inter-tilling should be done to enhance growth of root system. Irrigation and inter-tilling should be done to enhance growth of root system. Irrigation is done according to soil condition in order to meet water demands of plants.

Cold bed: Before over wintering, stock plants are transplanted into cold bed. Watering is necessary before severe winter comes and also the cold bed is required to be covered with straw mat or reed curtain at night and removed in the daytime to receive sunshine. In the early spring when the weather gets warm, more attention should be paid to prevent stock plants from exposure to heat. All mats on the bed must be removed and dried in the sun during the daytime. When the bed is covered with straw at night a small outlet is left. Otherwise, plants exposed to heat which leads to leaf drop and thereby affecting seed yield.

Storage hidden in the earth: This method is suitable for parents with flat round head and sharp round head types. In this method the use soil environments with rather stable temperature and moisture is made. In this way, cold damage to stock plant can be prevented for safe over wintering. Normally, a higher place is chosen to dig a ditch on the shadow side. The ditch, which should be under the frozen layer, is one meter wide, and the depth depends upon the local climate. All stock plants are placed in the ditch with top of their heads in a position between frozen layer and unfrozen soil. Spacing among plants helps to avoid heat damage to stock plants. As temperature goes down, more and more soil is covered to protect the stem base and root system from freezing damage. If the ditch is too long or too wide, it is important to put a bundle of corn straw in the middle at a certain interval in order to be well ventilate it and reduce temperature and moisture in the ditch. To meet the seedling size for vernalization, the stock plants to be buried should have small loose head. According to weather changes the covered soil is added or removed to allow stock plants safely over winter.

Pit storage: For this type of storage, all stock plants are put in the semi-underground tunnel with a skylight. Inside the tunnel there are 3-4 shelves, 50 cm wide and 150 cm high made of bamboo or wood sticks. Between shelves there is 0.8-1.0 m space for corridor. In each self, 3-5 layers of stock plants are stacked to be well ventilated and rearranged. All stock plants are harvested at 15 days before soil gets frozen and dried in the sun for 2-3 days. All old and infected leaves on the stock are removed. The stock plants are stacked in a way of putting lower parts with roots toward the center and head parts outwards. When the temperature falls below 0°C, the stock plants are moved into the tunnel and rearranged within each other for 15 days. Attention should be paid to ventilation and temperature control in the tunnel during storage. Stock plants for pit storage are required to be as big as that for buried storage. According to weather change, temperature should be controlled by ventilation, and freezing protection should be ensured. Furthermore, moisture should be monitored to prevent drying of the stock roots. Otherwise, re-rooting will be affected in the coming year.

Transplanting Methods and Management of F1 Hybrid Production Plots

Transplanting of stock plants can be done either in the open field or in the protected area. If the flowering of female and male parents is synchronous, it is better to produce F1 hybrids in the open field. If not, F1 hybrids should be produced in the protected cultivation such as cold bed, and mini-tunnel. The plant ratio of female to male is 1:1 and planted in alternate rows. If the plant vigor of both parents differs from each other greatly, ratio might be 2:2 which is in favor of insect pollination. For the parents which have different seed set, it is recommended that parent with better seed set should be planted more with maximum ratio of 2:1. For F1 seed collection, row spacing is 50 cm plant spacing is 33-37 cm. In the warm temperate region where F1 hybrids are produced in the open field normally stock plants are planted in the field before winter comes. After over winter it is not necessary to plant them again. In the early spring before stock plants turn green, old and infected leaves are removed. Proper watering and inter-tilling with fertilizer application are required to enhance early bolting and flowering. In the case of stock plants which are put in the cold bed, buried storage and pit storage. 60 t/ha of base manure should be applied before stock plants are planted in the field in plots of 10 m length with 1 m width are prepared after leveling the field. For ridge cultivation, ridge width is about 50 cm. Although the planting plots are prepared differently in different regions of China, rational density should be considered. For a combination of mid and late maturity 60,000 plants/ha is recommended. For early combination 67,500 plants/ha is suggested. Several aspects for planting stock in the early spring are: (1) plants in cold bed are transplanted with soil which can protect injury of roots, (2) if the soil in the plot has proper moisture, drip irrigation can be adopted to maintain soil temperature and quick rooting of stock plants, and (3) plastic mulch 10-15 days before planting is suggested to increase soil temperature to help quick rooting.

If the F1 hybrids are produced in the cold bed or mini-tunnel, stock plants should be planted 15 days before soil gets frozen. Generally 1.233 m width of cold bed for 4 rows and 1.66 m for 5 rows of plants are planted in a space 30-40 cm between plants. For the combination of parents flowering at the same time, the female and male are planted in the direction of south to north at a ratio of 1:1 of 2:2. If the two parents flower at different times, plants are grown in the direction of east to west; late flowering parent in the north of cold bed and early flowering parent in the south so that flowering time of both parents can be adjusted. The plant ratio is still 1:1 of 2:2. Before winter comes, watering and inter-tilling are necessary to keep moisture. In the coming spring before the stock plants turn green, watering can increase soil temperature and help plants to recover. Compound fertilizer can be applied before watering. At the time when stock plants start to bolt, aphid is a major problem and should be well controlled. When the stock plants are in early flowering, sufficient watering and fertilizer should be applied in order to meet the nutrition and water demands of bolting and flowering.

Special Considerations for F1 Hybrid Production Plots

Isolation should be strictly guaranteed: Other *Brassica* lines easily pollinate self-incompatible lines of cabbage. Therefore, hybrid production plot should be isolated by at least 2000 meters where no cauliflower, kohlrabi, broccoli, kale, brussels sprouts, etc., are planted. Especially early bolting plants of early cabbage, cauliflower and kohlrabi in the commercial production plot around F1 hybrid seed production plots should be prohibited.

Bee population should be large enough: To improve seed set it is necessary to have sufficient bee population for pollination. Based on experience, at least 15 boxes of bee must be placed in one hectare area for pollination, which is helpful to increase hybridization, and seed yield.

Build up framework to prevent lodging: The cabbage stock plants get easily lodged during flowering and pod setting. The lodging affects insect pollination and also causes pod mildew, which decreases seed yield and germination. Therefore, at early flowering stage, a framework should be built up with bamboo sticks and straw rope or stick stands beside each plant in order to prevent lodging.

Insect and disease control: Major pests are aphids, cabbage worm and the main diseases are black rot, viruses, and damping-off at seedling stage and powdery mildew at flowering stage, the control measures include spraying chemicals besides using resistant parents and avoiding continuous cropping with cabbage crops.

Adjustment of both parents' flowering time: If flowering time of both the parents does not coincide then adjustment of both parents' flowering time is necessary to ensure seed yield and quality. The most important issue is to select inbred lines with similar flowering characteristics as parents. If not possible, overlapped flowering time of both parents can be obtained by adjusting the sowing time, transplanting time and stock plant trimming which leads to synchronization of flowering.

Harvest of F1 Seed

If parents are self-incompatible lines, harvested seeds from both parents can be mixed up, but to improve the seed uniformity, it is recommended that F1 seeds of both the parents be harvested separately. When the pod gets waxy yellow and seeds at lower branches of plants becomes reddish brown, it is time to harvest. Plants are dried in time after harvest to prevent seeds from getting infected with mildew. If seeds are harvested prematurely and plants are stacked up too long after harvest, their germination will be adversely affected. Washing of seeds is not allowed if the cleanliness of newly harvested seeds is not good enough, otherwise, solute leakage from seed will reduce germination. After drying and cleaning, all new seeds should be stored in a dry and cool (0-4°C) room.

REFERENCES

Abdurashidov. DA. 1989. The effect of the preparation on the growth, development and seed production of white head cabbage. Turselektsiya I semenovodstvo ovoshchngkh I bakhchevgkh kul'tur, 46-51 Moscow, USSR.

Bannerrot, H., Loulidard, L. and Tempe, T. 1974. Cytoplasmic male sterility transfer from *Raphanus* to *Brassica*. In Eucarpia-Cruciferae Conference, pp. 52-54.

Chiang, M.S. and Jacob A. 1992. Inheritance of precocious seed germination in silique of cabbage. Canadian Journal of Plant science. 72(3): 911-913.

Cole, K. 1959. Inheritance male sterility in green sprouting broccoli (*B. oleracea* L.var. *italica*). Journal of American Society for Horticultural Science. 95(1): 13-14.

Dickson, M.H. 1997. A temperature sensitive gene in broccoli. Can. J. Genet. Cytol. 1: 203.

Earle, E. Stephenson, C. Walters T. and Dicksn, M. 1994. Cold-tolerant *Ogura* CMS *Brassica*. Vegetables for Horticultural Use. 16: 80-81.

Fang, Zh.Y., Sun, P.T., Liu, Y.M., Yang, L.M., Wang X.W. and Zhuang, M. 1997. A male sterile line with dominant gene (Ms) in cabbage and its utilization for hybrid seed production. Euphytica 97: 165-268.

Fang, Zh.Y., Sun, P.T. and Liu, Y.M. 1991. Cultivation Technology of Cabbage. Jindun Press, Beijing.

Fang, Zh.Y., Sun, P.T. and Liu, Y.M. 1983. Utilization of heterosis and some problems encountered in self-incompatible line selection in cabbage. Chinese Agricultural Science, 3: 51-61.

Kandic, B., Markovic, Z. Dinovic, I. Todorovic, V. and Stankovic, L. 1992. Manifestation of heterosis for various traits in F1 hybrids (*Brassica oleracea* L.). Savremena Poljoprivreda. 40: 154-157.

Len, H.N. 1988. Main obstacles in the utilization of male sterility in cabbage. Chinese Vegetables 4: 8-10.

Ludilov, V.A. 1995. Problems in the breeding and seed production of vegetable crops. Kartofel'I ovoshchi, No. 3, 29-30.

Nasrallah, M.E. and Wallace, D.H. 1968. The influence of modifier genes on the intensity and stability of self-compatibility in cabbage. Euphytica, 17: 493.

Nikanishi, T., Esashi, T. and Hinata, K. 1975. Self seed production by CO_2 gas treatment in self-incompatible cabbage. Euphytica, 24: 117.

Ogura, H. 1968. Studies on the new male sterility in Japanese radish with special reference to the utilization of this sterility towards the practical raising of hybrid seed. Mem Rac Agri Kagoshima University, 6(2): 39-78.

Pearson, O.H. 1972. Cytoplasmically inherited male sterility characters and flavor components form the species *Brassica nigra* (L.) Koch X B. *oleracea* L. J. Am. Soc. Horti. Sci., 97: 397-402.

Pearson, O.H. 1931. The influence of inbreeding upon the season of maturity in cabbage. Proc. Am. Soc. Hortic. Sci. 29: 359.

Pelletier, G., Primard, C., Vedel, F., Chetrit, P., Remy, R., Rousselle, P. and Renard, M. 1983. Intergenetic cytoplasmic hybridization in Crucifereae by protoplast fusion. Mol. Gen. Genet. 191: 244-250.

Polegaev, V.I. and Magomedov, I.R. 1990. Storability and seed production of head cabbage mother plants in relation to the time of retardant application. Vestiga Timirgazevskoi Sel'skokhozgaistvennoi Akademii. No. 6, 109-117.

Renard, M., Delourme, R., Mesquida, J., Pelletier, G., Primard, C., Boulidard, L., Dove, C., Ruffic, V., Herve, Y., and Morice, J. 1992. Male sterility and F1 hybrid in *Brassica*. In: Dattee, Y., Dumas, C., and Gallais, A. (eds.) Reproductive Biology and Plant Breeding. Springer, Berlin, Heidelberg, New York, pp. 107-109.

Roggen, H.P., and VanDijk, A.J. 1976. Thermally aided pollination: A new method of breaking self-incompatibility in *Brassica oleracea* L. Euphytica, 25: 643.

Roggen, H.P. and Van Dijk, A.J. 1973. Electric aided and bud pollination, which method to use for self-seeds on cole crops (*Brassica oleracea* L.). Euphytica, 22: 260.

Sampson, D.R. 1966. Genetic analysis of *Brassica oleracea* using genes from sprouting broccoli. Can. J. Genet. Cytol. 8: 404-413.

Sambandamurthi, S. and Sundaram K.S. 1998. A note on seed production in cabbage under Kodaikanal conditions. South Indian Horticulture. 37(3): 183-185.

Suteki, Shinohera, 1984. Vegetables Seed Production Technology of Japan. Tokyo Yohkendoh Co., Ltd.

Verma, L.R. and Partap, U. 1994. Foraging behavior of *Apis cerana* on cauliflower and cabbage and its impact in seed production. Journal of Apicultural Research 33(4): 231-236.

Xiaowu Wang and Zhiyuan Fang. 2000. An extend random primer amplified region (ERPAR) marker linked to a dominant male sterility gene in cabbage (*Brassica oleracea* var. *capitata*). Euphytica. 112: 267-273.

Current Researches in Hybrid Broccoli

Pritam Kalia
S. R. Sharma

SUMMARY. Broccoli is the most nutritious member of Cole group belonging to family Brassicaceae. Its marketable plant part is "head" composed of immature flowers or florets. It is very popular in the developed world because of its highly nutritional and medicinal properties. It is recommended for consumption as a measure to decrease the incidence of human cancer. Different types of broccolis are available but green types are more popular. Genetic resistance is also available for many diseases. Self-incompatibility and male sterility are also present in the crop. Commercial heterosis is reported by many workers. The use of self-incompatible and cross-compatible lines and genic and cytoplasmic male sterile lines helps in the seed production of hybrid varieties. Many hybrids have been released for commercial cultivation and hybrid seed production is successfully exploited. *[Article copies available for a fee from The Haworth Document Delivery Service: 1-800-HAWORTH. E-mail address: <docdelivery@haworthpress.com> Website: <http://www.HaworthPress.com> © 2004 by The Haworth Press, Inc. All rights reserved.]*

KEYWORDS. Broccoli, Brassicaceae, genetics, heterosis, self-incompatibility, male sterility, seed production, hybrid seed production

Pritam Kalia and S. R. Sharma are affiliated with the Division of Vegetable Crops, Indian Agricultural Research Institute, New Delhi 110 012, India.

[Haworth co-indexing entry note]: "Current Researches in Hybrid Broccoli." Kalia, Pritam, and S. R. Sharma. Co-published simultaneously in *Journal of New Seeds* (Food Products Press, an imprint of The Haworth Press. Inc.) Vol. 6, No. 2/3, 2004, pp. 109-134; and: *Hybrid Vegetable Development* (ed: P. K. Singh, S. K. Dasgupta, and S. K. Tripathi) Food Products Press, an imprint of The Haworth Press, Inc., 2004, pp. 109-134. Single or multiple copies of this article are available for a fee from The Haworth Document Delivery Service [1-800-HAWORTH, 9:00 a.m. - 5:00 p.m. (EST). E-mail address: docdelivery@haworthpress. com].

Digital Object Identifier: 10.1300/J153v06n02_06

INTRODUCTION

Broccoli (*Brassica oleracea* L. var. *italica* Plenck.) is an Italian word from the Latin brachium meaning an arm or branch (Boswell, 1949) and is used in Italy to refer to young edible floral shoots on brassicas, but commonly refers to sprouting and heading forms of *italica* group of *Brassica oleracea* L. (2n = 2 x = 18). It is a cool season crop and its most types are biennials. It is the most nutritious member of Cole group belonging to family Brassicaceae. Its marketable plant part is head composed of immature flowers or florets. There is considerable diversity in cultivated Coles, with different parts of the plant being used for consumption as vegetables. In broccoli and cauliflower, the developing inflorescence is the edible part and in this respect the two crops are very similar. Due to this apparent similarity, there is considerable confusion that existed in both scientific and lay nomenclature, since the two terms (broccoli and cauliflower) were being used interchangeably. A practical distinction between the two could, therefore, be made on their comparative ontogeny at marketable maturity (Gray, 1982; 1989) when cauliflower head is a curd or dome of tissue made up of a mass of proliferated floral meristems (Sadik, 1962), of which 90 percent or more abort prior to flowering (Crisp and Gray, 1979) whereas in broccoli the head and sprouts are actually composed of fully differentiated flower buds of which relatively few abort prior to flowering. The marketable stage in cauliflower is, therefore, ontogenetically at quite early stage than that of broccoli.

ORIGIN AND EVOLUTIONARY HISTORY

Broccoli is thought to be of European or Siberian ancestry like other members of Brassicaceae. Its wild forms are found along the Mediterranean Sea and it is thought to have originated as an abnormality from crosses between early forms of branching shrub kale and other botanical varieties followed by selection. Branching or sprouting may be indicative of a role played by Kales or *Brassica critica* at some stage during the evolution of broccoli (Snogerup, 1980). However, Song et al. (1990) supported monophyletic origin of cultivated morphotypes in *Brassica oleracea* based on nuclear restriction fragment length polymorphisms (RFLPs) data. Broccoli is believed to have originated in eastern Mediterranean with Italy as main center of its distribution. There are several forms within *italica* group (Table 1) that are grown for fresh markets but calabrese is also grown for the frozen food industry. Purple and white sprouting types and to a lesser extent purple cape broccoli are popular garden crops in Europe. The Romans grew a sprouting form in the first century. It was

TABLE 1. Broccolis belonging to the *italica* group*

Crop type	Characteristics features
Purple sprouting broccoli	Variable in nature with early and late types, branched purple sprouts, biennial.
Purple cape broccoli	Variable biennial purple heading type.
White sprouting broccoli	Biennial, branched, white sprouts early and late botrytis type.
Purple Sicilian broccoli (Cavolfiore violetto di Sicilia)	Pale-purple heading, also called 'purple cauliflower'.
Couve broccolo 'Roxo de Cabeca'	Purple sprouting, tall from Portugal.
Calabrese (Cavolo broccolo ramose verde di Calabria)	Calabrian green sprouting broccoli, shifting to heading type.
Black broccoli (broccolo neri)	Romanian annual, branched, dark green sprouts, sickle shaped leaves.

*Adapted from Gray, 1993

relatively unknown in England until about 1720 and was first mentioned in United States in 1806. The broccoli industry in the United states, the largest producer of broccoli in the world, is thought to be started establishing in 1923 when D'Aqrigo Brothers company began growing broccoli in California. The sprouting forms first develop terminal head and after harvest of this, a number of smaller axillary sprouts start appearing bearing smaller heads, which are sold in small bunches. During nineties, in order to satisfy changing horticultural requirements, heading crossing *botrytis* and *italica* groups evolved broccolis. The head in these are of green, yellowish green, purple, chocolate color. It is suitable both for fresh market and processing industry, especially freezing. Broccoli in grown as a specialist crop in most temperate and subtropical regions of the world, especially in USA, Japan, and Europe. In India, its sporadic cultivation began in the beginning of last decade of twentieth century. Due to the development of indigenous appropriate varieties in the middle of this decade and availability of seeds to the growers, its cultivation increased during the second half of this decade and it become quite popular with most Indian consumers because of its high nutritive value and anticancerous properties. Broccoli can be consumed as raw, boiled, in soup or in cooked from single or mixed with peas, etc. Besides, it can also be pickled. It has a bright future in India both from indigenous and export market point of view.

CYTOLOGY

Studies regarding secondary association of bivalents, pachytene chromosome morphology and pairing behavior in haploid revealed the basic chromosome number to be either 5 or 6. The first suggestion was Catcheside (1934, 1937) who studied the secondary association in *napus* and *oleracea* and concluded that 6 is the basic number for the genes. He was of the opinion that the primitive haploid number of Cruciferae is seven from which the fusion of two chromosome could give rise to the basic number 6.

Haga (1938), however, observed a frequent occurrence of 2 groups of two bivalent and four pair in *nigra* and 3 groups of two bivalents and three pair in *oleracea*. He proposed the basic number to be 6 and genomic formula AABBCCDEF for *oleracea*. Sikka (1940) has classified the genome of 9 according to size and there are 1 very long, 4 long, 3 medium and one small chromosomes. Studies on the secondary association in *oleracea* and other species of *Brassica* led Sikka to conclude that 5 is the basic number of the genus. He regarded *oleracea* as a modified amphidiploid from a cross between two primitive 5 chromosome species with subsequent loss of one chromosome.

Robbelen (1960) reported *oleracea* genome consist of 1 very long, 5 long, 2 medium and 1 short chromosome where the centromere of one chromosome was sub-telocentric, those of two were submetacentric and those of rest were metacentric. In general, chromosome 1(A) was satellite chromosome. Nucleoli were attached to chromosome 1 and 4 (A and C). He proposed genetic constitution of genome to be ABBCCDEEF which is triple tetrasomic for types B, C and E. These result clearly established that these species are secondary balanced polyploid with the basic chromosome number 6.

Sen (1957) while studying a claimed hybrid of broccoli and cauliflower observed a high degree of meiotic irregularities like univalent and multivalent formation which seems to be contrary to all the reports of *Brassica oleracea*.

GENETICS

The populations developed from crosses between heading and sprouting forms indicate that the heading versus sprouting character is controlled by several genes with segregating populations showing a wide range of intermediate types. Keyes and Honma (1986) reported that reduced development of lateral shoots, i.e., lateral suppression is dominant over branching. They proposed a two gene 9:7 epistatic model where $Ns_1 Ns_2$ governs lateral suppression from F_2 segregation ratio. While studying the inheritance of annual flowering, Bagget and Kean (1989) reported dominance for annual habit from F_1 crosses between annual Calabrese and other biennial Coles. They further observed that the segregation in F_2 generations confirmed strong dominance for the trait, though the ratio of annuals to biennials was influenced by the use of early or late maturing annual parents indicating some additive effects. While studying F_2 population of the crosses between cape broccoli (spring-maturing biennial) and purple Sicilian broccoli (autumn maturing early biennial), the segregation was towards the early parent, indicating dominance. Thus, annual habit is dominant over biennial and evidently controlled by several major genes with strong effects by modifiers from both the parents.

GENETIC RESOURCES

To conserve the genetic variation within the broccoli group, a comprehensive base collection of the cultivated forms of *Brassica oleracea*, including broccoli, has been established at the Vegetable Gene Bank of Horticultural Research International (HRI), Wellesbourne, Warwick (UK). Besides, gene banks have also been established at Instituut voor de veredeling van Tuinbouwgewssen, Wageningen, the Netherlands; the Instituto del Germoplasm, Bari, Italy, and National Bureau of plant Genetic Resources (NBPGR) at New Delhi, India for conservation of genetic resources. The International Board for Plant Genetic Resources (IBPGR) has recognized this as an international centre for conservation of *Brassica oleracea* (Anon, 1989 and Innes, 1985). In Britain, biennial sprouting broccoli and cape broccoli are readily available as unimproved seed stocks. Seed directly purchased from seed houses provides an effective way of collecting a range of cultivars, including many primitive ones, from several sources (Crisp and Ives, 1982). Seed is relatively cheap and usually of high germination, thereby avoiding the need to regenerate. The accessions are renewed at Wellesbourne (UK) gene bank by growing 30 plants in isolation cages or polythene tunnels (Astley et al., 1982 and 1985), and pollinated using blowflies (Innes, 1985). Each plant is individually harvested, and equal weights of seed are combined to provide the new population for long-term storage. This minimizes loss of seed viability. A source of genetic erosion is through losses of seed due to incorrect storage. Broccoli seeds are stored as dry, dried to 5 percent moisture content and sealed hermetically in foil laminate pouches, stored at $-20°C$ and monitored for viability (Anon. 1989).

Collections in gene banks are genetic resources generally available for research and breeding purposes. Crisp et al. (1989) proposed schematic screening of genetic resources for pests and diseases for various *Brassica oleracea* crops in relation to club root resistance. Broccoli germplasm has been evaluated for resistance to downy mildew. It is possible that similar schemes could be developed for all *Brassica* crops including broccoli. Crisp and Astley

(1985) reported that limited taxonomic work on cultivated vegetables is hindering the effective management of genetic resources.

Germplasm collections of broccoli are available in several research centers, particularly in Europe, the United States, India, and Japan. However, within the main Italian and secondary central and northern European gene centre, every effort should be made to preserve the genetic variability for future breeding purposes. Genetic erosion also is a real danger in all South East Asian countries, because local selections are being replaced by commercial cultivars from seed companies. Many old cultivars have already disappeared, representing considerable genetic erosion and probable loss of unique genes. The single head broccoli (calabrese) is particularly very narrow-based genetically (Van der Vossen, 1993).

GLUCOSINOLATES

Brassica vegetables are characterized by sulphur containing compounds glucosinolates (thioglucosides) (Table 2). Glucosinolates have a generalized structure (Figure 1). These are anions and occur in plants mostly as potassium salts (Tookey et al., 1980). Their break down products possess important sensory properties such as odor and flavor. Glucosinolates are readily hydrolyzed under moist conditions by the coexisting endogenous enzyme myrosinase (thioglucoside glucohydrolase, EC 3.2.3.1). The products of enzyme decomposition of glucosinolates are β-D-glucose, sulphate, and an organic aglucon moiety. Depending on conditions, such as the proteins present, pH, trace metals, etc., the aglucon can under go intermolecular rearrangement and/or fragmentation yielding products such as thiocyanates, isothiocyanates nitrites, cyanides, and ozazolidine-2-thiones (Tookey, 1980 and Van Etten and Daxenbichler, 1977).

The glucosinolate content of purple-headed broccoli was found to be in the range of 72-212 mg/100 g (Lewis et al., 1991). Grodrich et al. (1988) observed a number of quantitative changes in specific glucosinolate levels in broccoli harvested in municipal sewage sludge-amended soil, whereas the total glucosinolate content remained unchanged.

Broccoli has therapeutic properties. The vegetables of *Brassica* genus have been recommended for increased consumption, as a measure to decrease the incidence of human cancer (Anon., 1982). The reduced incidence of some forms of cancer has been found due primarily to glucosinolate derived products such as isothiocyanates and indoles (Birt,1988). Heaney and Fenwick (1984) proposed that rather than aiming for total elimination of glucosinolates, *Brassica* vegetable breeders should try to adjust the content of individual glucosinolates by reducing those responsible for goiter and undesirable flavor,

TABLE 2. Glucosinolates present in *Brassica* vegetables: Structure and trivial nomenclature.

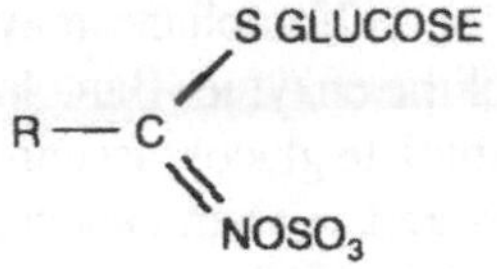

S GLUCOSE

R — C

NOSO$_3$

Glucosinolate skeleton

	R	Trivial names
I	Prop-2-enyl	Sinigrin
II	2-Hydroxybut-3-enyl	Progoitrin
III	2-Hydroxypent-4-enyl	Gluconapoleiferin
IV	3-Methylthiopropyl	Glocoiberverin
V	3-Methylthiobutyl	Glucoerucin
VI	3-Methylsulfinylpropyl	Glucoiberin
VII	4-Methylsulfinylbutyl	Glucoraphanin
VIII	2-Phenethyl	gluconasturtiin
IX	Indolyl-3-methyl	Glucobrassicin
X	4-Hydroxy indolyl-3-methyl	4-Hydroxyglucobrassicin
XI	2-Methoxyindolyl-3-methyl	4-Methoxyglucobrassicin
XII	1-Methoxyindolyl-3-methyl	Neoglucobrassicin

Adapted from Lewis et al. (1991)

FIGURE 1. Structure of glucosinolate

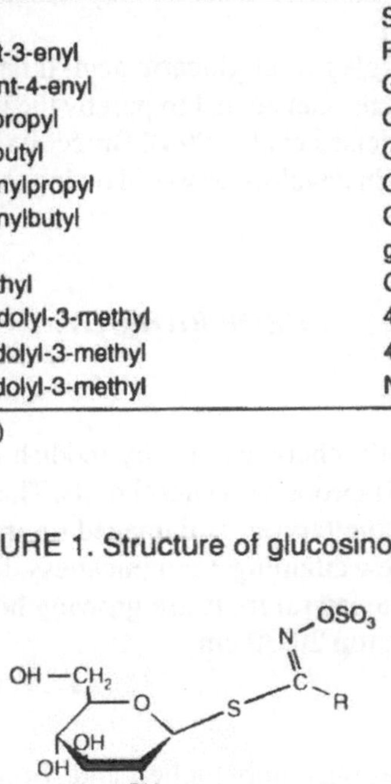

while maintaining or even enhancing those responsible for desirable flavor and anticarcinogenic effects.

Wattenberg (1983) suggested that tumorigenesis is inhibited by carcinogen-metabolizing systems induced by compounds in *Brassica* plants. Sulphur containing photochemicals of two different kinds are present in broccoli and other cruciferous vegetables: glucosinolates and S-metholcysteine sulfoxide. Numerous studies have indicated that the hydrolytic products of some of the glucosinolates have anticarcinogenic activity (Stoewsand, 1995).

Sulforaphane [1-isothiocyanato-4-(methyl-sulfinyl) butane] isolated from broccoli was found to inhibit the phase 1 enzyme cytochrome P-450 iso-enzyme 2 E1, which is responsible for activation of several carcinogens, including diakylnitrosamines. Inhibition of this enzyme by sulforaphane may offer chemo-protection against carcinogenic substrates of the enzyme (Barcelo et al., 1996). Indole-3-carbinol, a metabolite of the glucosinolate glucobrassicin, has shown inhibitory effects in studies of human breast and ovarian cancers (Stoewsand, 1995). Goodman et al. (1992) studied the effect of various vegetables in the diet on the survival of lung cancer patients and observed that increased consumption of broccoli, particularly among women, appeared to improve survival in lung cancer. Broccoli may also play a role in reducing levels of serum cholesterol.

It contains about 3.5 g/kg of D-glucaric acid. It has been suggested that dietary calcium requirements can be met in part by the intake of broccoli, among other vegetables (Michaelsen et al., 1994). Broccoli should be encouraged as a constituent of daily diet in developing world owing to its therapeutic and nutritional properties.

CROP BIOLOGY

Root System

The seedling generally characterized by reddish colored hypocotyls, two notched cotyledons and taproot with lateral roots. There are many adventitious roots that emerge after the taproot is damaged on transplanting. Most of the roots are 0.5 mm with few attaining 1 cm thickness. In the beginning the roots are quite shallow and the lateral roots are growing horizontally. The majority of the roots occur in the top 20-30 cm.

Stem

The stem is waxy, usually unbranched and, from it arise the leaves and flower heads.

Leaves

The leaves are simple, alternate and without stipules. Many times they are pinnately lobed.

Flower

Branched flower clusters form on 2-2 1/2 tall plants. The flowers are bright yellow. There are four sepals, six stamens, two carpel and four petals. Broccoli flowers have a superior ovary. The buds are dark green and tightly packed on top of the plant. Broccoli exposed to 40°F will initiate flower primordia much quicker than plants grown in high temperatures. Mostly bees pollinate the flowers.

Seed

The fruit of broccoli is a glabrous silique. There are between 10-30 seed per silique. About 325 seed will constitute a gram.

Other Descriptions

In single headed broccoli or calabrese leaves are more divided and petiolate, the main head consists of clusters of fully differentiated flower buds, green to purple, less densely arranged and with longer peduncles; axillary shoots with smaller flower heads usually develop after removal of the dominant terminal shoot. The flower head is fully exposed form an early stage of development. Broccoli plants carry inflorescence from lateral branches too. The sprouting broccoli bears many, more or less uniform, relatively small flower heads instead of one large head as in the calabrese type.

The distinction between different types in broccoli is that sprouting types are multi-branched with relatively small heads with green, purple or even white flower buds, e.g., 'Purple sprouting', whereas heading types or calabrese have a mass of flower buds in one large main head on a short central stalk; smaller heads may develop on lateral branches after removal of the main head; the color of the head can be dark purple (e.g., purple cape; 'Sicilian purple'), brownish (e.g., 'brown headed'), yellowish-green (e.g., 'sulphur colored'), dark green (e.g., 'chartreuse'), or white.

FLORAL BIOLOGY

Broccoli has a typical cruciferae flower, during differentiation of which the successive development of four sepals, six stamens, two carpels, and four symmetrical equal petals takes place, which may be yellow or white. The carpels form a superior ovary with a 'false' septum and two rows of campylotropous ovules. The androecium is tetradynamous, i.e., there are two short and four long stamens.

The flowers are borne is racemes on the main stem and its branches. The buds open under the pressure of the rapidly growing petals. This process starts in the afternoon and, usually, the flowers become fully expanded during the following morning. The anthers open a few hours later being slightly protogynous. The anthesis starts at 8 a.m. and continues up to 11.30 a.m., under normal conditions. Peak hours of anthesis are between 8 to 9 a.m. The anther dehiscence begins at the time of anthesis, however maximum dehiscence occurs between 10 a.m. and 12 noon. The pollen fertility is considerably high at the time of anthesis. The stigma becomes receptive two days before and remains up to two days after anthesis. The pollination is carried out by bees, which collect pollen and nectar. It is secreted by two nectaries situated between bases of the short stamens and the ovary. An additional nectary is situated outside the bases of each of the two pairs of long stamens, but these two nectaries are not active.

SELFING AND HYBRIDIZATION TECHNIQUES

Brushing or shaking the open flowers, if the plant is self-compatible can carry out self-pollination in broccoli. In general, the plants are self-incompatible, therefore selfing can be accomplished by opening the buds 1-4 days before they will open naturally and bud pollinate them. Applying their own fresh pollen from previously bagged flowers of the same plant does pollination. The largest unopened bud is probably too old for successful bud pollination as the incompatibility factor will be biosynthesized by that stage. Hence, 6 to 8 buds can be opened at one time with forceps and pollen from already covered older open flower can be transferred to the stigma with a camel hairbrush. The genotypes can be maintained by using pooled pollen from selected plants for pollination. Emasculation is not necessary for self- and sib bud-pollination. However, after pollination the buds are enclosed in butter paper bags to avoid contamination from foreign pollen.

In case of hybridization program, the buds of the female parent are emasculated one or two days prior to opening and are pollinated by pollen collected from the flowers of the desired male parent also bagged before opening. The pollination is done by hand by shaking the pollen over the stigma directly from the freshly opened but previously bagged buds of the male parent. The pollen can be transferred with the help of camel hairbrush. It is desirable to enclose 1 or 2 flowering branches in a muslin or butter paper bag and use individual flower buds or flowers for bud pollination or crossing as and when they become ready for the purpose. Each branch containing pollinated buds or flower is tagged with a small tag label. Usually 8 to 10 flower buds or flowers are bud pollinated or closed in each inflorescence branch to ensure better seed set and the unopened young flower buds at the terminal end are removed.

For obtaining large quantity of crossed seed form male and female parental lines, homozygous for self-incompatibility but are cross compatible, these are planted in alternate rows spaced 45 or 60 cm apart. The plants are covered under an insect proof wire net or plastic cage of 22 or 24 mesh 3 to 4 days before opening of buds. A small honeybee colony is sufficient for a cage covering four plants, 2 each of male and female parent. This procedure can also be used for producing sib-mated seeds for maintenance of varieties under insect-proof cages. However, in this case 20-30 selected plants are required to be placed under a cage to avoid inbreeding depression. A wire-net or plastic-net cage of dimensions 3 m length $\times$ 3 m width $\times$ 2.5 m height with a small door on one side is suitable for this purpose.

BREEDING OBJECTIVES

The main breeding objectives include head shape, size, firmness, weight, dark green color, no physiological disorders, high yield, earliness, stand ability, heat tolerance, resistance to diseases and insect pests.

Heterosis and F_1 Hybrid Breeding

Based on single crosses between inbred lines, breeding F_1 hybrid are now the main goal of most breeding programmes in broccoli and cauliflower. Seed production depends on the system of sporophytic self-incompatibility present in most types. S-alleles have been re-introduced in the self-compatible annual cauliflower of northern Europe. Self-fertilization, necessary to develop homozygous inbred lines, is effected by bud pollination or treatment with CO_2 (2-6%) before bee pollination, to temporarily break the self-incompatibility. Inbred lines have also been developed from di-haploid plants regenerated from anther or microspore cultures.

Phenotypic uniformity of an F_1 hybrid depends upon genetic homozygosity in the inbred parents. Out-breeding crops are more or less intolerant of inbreeding, which exposes and fixes unfavourable recessive alleles and leads to inbreeding depression. Cole crops permit inbreeding to I_5, albeit with low seed production. This loss of vigor results in low yield of hybrid seed and difficulties in maintaining the inbreds. Inbreeding depression may rapidly reduce the number of surviving lines unless compromises are made in the degree to which inbreeding is enforced. Phenotypic selection among second and third generation inbreds may be followed by mass pollination between the best of them to give an improved population for a further cycle of selfing. At some point the general and specific combining abilities of selected inbreds will be determined. General combining ability (GCA) can be estimated in top crosses. This involves crossing each of the inbred to a widely based tester stock, usually an OP variety. Identification of outstanding combinations of inbreds, those depending upon specific combining ability (SCA) on top of GCA, is the next important step (Simmonds, 1979). Honma (1965) which developing and maintaining uniformity in broccoli during early 1950s, he was looking for a cheap method of obtaining uniformity in calabrese in USA at the time of mechanical harvesting aids advent. Open pollinated cultivars were then in use and commercial F_1 hybrids were not known. Occurrences of self-compatibility in calabrese suggested that lines could be developed as cultivars (Moore and Anstey, 1954; Ockenden et al., 1979). The purple heading cultivar 'Rosalind' was developed in a similar way on an SC inbred line (Crisp and Gray, 1984). The fixing of additive genetic variation leads to development of inbred cultivars, which is evident for most economically important characters (Leg and Lippert, 1966). Honma (1965) devised a system based on developing matching inbred lines, each self-incompatible but cross-compatible. Two closely related cross compatible lines were mass pollinated and this led to the development of the synthetic cultivar 'Spartan early' (Honma, 1960).

Improvement of cross-pollinating crops like broccoli has traditionally involved mass selection or recurrent selection to increase gene frequencies for desired traits. Improved population can be used as sources of new open-pollinated cultivars or for the extraction of superior inbreds for use in synthetic population or as parents of F_1 hybrids (Coyne, 1980). F_1 hybrids have become commercially important in calabrese in the developed world where uniformity and early maturity are at a premium. Breeding of F_1 hybrids exploits the expression of heterosis when inbreds of diverse origin are crossed together (Legg and Souther, 1968). The F_1 hybrids also have a commercial advantage arising from ownership and maintenance of parental lines, providing on inbuilt exclusivity (Riggs, 1988). For most economical characters, additive genetic variation accounts for a greater proportion of the total genetic variance (Legg and Lippert, 1966). Dominance contributes to variance for earliness. Hulbert and Orton (1984), and Legg and Souther (1968) observed expression of heterosis in increased head weight and early maturity. Generally, there is a high correlation between mean maturity of hybrids and the mean maturity of their parents. Earliest hybrids are usually products of crosses involving early parents. This positive relationship, however does not appear to exist between uniformity of maturity in broccoli hybrids and in their parents. The tendency often is for hybrids to be more uniform than their mid-parent and more uniform than the most uniform parent of crosses (Hulbert and Orton, 1984). Crosses involving broccolis belonging to different seasonal maturity groups, or with cauliflower, have been used to broaden the genetic base and select new horticultural forms free from cosmetic defects in calabrese. Developing varieties of broccolis for specific season by involving populations from different maturity groups is a type of breeding of which the approach is based on the high heritability of heading characters, head color and time of maturity (Crisp and Gray, 1984).

PROSPECTS

The importance of broccoli is likely to increase further is South-East Asia. In India, its commercial cultivation started about a decade ago and now it is increasing very fast as the indigenous cultivars have become available to the growers (Kalia, 2002). Heat tolerant cultivars enable cultivation at low altitudes, but market gardening will continue to prevail in high lands because of higher yields, better head quality and fewer disease and pest problems. Considerable progress is being made with effective methods of integrated pest management, as in cabbage, and this will reduce pesticide use. Cultivars resistant to *Fusarium*, black rot and downy mildew will gradually become available, but durable resistance to club root is very hard to realize.

The development of DNA markers for more precise indirect screening for resistance to diseases, pests and other important traits can increase breeding efficiency. Cytoplasmic male sterility might, in the long run, replace self-incompatibility as a more reliable technique for hybrid seed production.

REPRODUCTIVE BIOLOGY

Pollination Control Mechanisms

Large-scale production of hybrid seed requires that there be some mechanisms for ensuring that cross-pollination occurs between the parental lines with a minimum of self-fertilization. Hand emasculation is considerably more difficult in species with hermaphrodite flowers. The economics of hand emasculation will depend upon the ease with which it can be carried out, the return of seeds per pollination, the price of hybrid seed in relation to the cost of production, and the seed rate (Frankel, 1973a).

There are two naturally occurring mechanisms for ensuring cross-fertilization of hermaphrodite species like broccoli are self-incompatibility and male sterility. The use of both these mechanisms in hybrid seed production was reviewed by Frankel (1973a, 1973b). Apart from the need to prevent selfing of the female line, it is necessary that pollen of the male line be effectively transferred to the female line. This requires either artificial pollination or an efficient pollen dispersal mechanism.

Self-Incompatibility(SI)

Self-incompatibility (SI) systems are found in many species of flowering plants and there are two systems of self-incompatibility, viz., gametophytic and sporophytic. Gametophytic self-incompatibility is commonly associated with binucleate pollen and pollen tube inhibition that usually occurs in the style; sporophytic self-incompatibility is associated with trinucleate pollen and inhibition of pollen at stigmatic surface (Hodgkin et al., 1988). In the gametophytic self-incompatibility system, the pollen reaction is determined by the genotype of the gametophyte whereas in the sporophytic system pollen reaction is determined by the genome of the somatic tissue (of the sporophyte) in which this pollen developed. In other words, pollen reaction is controlled by the genotypes of the sporophyte. *Brassica oleracea* crops have a sporophytic self-incompatibility system. So is with broccoli in which this system is controlled by a series of multiple alleles at a single locus (Bateman, 1955). The sporophytic system of self-incompatibility can be either heteromorphic (heterostyly) or homomorphic. Heterostylic sporophytic incompatibility is controlled by one gene with two alleles in distyly and by two genes, each with two alleles in tristyly (Kalloo, 1988). Homomorphic sporophytic incompatibility is controlled by one locus with multiple alleles, and occurs in *Brassica* vegetables. The genetic features of the sporophytic systems of self-incompatibility (Frankel and Galun, 1977) are:

1. Incompatibility is controlled by one S locus having several alleles over 50 to 70 alleles were found in *B. oleracea*.
2. The reaction of the pollen is determined by the genotype of the sporophytic tissue in which the pollen was formed, and therefore controlled by two S alleles.
3. All the pollen of a plant have the same incompatibility reaction.
4. The two S alleles may react as codominance (independent) or they may interact by one being dominant over the other.
5. Active allele identity in both pollen and pistil leads to incompatibility.
6. The dominance/independence relationships of the S alleles in the pollen and in the pistil may differ.

Wallace (1979) presented detailed procedures for identifying the S allele genotypes of *Brassica*, and Nasrallah (1989), and Nasrallah and Nasrallah (1986) reported on *Brassica oleracea* self-incompatibility at the cellular and molecular levels. In a self-incompatible pollination, the papilla cells of the stigma produce callose which inhibits pollen germination and/or pollen tube growth (Kanno and Hinata, 1969).

Thompson and Taylor (1966) documented self-incompatibility for Kales and also for horticultural brassicas including broccoli. Ockendon (1980) surveyed the distribution of S-alleles in broccoli and breeding structure in purple cape and purple Sicilian broccolis. Similarly, Voss stern et al. (1982) surveyed various hybrids and breeding lines of calabrese (Table 3). According to Gray and Crisp (1985), there is limited investigation of S-allele composition in pop-

TABLE 3. S-alleles in some broccoli types

S-allele	Calabrese	Purple Sicilian broccoli (autumn heading)	Purple cape broccoli (spring heading)
S_2	+	+	+
S_8		+	
S_{12}		+	
S_{13}	+		
S_{14}		+	
S_{15}	+	+	+
S_{16}			+
S_{18}	+	+	
S_{20}		+	
S_{23}		+	
S_{24}	+		+
S_{25}		+	+
S_{26}		+	
S_{29}		+	+
S_{36}		+	+
S_{39}	+	+	
S_{45}			+
S_{51}	+	+	
S_{52}		+	
S_{57}			+
S_{58}			+

ulations of sprouting broccoli, although purple sprouting broccoli is known to contain strong self-incompatibility.

Relatively large number of S-alleles per 'cultivar' (totaling 20) and very few self-compatible (SC) individuals were found in populations of cape broccolis and purple sicilian broccoli, indicating as high level of out crossing not expected in an undeveloped crop (Ockendon, 1980). Purple and white sprouting broccoli surveyed by Ockendon (1982) showed relatively fewer S-alleles. Seven S-alleles were identified by Voss stern et al. (1982) in calabrese, indicating a narrow genetic base. Self-compatible individuals occasionally appeared and in two open pollinated cultivars many plants were partially or strongly self-compatible (Ockendon et al., 1979). There is a degree of natural self-pollination in open pollinated populations of Calabrese, and random populations, therefore, are likely to contain both SI and SC individuals (Anstey, 1954 and Moore and Anstey, 1954).

The sporophytic SI system, operative in all types of *Brassica oleracea*, varies in effectiveness in the different crop types, being strongest in Kale and weakest in summer cauliflower (Watts, 1980), but there are many factors, genetic and environmental, which affect the expression (Ockendon, 1971). The S-allele system is complex, with about 50 alleles at a single locus, forming a dominance series in which, generally speaking, the dominant alleles are strong and the recessive alleles weak. A high level of self-incompatibility was noticed in Indian genotypes of green sprouting broccoli as most of their plants were in 0-10% compatibility grade (Kalia and Yadav, 2000) and Type IV S-allele interaction found prevalent in the four Indian genotypes has potential use in three- and four-way F_1 hybrid breeding (Yadav, 1998 and Kalia and Yadav, 2002). The very first objective in F_1 hybrid breeding is to produce inbred lines as parents. Therefore, the incompatibility reaction has to be overcome to allow self-pollination. This can be done in some species by pollinating the flowers in the bud stage before the Self-Incompatability (SI) mechanism is operative. The glycoproteins play an important role in sporophytic SI and the increased rate of synthesis of these S-locus specific glycoproteins in the developing stigma correlates with the onset of the incompatibility reaction (Nasrallah et al., 1985). Since commercial production of F_1 hybrid seed in vegetable brassicas including broccoli depends on crossing together two inbred lines, each homozygous for a different S-allele, therefore, each line is required to be both sib and self-incompatible but the two lines must be cross-compatible. Toriyama et al. (1991) transformed Green Comet and Kairan hybrids of broccoli with an SLG gene isolated from the *Brassica campestris* S8 homozygote. The mutant phenotypes generated from the introduction of the S8 transgene provide direct evidence that the pollen-stigma interaction of incompatibility is influenced by the expression of the SLG gene. Camargo et al. (1997) located a single locus controlling SI on RFLP and RAPD linkage map of *Brassica oleracea* by assaying for SI phenotype and segregation of an RFLP locus detected by the cloned SI gene SLG6.

Male-Sterility

This is of two types, genic and cytoplasmic.

Genic Male Sterility. This has been isolated from calabrese and purple heading broccoli within broccoli group (Borchers, 1971 and Cole, 1959). Dunemann and Grunewaldt (1987) induced male sterility in broccoli by *in vitro* mutagenesis. This has possible practical value in hybrid seed production, particularly because self-incompatibility (SI) in some broccolis and cauliflowers is often inadequate for reliable hybrid production. Genic male sterility is conditioned by recessive male sterility (ms) genes, of which several designated as ms_1, ms_2, ms_4, ms_5, and ms_6 have been identified (Dickson, 1970), which are non-allelic. This sterility can be responsive to temperature, with low temperature, i.e., 10-11°C (continuous), or 15°C (day) and 7°C (night), restoring male fertility in individuals carrying the ms_6 gene, with ms_6, ms_6 being highly male fertile. This conversion from complete male sterility to male fertility takes about 30 days at 10-11°C, the control mechanism apparently operating early in the buds initials (Dickson, 1970). Borchers (1966) recognized male steriles by their small flowers, short stamens and shriveled anthers. Their pollen grains were characteristically small, abortive, non-staining and often clump together. Borchers (1971) studied hybrid broccoli seed production utilizing the ms_6 gene for male sterility and observed that the use of temperature sensitive male sterility such as that conferred by ms_6 gene could enable hybrid seed to be produced at relatively high temperature with few sibs, while Dickson (1970) reported that low temperature could be used to maintain homozygous male sterile lines by self-pollination.

Cytoplasmic Male Sterility (CMS)

This sterility was not known in *Brassica oleracea* until Pearson (1972) crossed *B. nigra* (L.) Koch, a wild mustard, with broccoli. By backcrossing the 4x amphidiploid of *Brassica nigra* × *Brassica oleracea* (as broccoli) with broccoli pollen for three generations, the cytoplasm of *B. nigra* was introgressed with the genome of *B. oleracea*. The backcrosses with 'Green Globe' cabbage established two systems of cytoplasmically inherited male sterility, one based on petaloid where flowers have no nectaries and the other on vestigial anther male sterility. Dickson (1975) developed cytoplasmic male sterile inbreds in broccoli from the cabbage stocks of Pearson. While looking for maintainers, male sterility genes were introduced to *oleracea* cytoplasm in Waltham 29 broccoli by backcrossing. It was also observed by Dickson (1975) that the necessity to select for self-fertility in maintainers, sib fertility in back crosses and high female fertility in petaloid steriles concurrently, probably contributed to fewer homozygous steriles (all petaloid) than expected. This CMS system had affected a well worked out mechanism for hybrid seed production in *Brassica* vegetables.

CMS from *Raphanus sativus* (Ogura, 1968) is a vestigial anther type with normal flowers and full seed fertility. Nectary development may not be normal but there seems to be scope for selection. Initial attempts to introduce male sterility to *Brassica oleracea* by transferring *oleracea* genome to radish cytoplasm via the amphidiploid were difficult due to the persistence of radish chromosomes in back crosses. Embryo rescue effectively enabled plants with low numbers of radish chromosomes to be recovered, as did the use of 4x broccoli (as pollen) in back crosses with out embryo rescue (Bannerot et al., 1975 and McCollum, 1981). Also plants when grown at low temperature, they exhibit chlorosis.

Petaloidy CMS has a wide range of expression as in its extreme, its all stamens are converted into narrow incurved petals. Pistils are enlarged and malformed, and nectaries are generally absent. Pollinators are less attracted to these flowers. Female fertility varies broccoli petaloid male steriles but can be improved by selection. Dickson (1975) selected CMS broccoli lines that set seed freely had the largest nectaries, although these were still reduced when compared with those of the male fertile maintainers. Pearson (1972) also stated that in vestigial anther male sterility, the flowers are normal in appearance but are smaller, with reduced anthers but apparently functional nectaries, and are therefore more attractive to pollinating insects. Dickson (1975) reported that homozygous vestigial anther lines were, however, not recovered after six generations of back crossing and testing in broccoli. Lin-BiYing et al. (1997) developed CMS line 5A of broccoli after a single crossing generation and four generations of backcrossing using male-sterile cauliflower line MSA as CMS source. This line is genetically stable and has excellent horticultural characteristics. Sharma and Vinod (2002) developed CMS lines in broccoli suitable for growing at a temperature range of 20-27°C using male sterile backcross progeny (Kale EC 173419 × G-SB^2 × 005559^2) as a source after three backcrosses. These showed normal or near normal floral characteristics with good seed setting ability. Honeybees were frequent visitors as these had functional nectaries.

The *Brassica oleracea* CMS lines based on radish cytoplasm develop chlorosis under low temperatures (below 12°C) (Bannerot et al., 1977). The intensity of chlorosis depends upon the recurrent pollen parent used (McCollum, 1981). Hoser-Krauze (1987) found that chlorosis in seedlings and young leaves delays maturity, as observed in early lines of cauliflower developed using CMS broccoli lines based on radish cytoplasm.

When *B. nigra* is used as sources of CMS, the derived broccoli lines are considerably altered in flavor and the characteristic broccoli flavor being replaced by a sharper, maternally inherited mustard flavor (Pearson, 1972). These problems slowed the pace of utilization of CMS system in production of commercial F$_1$ hybrid broccoli.

Somatic hybridization by protoplast fusion could provide a means of directly combining a male sterile cytoplasm with nuclear genome of an improved inbred, thus avoiding the need for lengthy backcrossing. This technique could also be used to create unique and possibly heterotic material by combining different cytoplasm to form cybrids as reported by Pelletier et al. (1983) in which the male sterile cytoplasm from *R. sativus* has been combined with the chloroplasts of *B. napus*. Recombination between mitrochondria is useful in separating CMS from 'defective nectary,' as both characters are mitochondrially determined. A lack of suitable restorers prevents the use of this source of CMS in rape (*B. napus*) but not in vegetable brassicas where restoration is unnecessary. Conventional sexual hybrids contain only maternally derived cytoplasmic factors but somatic fusion combines both parental cytoplasms.

Subsequent cytoplasmic segregation results in the elimination of one or the other cytoplasm (Aviv et al., 1980) but evidence from restriction analysis of mitochondrial DNA suggests that recombination occurs between parental mitrochondrial DNAs prior to cytoplasmic segregation (Belliard et al., 1979). Thus, protoplast fusion can give rise to unique nuclear-cytoplasmic combinations, not available using conventional methods (Sharp et al., 1984).

The cloning of DNA from the S_6 allele of *Brassica oleracea* (Nasrallah et al., 1985) have opened up the use of such clones as probes which may be more convenient way to characterize plants for their S-locus genotype than using genetic and electrophoretic methods (Charlesworth, 1985).

MAINTENANCE OF INBREDS

For multiplication of parental inbreds, bud pollination is often used but this process requires a large poll of temporary skilled labour. Insect proofing to a high degree is required as small amounts of foreign pollen will cause disproportionate amounts of contamination.

Nakanishi and Hinata (1975) used CO_2 gas treatment to over come self-incompatibility systems in order to produce self seed in *Brassica oleracea*. Fly or bee pollination can then be used even on high self-incompatible inbreds. In this case capital costs are high but operational costs are much lower than using bud pollination. High levels of CO_2 (5%) are required and glass houses have to be sealed in such a way that there is no leakage. Strict safety precautions are also necessary to prevent accidental entry as 5 percent CO_2 is not breathable, although blow flies show little apparent ill effect. Spraying of 5 percent NaCl at initiation of flowering as well as at peak flowering increase self-fertility in *Brassica oleracea*. Carafa and Carratu (1997) treated *B. oleracea* stigmas with a drop of salt solution and 24 hour after pollination the pollen tube growth was determined. It was found that pollen tubes penetrated in to the ovary, which led to the conclusion that salt treatment alters the conformation of S-locus specific glycoproteins.

COMMERCIAL F_1 HYBRID SEED PRODUCTION

F_1 hybrid seed can be produced commercially out doors in topographically isolated areas, especially in hilly and mountainous regions, although some has been produced under glass, and Faulkner and Hinton (1980) advocated the use of polytunnels as being the most reliable environment when blow flies were used as pollinating agents. Although protected cropping is more reliable than field production, but it may not be more economic.

The problem of sib production in F_1 seed of *Brassica oleracea* was high lighted by Hodgkin (1981) to the extent of 50 percent. The factors affecting the sib rate in field-scale production are numerous, e.g., bee behavior, coincidence of flowering and plant height (Faulkner, 1974), but failure of self-incompatibility to operate efficiently is the most important.

If both the inbred parents are to produce seed then they are planted in 1:1 ratio. But when one parent is more vigorous a 2:1 ratio is adopted. In case seed is harvested from one dependable self-incompatible line, higher planting ratios can be contemplated. In this case the pollinator line is destroyed immediately after flowering. If both the parents are allowed to mature seed then plants of each parent line are eventually harvested separately so that two seed lots representing F_1 hybrid seed from each parent line are obtained. Occurrence of sibs (seed derived from self or sib pollination of parent lines) in F_1 hybrid seed lot is inevitable. It is not feasible to remove sib seed but determination of sib rates in commercial seed is now standard practice using isozyme analysis by polyacrylamide gel electrophoresis. The acid phosphatase system was first suggested by Te Nijenhuis (1968) and developed by Woods and Thurman (1976). Sib levels greater than 5 percent make the bulk seed lots virtually unsaleable.

SEED PRODUCTION

Broccolis are relatively easy to transplant at maturity. The individual plants can be harvested and recorded for selection purposes and even after that produce sufficient flowering shoots to give 10-50 g of seed (Crisp et al., 1985). For producing seeds of open pollinated varieties on a large scale in open in tropical and sub tropical areas where winters are not harsh, the off type plants are rouged at marketable maturity and ensured that isolation of at least 1600 m is maintained and/or physical barriers in mountainous valley areas are ear marked for different varieties of broccoli and/or other related crops like cauliflower, cabbage, Brussels sprout, knol khol, and kale. For ensuring better sub set, honeybee boxes are installed in close vicinity. In a breeding program also in order to make progress, it is necessary to be able to obtain sufficient seed in any generation from selected parents.

So for this purpose, individual plants lifted from the field population are transferred to large (30 cm) pots accommodated in isolation cages in a glass house or polythene tunnel. Following re-establishment, flowering takes place about 3-4 weeks later, when blowflies and/or bees are introduced as pollinators. Seed harvested as separate progenies are used for assessment in a trial to identify the best parental material for the next cycle of pollination. For developing self-compatible inbred lines as cultivars from self-compatible selections, small number of blowfly pupae can be introduced into uncoated cellophane bags or finely perforated bread bags enclosing the inflorescences of the selected parent. Using Remay nylon bags in place of cellophane, this method has been successfully adapted for seeding selected plants of broccoli *in situ* under field conditions (Crisp et al., 1981).

The self-incompatibility in brasicas can be over come by pollen transplantation (Kroh, 1966), stigma mutilation (Linskens and Kroh, 1967), chemical treatment (Tatebe, 1968), carbon dioxide treatment (Nakanishi et al., 1969), high temperature (Gonai and Hinata, 1971), and electrical aids (Roggen et al., 1972).

For quick propagation of parent plants of desirable self-incompatible crosses in broccoli, micropropagation method, using flower buds, have been devised and extended to large scale F_1 hybrid seed production (Anderson and Carstens, 1977). Such methods eliminate any bud pollination needed to maintain SI parental lines.

REFERENCES

Anderson, W.C. and Carstens, J.B. (1977). Tissue culture propagation of broccoli *Brassica oleracea* (*italica* group) for use in F_1 hybrid seed production. *J. Am. Soc. Hortic. Sci.*, 102: 69.

Anonymous (1989). Vegetable Gene Bank. Published by AFRC Institute of Horticultural Research with the assistance of the Vegetable Research Trust, UK.

Anonymous, (1982). National research council, Inhibitors of carcinogenesis. In: Diet, Nutrition, and Cancer, National Academic Press, Washington, DC. pp. 358.

Anstey, T.H. (1954). Self-incompatibility in green sprouting broccoli (*Brassica oleracea* L. var. *italica* Plenck.) I. Its occurrence and possible use in a breeding programme. *Canadian J. Agric. Sci.* 34: 59.

Astley, D., Cleeton, A.E. and Lockelt, A.H. (1982). Vegetable Gene Bank. *Rep. Natl. Veg. Res. Stn. for 1981*, pp. 57.

Astley, D., Pinnegar, A.E., Collett, M.B. and Lockelt, A.H. (1985). Vegetable Gene Bank. *Rep. Natl. Veg. Res. Stn. for 1984.* pp. 62.

Aviv, D., Fluhr, R., Edelman, M. and Galun, E. (1980). Progeny analysis of the inter specific somatic hybrids: *Nicotiana tabacum* (CMS) + *Nicotiana sylvestris* with respect to nuclear and chloroplast markers. *Theor. Appl. Genet.* 56: 145-50.

Bagget, J.R. and Kean, D. (1989). Inheritance of annual flowering in *Brassica oleracea. Hort. Sci.* 24: 662.

Bannerot, H., Boulidard, L. and Chupeau, Y. (1977). Unexpected difficulties met with the radish cytoplasm in *Brassica oleracea. Eucarpia Cruciferac Newsletter*, 2: 16.

Bannerot, H., Boulidard, L., Cauderon, Y. and Temp, J. (1975). Transfer of cytoplasmic male sterility from *Raphomus sativus* to *Brassica oleracea*, pp. 52-54. In: Proceedings of Eucarpia meeting, Cruciferae 1974, SCRI, Dundee, UK.

Barcelo, S., Gardiner, J.M., Gescher, A. and Chipman, J.K. (1996). CYP 2E1-mediated mechanism of anti-genotoxocityc of the broccoli constituent sulforaphane. *Carcinogeneins* 17: 277.

Bateman. A.J. (1955). Self-incompatibility systems in angiosperm. III. Cruciferae. *Heredity* 9: 53.

Belliard, G., Vedel, F. and Pelletier, G. (1979). Mitochondrial recombination in cytoplasmic hybrids of *Nictiana tabacum* by protoplast fusion. *Nature, UK* 281: 401-403.

Birt, F.D. (1988). Anticarcinogenic factors in cruciferous vegetables. In: Horticulture and Human Health (Quebedeaux, B. and Bliss, F.A., eds.), p. 160.

Borchers, E.A. (1966). Characteristics of a male sterile mutant in purple cauliflower (*Brassica oleracea* L.). *Proc. Am. Soc. Hortic. Sci.* 88: 406.

Borchers. E.A. (1971). Hybrid broccoli seed production utilizing the ms$_6$ gene for male sterility. *J. Am. Soc. Hortic. Sci.* 96: 542.

Boswell, V.R. (1949). Our vegetable travelers. *Natl. Geogr. Mag.* 96: 145-217.

Camargo, L.E.A., Savides, L., Jung, G., Nienhuis, J. and Osborn, T.C. (1997). Location of the self-incompatibility Locus in an RFLP and RAPD Map of *Brassica oleracea*. *The Journal of Heredity* 88(1): 57-59.

Carafa, A.M. and Carratu. G. (1997). Stigma treatment with saline solutions: A new method to overcome self-incompatibility in *Brassica oleracea* L. *J. Hortic. Sci.* 72(4): 531-535.

Catcheside, D.G. (1934). The chromosomal relationships in the swede and turnip groups of *Brassica*. *Annals of Botany* 48: 601-633.

Catcheside. D.G. (1937). Secondary pairing in *Brassica oleracea*. *Cytologia, Fuji Jubilee*, Vol. I: 366-378.

Charlesworth, D. (1985). Molecular view of pollen rejection. *Nature, UK* 318: 231-232.

Cole, K. (1959). Inheritance of male sterility in green sprouting broccoli. *Canadian J. Genet. Cytol.* 1: 203.

Coyne, D.P. (1980). The role of genetics in vegetable improvement. *Scientific Hort.*, 31: 74-88.

Crisp, P. and Astley, D. (1985). Genetic resources in vegetables. In: Progress in Plant Breeding (Russell, G.E., ed.), Butterworths, London, p. 281.

Crisp, P. and Gray, A.R. (1979). Successful selection for curd quality in cauliflower, using tissue culture. *Hort. Res.* 19: 49-53.

Crisp, P. and Gray, A.R. (1984). Breeding old and new forms of purple heading broccoli. *Eucarpia Cruciferae Newslett.* 9: 17.

Crisp, P. and Ives, S.J. (1982). Genetic conservation of crucifers. *Rep. Natl. Veg. Res. Stn. for 1981*, p. 58.

Crisp, P., Crute, I.R., Sutherland, R.A., Angell, S.M., Bloor, K., Burgess, H. and Gordon, P.L. (1989). The exploitation of genetic resources of *Brassica oleracea* in breeding for resistance to club-root (*Plasmodiophora brassicae*). *Euphytica* 42: 215.

Crisp, P., Gray, A.R., James, H., Dudley, J.M. and Angell, S.M. (1981). Broccoli Rep. *Natl. Veg. Res. Stn. for 1980*, p. 56.

Crisp, P., Gray, A.R., James, H., Ives, S.J. and Angell, S.M. (1985). Improving purple sprouting broccoli by breeding. *J. Hortic. Sci.* 60: 325.

Dickson, M.H. (1970). A temperature sensitive male sterile gene in broccoli (*Brassica oleracea* L. var. *italica* Plenck.). *J. Am. Soc. Hortic. Sci.* 95: 13.

Dickson, M.H. (1975). G1117A, G1102A and G1106–A cytosterile broccoli inbreds. *Hortsci.* 10: 535.

Dunemann, F. and Grunewaldt, J. (1987). Male sterile broccoli (*Brassica oleracea* var. *italica*) induced by *in vitro* mutagenesis. *Eucarpia Cruciferae Newsletter* 12:50.

Faulkner, G.J. (1974). Factors affecting field scale production of F_1 hybrid Brussels sprouts. *Ann. Appl. Biol.* 77: 181.

Faulkner, G.J. and Hinton, W.L. (1980). F_1 hybrid Brussels sprout seed: an assessment of production methods and their economic viability. *Hort. Res.* 20: 49.

Frankel, R. (1973a). The use of male-sterility in hybrid seed production. In: Agricultural Genetics: Selected Topics (Moav, R., Ed.) John Wiley and Sons, New York, 85-94.

Frankel, R. (1973b). Utilisation of self-incompatibility as on out-breeding mechanisms in hybrid seed production. In: Agricultural Genetics: Selected Topics (Moav, R., ed). John Willey and Sons, New York, 95-107.

Frankel, R. and Galun, E. (1977). Pollination mechanisms, reproduction and plant breeding. In: Monograph on Theoretical and Applied Genetics 2, Springer-Verlag, Berlin Heidelberg, New York.

Gonai, H. and Hinata, K. (1971). Effect of temperature on pistil growth and phenotypic expression of self-incompatibility in *Brassica oleracea* L. Japanese *J. Breed.* 21: 195.

Goodman, M.T., Kolonel, L.N., Wilkens, L.R., Yoshizawa, C.N., Le-Marchand, L. and Hankin, J.H. (1992). Dictary factors in lung cancer prognosis. *European J. Cancer* 28: 495.

Gray, A.R. (1982). Taxonomy and evolution of broccoli (*Brassica oleracea* L. var. *italica*). *Econ. Bot.*, 36(4): 397-410.

Gray, A.R. (1989). Taxonomy and evolution of broccolis and cauliflowers. *Bailey*, 23(1): 28-46.

Gray, A.R. (1993). Broccoli (*Brassica oleracea* L. *italica* group). In: Genetic Improvement of Vegetable Crops (Kalloo, G. and Bergh, B.O., eds.), Pergamon Press, pp. 61-86.

Gray, A.R. and Crisp, P. (1985). The genetic potential of purple sprouting broccoli. Proc. of conference 'Better Brassicas 84', Dundee, UK, Scot. Crop Res. Inst. Sept. 1984.

Grodrich, R.M., Parker, R.S., Donald, J.C. and Stoewsand, G.S. (1988). Glucosinolate, carotene and cadmium content of *Brassica oleracea* grown on municipal sewage sludge. *Food Chem.* 27: 141.

Haga, T. (1938). Relationship of genome to secondary pairing in brassica (A preliminary note). *Japanese J. Genet.*, 13: 277.

Heaney, R.K. and Fenwick, G. R. (1984). Glucosinolates–should the breeder be concerned? Proceeding of 'Better Brassicas 84' conference. p. 175.

Hodgkin, T. (1981). Some aspects of sib production in F_1 cultivars of *Brassica oleracea*. *Acta Hortic.* 111: 17.

Hodgkin, T., Lyon, G.D. and Dickinson, H.D. (1988). Recognition in flowering plants: A comparison of the *Brassica* self-incompatibility system and plant pathogen interaction. *New Phytol.* 110: 557.

Honma, S. (1960). Spartan Early–a new variety of broccoli. *Quart. Bull. Mich. St. Univ. Agric. Expt. Stn.* 42: 470.

Honma, S. (1965). Development and maintenance in broccoli (*Brassica oleracea* L. var. *italica*). *Proc. Am. Soc. Hort. Sci.* 87: 295.

Hoser-Krauze, J. (1987). Influence of cytoplasmic male sterility source on same characters of cauliflower (*Brassica oleracea* var. *botrytis* L.). *Genet. Pol.* 28: 101.

Hulbert, S.H. and Orton, T.J. (1984). Genetic and environmental effects on mean maturity date and uniformity in broccoli. *J. Am. Soc. Hort. Sci.* 109: 487.

Innes, N.L. (1985). The work of the NVRS in conservation and breeding of vegetables. *The Garden* 110: 57.

Kalia, P. (2002). Breeding broccoli genotypes for Indian conditions–a success story. Int. Conf. on vegetables for sustainable food and nutritional security in the new millennium held at Bangalore, India. November 11-14, 2002. Abstr. II-6-P, p. 54.

Kalia, P. and Yadav, S.M. (2000). Level of self-incompatibility in green sprouting broccoli (*Brassica oleracea* L. var. *Italia* Plenck.). *Crucifereae Newsletter No.* 22: 33-34.

Kalia, P. and Yadav, S.M. (2002). Inheritance of self-incompatibility and S-allele interactions in Indian genotypes of green sprouting broccoli (*Brassica oleracia* L. var. *italica* Plenck.). XXVIth Int. Hortic. Cong. Held at Ontario, Toronto, Canada, August 11-17, 2002. Abstr. S17-P-57A, pp. 26-27.

Kalloo, G. (1988). Vegetable Breeding Vol. I. CRC Press Inc. Boca Raton, Florida. p. 239.

Kanno, T. and Hinata, K. (1969). An electron microscopy study of the barrier against pollen tube growth in self-incompatible crucifereae. *Plant Cell Physiol.* 10: 213.

Keyes, K.A. and Honma, S. (1986). Inheritance of lateral suppression and leaf number in broccoli (*Brassica oleracea* L. *italica* group). *Eucarpia Cruciferae Newsletter* 11: 43.

Kroh, M. (1966). Reaction of pollen after transfer from one stigma to another (contribution to characterisation of the incompatibility mechanisms with Cruciferae). *Zuchter* 36: 185.

Legg, P.D. and Lippert, L.F. (1966). Genetic veriation in open pollinated varieties of broccoli (*Brassica oleracea* L. var. *italica*). *Proc. Am. Soc. Hort. Sci.* 88: 411.

Legg, P.D. and Souther, F.D. (1968). Heterosis in intervarietal crosses in broccoli (*Brassica oleracea* L. var. *italica*). *Proc. Am. Soc. Hort. Soc.* 92: 432.

Lewis, J.A., Fenwick, G.R. and Gray, A.R. (1991). Glucosinolates in *Brassica* vegetables. Green curded cauliflowers (*Brassica oleracea* L. Botrytis group) and purple headed broccoli (*Brassica oleracea* L. *italica* group), Lebensm. *Wiss,* 24: 361.

Lin-BiYing, Wei-Wenlin, and Gao-shan. (1997). The breeding of a cytoplasmic male sterile line of broccoli. *J. Fujian Acad. Agric. Sci.* 12(1):24-26.

Linskens, H.F. and Kroh, M. (1967). Inkompatibilitat der Phanerogamen. *Handb. Pflanzen-Physiol.* 18: 506.

McCollum, G.D. (1981). Induction of an alloplasmic male sterile *Brassica oleracea* by substituting cytoplasm from 'Early Scarlet Globe' radish (*Raphanus sativus*). *Euphytica* 30: 855.

Michaelsen, K.F., Astrup, A.V., Mosekilde, L., Richelsen, B., Schroll, M. and Sorensen, O.H. (1994). The importance of nutrition for the prevention of osteoporosis. *Ugeskr. Laeger,* 156: 958.

Moore, J.F. and Anstey, T.H. (1954). A study of the degree of natural selfing in green sprouting broccoli (*Brassica oleracea* L. var. *italica* Plenck.) a normally cross pollinated crop. *Proc. Am. Soc. Hort. Sci.* 63: 440.

Moore, J.F. and Anstey, T.H. (1954). A study of the degree of natural selfing in green sprouting broccoli (*Brassica oleracea* L. var. *italica* Plenck.), a normally cross-pollinated crop. *Proc. Am. Soc. Hortic. Sci.* 63: 440.

Nakanishi, T. and Hinata, K. (1975). Self seed production by CO_2 gas treatment in self-incompatible cabbage. *Euphytica* 24: 117.

Nakanishii, T., Esashi, Y. and Hinata, K. (1969). Control of self-incompatibility by CO_2 gas in *Brassica. Plant cell Physiol.* 10: 925.

Nasrallah, J.B. (1989). Molecular genetics of self-incompatibility in *Brassica*. In: Plant Reproduction for Floral Induction to Pollination (Lord, E. and Bemier, G., eds.), *The Am. Soc. Plant Physiol. Symp. Series* 1: 156.

Nasrallah, J.B., Kao, T.H., Gold berg, M.L. and Nasrallah, M.E. (1985). A cDNA clone encoding on S-locus specific glycoprotein from *Brassica oleracea. Nature, UK,* 318: 263-67.

Nasrallah, M.E. and Nasrallah, J.B. (1986). Molecular biology of self-incompatibility in plants. *Trends in Genetics* 2: 239.

Ockendon, D.J. (1971). Incompatibility studies report of the National Vegetable Research Station for 1971, 28-29.

Ockendon, D.J. (1980). Distribution of S-alleles and breeding structure of cape broccoli (*Brassica oleracea* var. *italica*). *Theor. Appl. Genet.* 58: 11.

Ockendon, D.J. (1982). Breeding systems and interspecific hybrids, Brassica, Rep. *Nalt. Veg. Res. Stn. for 1981*, p. 46.

Ockendon, D.J., Currah, L. and Voss, G.A. (1979). Breeding systems–*Brassica*. Rep. *Natl. Veg. Res. Stn. for 1978*, p. 54.

Ogura, H. (1968). Studies on the new male sterility in Japanese radish with special reference to the utilization of this sterility towards the practical raising of hybrid seeds. *Mem. Fac. Agric. Kagoshima Univ.* 6: 39.

Pearson, O.H. (1972). Cytoplasmically inherited male sterility characters and flavour components from the species cross *Brassica nigra x B. oleracea. J. Am. Soc. Hortic. Sci.* 97: 397-402.

Pelletier, G., Primard, C., Vedel, F., Chetrit, P., Remy, R., Rouselle, P. and Renard, M. (1983). Intergeneric cytoplasmic hybridization in cruciferac by protoplast fusion. *Molecular and General Genetics* 191: 244-50.

Rigg, T.J. (1988). Breeding F_1 hybrid varieties of vegetables. *J. Hort. Sci.* 63: 369.

Robbelen, G. (1960). Beikage zur analyze des brassica genomes. Chromosoma II, 205.

Roggen, H.P.J.R., Dijk, A.J. Van and Dorsman, C. (1972). 'Electric aided' pollination: A method of breaking incompatibility in *Brassica oleracea* L. *Euphytica* 21: 181.

Sadik, S. (1962). Morphology of the curd of cauliflower. *Am. J. Bot.* 49: 290-297.

Sen, S. (1957). Cytological analysis of a claimed varietal hybrid of *Brassica oleracea* L. *Bull. Bot. Soc. Bengal* 9: 17.

Sharma, S.R. and Vinod (2002). Breeding for cyploplasmic male sterility in broccoli (*Brassica oleracea* L. var. *italica* Plenck.). *Indian J. Genet.* 62(2): 165-166.

Sharp, W.R., Evans, D.A. and Ammirato, P.V. (1984). Plant genetic engineering: Designing crops to meet food industry specifications. *Food Technology, February,* 1984, 112-119.

Sikka, S.M. (1940). Cytogenetics of brassica hybrids and species. *J. Genet.* 40: 441.

Simmonds, N.W. (1979). Principles of Crop Improvement. Longman, London.

Snogerup, S. (1980). The wild forms of the *Brassica oleracea* group (2n = 18) and their possible relations to the cultivated ones. In: Brassica Crops and Wild Allies (Eds. Tsunoda, S., Hinata, K. and Gomez-Campo, C.), Japan Scientific Societies Press, Tokyo, pp. 121-132.

Song, K., Osborne, T.C. and Willams, P.H. (1990). Brassica taxonomy based on nuclear restriction fragment length polymorphism (RFLPs). 3. Genome relationships in Brassica and related genera and the origin of *B. oleracea* and *B. rapa* (Syn. *campestris*). *Theor. Appl. Genet.* 79: 497.

Stoewsand, G.S. (1995). Bioactive organosulphur phytochemicals in *Brassica oleracea* vegetables–a review. *Food Chem. Toxicol.* 33: 537.

Tatebe, T. (1968). Studies on the physiological mechanism of self-incompatibility in Japanese radish. II. Breakdown of self-incompatibility by chemical treatment. *J. Japanese. Soc. Hortic. Sci.* 37: 227.

Te Nijenhuis, B. (1968). The use of isozyme patterns of *brassica* seeds in hybrid breeding. In: Brassica Meeting of Eucarpia (Dixon, G.E., ed.), Wellesbourne, 42.

Thompson, K.F. and Taylor, J.P. (1966). Non-linear dominance relationships between S-allels. *Heredity* 21: 345.

Tookey, H.L., Van Etten, C.H. and Daxenbichler, M.E. (1980). Glcoosinolates. In: Toxic Constituents of Plant Foods, 2nd ed. (I.E. Liener, ed.) Academic Press, New York.

Toriyama, K., Stein, J.C., Nasrallah, M.E. and Nasrallah, J.B. (1991). Transformation of *Brassica oleracea* with an S-locus gene from *B. campestris* changes the self-incompatibility phenotype. *Theor. Appl.Genet.* 81: 769-776.

Van der Vossen, H.A.M. (1993). *Brassica oleracea* L. cv. groups cauliflower and broccoli. In: Plant Resources of South-East Asian Vegetables (J.S. Siemonsma and Kasem Piluek, eds.), Published by Pudoc Scientific Publishers, Wageningen, No. 8: pp. 111-115.

Van Etten, C.H. and Daxenbichler, M.E. (1977). Glucosinolates and derived products in cruciferous vegetables. Gas liquid chromatographic determination of the aglucone derivatives from cabbage. *J. Assoc. Anal. Chem.* 60: 950.

Voss Stern, G.A., Ockendon, D.J., Gabrielson, R.L. and Maguire, J.D. (1982). Self-incompatibility alleles in broccoli. *Hort. Sci.* 17: 748.

Wallace, D.H. (1979). Procedures for identifying S-allele genotype of *Brassica*. *Theor. Appl. Genet.* 54: 249-265.

Wattenberg, L.W. (1983). Inhibition of neoplasia by minor dietary constituents. *Cancer Res.* 43: 2448s.

Watts, L. (1980). Flower and Vegetable Plant Breeding. Grower Books, London.

Woods, S. and Thurman, D.A. (1976). The use of seed acid phosphatases in the determination of purity of F_1 hybrid Brussels sprout seeds. *Euphytica* 25: 707.

Yadav, S.M. (1998). Self-incompatibility studies in broccoli (*Brassica oleracea* L. var. *Italica* Plenck.). MSc thesis, HP Agric. Univ., Palampur, HP 176 062, India.

An Overview
of Hybrid Khol Rabi Breeding

T. S. Verma

S. C. Sharma

SUMMARY. Khol rabi is a member of family Brassicaceae, is a popular vegetable of the German speaking countries of Europe. It is also popular in Asian countries. The edible portion of this vegetable is called "Knob" resembles an above ground turnip. It is a rich source of minerals. Hybrid varieties of this crop are also becoming popular among the farmers because of early maturity, uniform knobs and delayed fiber development. Producing high quality seed requires careful management of the parental lines, skilled labor and proper processing of the resulting seed. *[Article copies available for a fee from The Haworth Document Delivery Service: 1-800-HAWORTH. E-mail address: <docdelivery@haworthpress.com> Website: <http://www.HaworthPress.com> © 2004 by The Haworth Press, Inc. All rights reserved.]*

KEYWORDS. Khol rabi, crop biology, pollination control mechanisms, isolation, heterosis, inbred lines, single cross hybrid, double cross hybrids

---

T. S. Verma and S. C. Sharma are affiliated with the Indian Agricultural Research Institute, Kullu 175129, H.P., India.

[Haworth co-indexing entry note]: "An Overview of Hybrid Khol Rabi Breeding." Verma, T. S., and S. C. Sharma. Co-published simultaneously in *Journal of New Seeds* (Food Products Press, an imprint of The Haworth Press, Inc.) Vol. 6, No. 2/3, 2004, pp. 135-150; and: *Hybrid Vegetable Development* (ed: P. K. Singh, S. K. Dasgupta, and S. K. Tripathi) Food Products Press, an imprint of The Haworth Press, Inc., 2004, pp. 135-150. Single or multiple copies of this article are available for a fee from The Haworth Document Delivery Service [1-800-HAWORTH, 9:00 a.m. - 5:00 p.m. (EST). E-mail address: docdelivery@haworthpress.com].

Digital Object Identifier: 10.1300/J153v06n02_07

INTRODUCTION

Khol rabi (*Brassica oleracea* var. *gongylodes* L. or *Brassica caulorapa* L.) is a member of cole group of vegetables belonging to the family Cruciferae with chromosome number (2n) = 18. 'Khol' has been derived from German word 'Kohl' meaning cabbage turnip which resembles an above ground turnip. The fleshy enlargement of the stem called knob, a rounded lump, develops entirely above the ground and is used as a vegetable before it becomes tough and fibrous, though in early strains the young leaves may also be cooked.

According to Encyclopedia Britannica, knol khol was first described in 18th century as another cabbage of western Europe and is not cultivated on a large scale in the world. It is reported to have originated in the middle of Mediterranean region (Gates, 1953). It has assumed importance in some of the German speaking countries of Europe. In India, it is popular in Kashmir, West Bengal, and some parts of the south although its cultivation is not done on a commercial scale.

Knol khol is a rich source of minerals like calcium, magnesium, potassium, phosphorous, sodium, and sulphur besides vitamins A and C. Schwerdtfeger (1976) reported the influence of genetical factors on chemical composition. European varieties were found to have a higher content than Asiatic ones. Varieties of knol khol vary widely in vitamin C content.

BIOLOGY

The life cycle of khol rabi is marked by two distinct phases, vegetative and reproductive, and thus its biology can be classified into two categories

Vegetative Biology

The seedlings of khol rabi comprise of a hypocotyl a few centimeters long, two notched cotyledons and a taproot with lateral roots. The first true leaves are petiolate. The foliage is glabrous coated with a layer of wax. After some period, there is a deviation in primary growth with thickening in the pith tissues. Increase in thickness starts from just below the growing-point downwards and causes the stem to increase markedly in size. The strong primary growth in thickness causes a severe check to the longitudinal growth of the stem leading to the development of a tuber above the ground. Translocation in the tuber takes place by a network of medullar vascular bundles where in the formation of sclerenchyma causes the tubers (knobs) to become woody. The usually petiolate small leaves are arranged in a compressed spiral on the tuberous part. The tubers may be green or violet and round to flat round in shape.

Reproductive Biology

Development of flower, fruit and seed constitute reproductive biology and its knowledge is essential for the breeders in hybridization programs.

Flower development: Differentiation of flower bud results in successive development of four sepals, four petals, six stamens and two carpels. The sepals are green and erect and form the outer protective layer. In the variety White Vienna of khol rabi the petals are bright yellow and are placed at right angle to each other with a flower cross-section of approximately 2.5 cm (Verma and Sharma, 1999).

The androecium is tetradynamous consisting of two short and four long stamens. The pollen mother cells in the anthers after meiosis give rise to pollen grains which are 30-40 in diameter. The carpels form a superior ovary with a false septum and two rows of campylotropous ovules.

The pressure of the rapidly growing petals causes the buds to open. This process starts in the afternoon and usually the flowers become fully expanded during the following morning. The anthesis of flowers occurs before 10 a.m. (Brar, 1975). Since the flowers are protogynous, the anthers open a few hours later. The anthers start dehiscing at the time of opening and pollen fertility and stigma receptivity is maximum at the time of anthesis. However the stigma begins to be receptive two to three days before the flower opens thereby facilitating selfing by bud pollination even in case of self incompatibility. Insects, particularly honeybees when they visit the flower for pollen and nectar, affect pollination. The two nectaries located between the bases of the short stamens and the ovary are nectar secreting while the other two situated outside the bases of the pairs of long stamens do not secret nectar.

The inflorescence on the main stem and its branches is raceme type which in the bud stage appears like a head with the oldest buds at the periphery and the youngest at the centre. Normally 3-4 buds open in one day. Average length of flower stalk and average number of branches on the main flower stalk are 135 cm and 18, respectively, in variety White Vienna of khol rabi (Verma and Sharma, 1999).

Fruit development: The fruits (pods) of khol rabi is called silique. The siliques are green and glabrous turning yellow to grey brown on maturity (drying). These are 10-12 cm long, approximately 0.4 cm wide, having two rows of seeds located along the edges of false septum. On ripening, dehiscence takes place from below upwards.

Seed development: According to Nieuwhof (1969) in *Brassica* species, the endosperm develops immediately after fertilization while the growth of the embryo does not start for some days and even after about two weeks the embryo remains still small but after from three to five weeks it fills most of the seed with the endosperm having been almost completely absorbed. The reserve food material is stored in the cotyledons which are folded together with the radical lying between them. The seed is globular to slightly oval, 2-3 mm is diameter, light brown initially but turning greyish-black to red-brown later on. Verma and Sharma (1999) reported average number of seeds per siliqua to be 25 in variety White Vienna with 1000 seeds weighing 3.4 g.

CLIMATE

Khol rabi thrives well in cool and moist conditions while in warm conditions the knobs soon tend to become fibrous and woody. It can withstand extreme cold and frost better than other cool season crops. For good growth of Khol rabi temperatures of 18°-25°C are favorable with an optimum of 22°C. The short-term temperature changes showed, with few exceptions, no significant differences for average values of measured rates of photosynthesis, respiration and tuber growth as compared to treatments with constant temperatures (Fink and Krug, 1998).

According to Nieuwhof (1969) that early khol rabi varieties exhibit the problem of premature bolting. The varieties which are susceptible to bolting lack a juvenile phase and become generative if exposed after germination to temperatures below 10°C. Even one week of low temperatures may cause bolting. The effect of low night temperatures on the seedlings can be contracted by high day temperatures, while high temperature after transplanting can delay the bolting of plants that have been vernalized in the seed bed. The maximum temperature at which the khol rabi varieties become generative is not accurately known, but it has been observed that very susceptible varieties become generative at 12°-14°C and for their successful cultivation a minimum temperature of 15°C is required, while the less susceptible varieties do not bolt when raised at low temperatures of 12°-14°C. In slow-bolting cultivars, aged plants are more responsive to low temperatures and get vernalized sooner than the young plants. However, the correlation between earliness and susceptibility to bolting is not well established as some exceptions also exist. Purple varieties often bolt more rapidly as compared to white ones. Even, the different strains of the same variety show considerable differences for bolting.

Nieuwhof (1969) also mentioned that the duration of vernalizing period also effects bolting and it has been established that the vernalizing effect decreases when this period lasts longer than a certain number of days. Thus the areas with very long winters are not suitable for the production of khol rabi seed. For the susceptible varieties this period is 42 days, while for the resistant ones, it is 56 days. The time of bolting depends on the growth stage at which low temperatures occur. The late varieties are not responsive to low temperature in the early stages of growth and behave in the same way as cabbage and other biennial cole crops. It is only after having passed juvenile stage, these respond to vernalization only at a particular stage of growth. The day length, though, has no strong effect on the growth of khol rabi but a long day possibly, may have a slightly stimulating effect on bolting. Based on the results of growth chamber experiments, a simulation model for the prediction of bolting in khol rabi as a function of vernalization and devernalization has been developed (Gross, 1995).

Khol rabi has one typical feature because it is the only cole crop of which the pre-soaked seeds can be vernalized especially in quick bolting varieties. In these varieties, the young seedlings can also be vernalized by placing them at temperatures below 10°C for about 60 days. Seed vernalization makes it possible to grow knol khol as an annual instead of a biennial seed-crop (Nieuwhof, 1969). Due to vernalization requirement for conversion of vegetative phase into reproductive phase, the seed production of khol rabi is possible in the temperate regions experiencing a low temperature of 4°-10°C for 60 days (Singh et al., 1959; Verma and Sharma, 2000).

POLLINATION CONTROL MECHANISMS

For hybridization the knowledge of pollination behavior is essential. The pollination control mechanisms make it easier and economize the development of hybrids and their seed production. Among the different mechanisms, self-incompatibility is an important one which restricts self-pollination and encourages cross-pollination because functional pollen is unable to set seed after self-pollination thus favoring heterozygosity in the population. It is prevalent in *Brassica oleracea* of which khol rabi is one of the members. In khol rabi, homomorphic sporophytic system of self-incompatibility has been reported (Thomson and Taylor, 1966). In this the incompatibility reaction of pollen is controlled by the genotype of the plant on which it is produced and is determined by one locus with multiple S-alleles. There is dominance of one allele over the other in the pollen or style or the two alleles may be active on both pollen and style. An effective system of self-incompatibility has been reported in khol rabi (Thompson and Taylor, 1966).

Khol rabi is highly cross-pollinated due to the mechanism of self-incompatibility and the extent of natural cross-pollination has been reported to be 91.0% (Watts, 1968). The incompatibility is due to the fact that the germination of the pollen on the stigma of the flowers borne on the same plant is not proper and the pollen tubes formed are twisted and rudimentary. The self-incompatibility is strongest in freshly opened flowers. When young buds are selfed, seed setting is much better than at later stages and is comparable to cross-pollination.

In breeding, the inbred lines are produced by using the method of bud-pollination. In the bud stage (5-8 mm bud length) about two to three days before the
flowers open, the S-alleles are inactive thus good seed setting is obtained on
self-pollination of buds. Self-pollination of very young buds decreases seed-
setting due to (i) low level of fertility of flowers and (ii) mechanical injury to
these buds during pollination. In older buds, the incompatibility gradually increases and is maximum at the time of opening of flowers because the activity
of S-alleles reaches at its peak by this stage and no seed-setting is obtained on
self-pollination. It decreases again after the opening of the flowers and selfing
of two to four days old flowers gives the same seed setting as is obtained in the
proper bud stage.

ISOLATION

For maintaining the purity of the seed of knol khol (*Brassica oleracea* var.
gongylodes), it needs to be isolated not only from its varieties but also from all
other botanical varieties of *Brassica oleracea* namely, *B. oleracea* var. *capitata*
(cabbage), *B. oleracea* var. *botrytis* (cauliflower), *B. oleracea* var. *italica* (sprouting broccoli), *B. oleracea* var. *gemmifera* (Brussels sprout) and *B. oleracea*
var. *acephala* (kale and collards). All these crops intercross freely among
themselves, therefore, proper isolation distance should be maintained between
two varieties of khol rabi or between khol rabi and any of the botanical varieties of *Brassica oleracea* when grown for seed at the same location. At least
1600 m isolation is recommended for the production of hybrid seed of khol
rabi (Tunwar and Singh, 1988).

HETEROSIS

Heterosis has been exploited for the improvement of vegetable crops in different aspects, viz., earliness, winter-hardiness, resistance to biotic and abiotic
factors, uniformity and higher yields. Among the cross-pollination mechanisms, self-incompatibility has been utilized successfully to economize the
hybrid seed production in cabbage, cauliflower, and broccoli and can be exploited in other cole crops also. The main advantage of using self-incompatibility is that it is possible to obtain F_1 hybrid seed from both the male and
female parental inbred lines. On the contrary, the hand emasculation and hand
pollination of individual flowers of khol rabi is not economical due to less
number of seeds per pod resulting from a single pollination. Disease-resistant
genes can also be incorporated to produce multiple disease-resistant hybrids.
Khol rabi, being consumed in some areas of the world, is not a cosmopolitan
crop as are cabbage and cauliflower. Therefore, the progress in the exploita-

tion of heterosis in this crop has been limited. Keeping in view the heterotic advantages, there is a need to make a concerted approach in this crop.

Rubatzky and Hang (1996) reported significant heterosis in weight and width of the enlarged stem and height and width of the plant of khol rabi with increased yields of 58.7-95.3% in the heterotic combinations.

At IARI, Regional Station, Katrain (Kullu Valley), Himachal Pradesh (India), one F_1 hybrid KKH-1 of khol rabi developed by hand emasculation and hand pollination method has performed well with respect to earliness, shape and size of knobs and higher yield and possessing no fibre even at full maturity. To economize hybrid seed production, development of cytoplasmic male sterile (CMS) line of its female parent is in progress (unpublished).

METHOD OF PRODUCTION OF HYBRID SEED

For commercial hybrid seed production of sexually propagated crops, inbred lines are highly desirable in comparison to open-pollinated varieties having a broad genetic base because (i) inbreds can be maintained in homozygous state while the genetic constitution of open-pollinated varieties may be modified by the evolutionary forces, (ii) the hybrids derived from inbreds are homogeneous year after year, and (iii) homogeneity of inbred-derived hybrids is desirable from the viewpoint of uniform quality of the produce. The different operations involved in the development of hybrid varieties are (i) Development of inbred lines, (ii) Testing of combining ability, (iii) Improvement of inbred lines, and (iv) Production of hybrid seed.

Development of Inbred Lines

Inbred lines are developed by continued inbreeding of selected plants from a genetically variable population. Inbreeding results in a large number of lethal week forms and a general decline in vigor, fertility, and fecundity. It increases homozygosity and decreases heterozygosity in the population. In khol rabi and in all cole crops selfing for inbreeding is done artificially by bud-pollination (Pearson, 1929). Other methods like electrically aided pollination (Roggen et al., 1972; Roggen and Van Dijk, 1973) and thermally aided pollination (Roggen and Van Dijk, 1976) can also be used for selfing. The number of generations of inbreeding/selfing depends on the degree of inbreeding depression. In knol khol, severe degree of inbreeding depression has been reported (Fabig, 1973), therefore five generations of inbreeding are required (Dorsman, 1976). Uniformity of inbred lines should be tested before their exploitation in hybrid development since the uniformity of the hybrids depends entirely upon the homozygosity of the parental inbred lines.

For the production of hybrid seed, the utilization of homozygous self-in-incompatible lines is commercially accepted economical technique. Khol rabi being self-incompatible has all the plants heterozygous for 'S' locus. A randomly selected S-allele heterozygous plant (Io) on the basis of horticultural, disease resistance or other characteristics is selfed by adopting bud pollination technique. The self seeds are grown as I_1 progeny in which, at the time of flowering, reciprocal crosses among I_1 sibling plants are made. If there is no or very poor seed-set, the line is homozygous for S-allele. The proportion of homozygous and heterozygous lines is 50:50. The type of S-alleles in different homozygous lines can be known by reciprocal differences as per procedure outlined by Wallace (1979 b). Alternatively, a procedure for identifying S-allele homozygous lines can be adopted by making reciprocal crosses between the Io parent and the I_1 siblings (Wallace, 1979 b). According to him, fluorescent microscopy can also be used to identify homozygous lines on the basis of pollen tube growth in the style. Such homozygous self-incompatible lines can be exploited for the production of single-, double- and triple-cross hybrid seeds (Odland and Noll, 1950; Thompson, 1964; Wallace, 1979 b; Bauch, 1983). A combined approach based on seed set and pollen tube growth is advantageous for the identification of homozygous self-incompatible but cross compatible lines.

The S-allele interactions often approach one of the two extremes (Table 1): (a) codominance, i.e., simultaneous full activity of both S-alleles of the heterozygote, (b) dominance, i.e., full activity of one S-allele of the heterozygote with complete inactivity of the other (Wallace, 1979a). Four classes of incompatibility in *Brassica* were described and designated as mode of inheritance for S-allele interaction Types, I, II, III, and IV (Thompson and Howard, 1959; Haruta, 1962). These four types are derived from two S-allele interactions (codominance and dominance) in factorial combination with the two sexual organs: pollen and stigma (Table 1). To recognize these factorial components, the types are redesignated as sexual-organ × S-allele-interaction Types I, II, III, and IV. The possibility of either codominant or dominant S-allele interaction occurring in the pollen and independently of the same alternative codominant or dominant S-allele interactions occurring in the stigma have been used to determine the incompatibility, compatibility and reciprocal difference expectations for reciprocal crosses among the three I_1 genotypes and these expectations are given (Table 2) for each of the sexual organ × S-allele interaction Types I, II, III, and IV (Wallace, 1979b). These expectations are used to interpret results from reciprocal crosses among the unknown genotypes of the I_1 sibling population and/or to interpret results for reciprocal crosses between the Io parent and the I_1 plants. This facilitates simultaneous determinations of (a) codominance or dominance that occurs in the pollen and

TABLE 1. Sexual organ × S-allele interaction Types I, II, III, IV as derived from the two S-allele interactions (dominance[a] and codominance[a]) in factorial combination with two sexual tissues (pollen and stigma)

	POLLEN	STIGMA		POLLEN	STIGMA
I	$S_a < S_b$	$S_a < S_b$	III	$S_a = S_b$	$S_a < S_b$
II	$S_a < S_b$	$S_a = S_b$	IV	$S_a = S_b$	$S_a = S_b$

[a] the allele to the open side of < is dominant while that to the closed side is recessive, and = indicates codominance, i.e., strong and near equal activity of both alleles.

TABLE 2. Expected compatible and incompatible interpretations and reciprocal difference interpretations[a] from reciprocal intercrosses among plants of the two homozygous and one heterozygous S-allele for sexual organ × S-allele interaction Types I, II, III, IV

Sexual organ S-allele interaction	S-allele interaction		Female genotype and phenotype	Male genotype and phenotype		
	Stigma	Pollen				
				$S_a\ S_a$	$S_a < S_b$	$S_b\ S_b$
			$S_a\ S_a$	Inc	Com	Com
I	DOMINANCE	DOMINANCE	$S_a < S_b$	Com	Inc	Inc
			$S_b\ S_b$	Com	Inc	Inc
				$S_a\ S_a$	$S_a < S_b$	$S_b\ S_b$
			$S_a\ S_a$	Inc	Com	Com
II	CODOMINANCE	DOMINANCE	$S_a = S_b$	Inc	Inc	Inc
			$S_b\ S_b$	Com	Inc	Inc
				$S_a\ S_a$	$S_a = S_b$	$S_b\ S_b$
			$S_a\ S_a$	Inc	Inc	Com
III	DOMINANCE	CODOMINANCE	$S_a < S_b$	Com	Inc	Inc
			$S_b\ S_b$	Com	Inc	Inc
				$S_a\ S_a$	$S_a = S_b$	$S_b\ S_b$
			$S_a\ S_a$	Inc	Inc	Com
IV	CODOMINANCE	CODOMINANCE	$S_a = S_b$	Inc	Inc	Inc
			$S_b\ S_b$	Com	Inc	Inc

[a] Reciprocal crosses with a reciprocal difference is indicated by ◀――――▶ and reciprocal crosses without a reciprocal difference by ◀――▶

[b] The S allele phenotype of heterozygotes is indicated by a the symbols < and = where < specifies recessive vs. dominant and = indicates codominance. These phenotypes correspond with the described S-allele iteractions in the stigma and pollen. The phenotypes for homozygotes always correspond with the genotype.

in the stigma of the S-allele heterozygote, (b) the sexual-organ × S-allele-interaction types, and (c) the S-allele genotypes of the I_1 plants.

Testing of Combining Ability

The success in exploitation of heterosis depends upon the combining ability of the parental inbred lines and the technique utilized to economize hybrid seed production. The selection of inbred lines for the development of hybrid is done on the basis of general combining ability (gca) of parental inbred lines and specific combining ability (sca) of the crosses, the latter being more reliable for the selection of parents for hybrid production. General combining ability represents additive type of gene action while in specific combining ability, dominance and epistatic gene actions are involved. Combining ability is tested by different mating designs. Polycross is adopted to study gca; single- and three-way crosses for sca, and top and diallel crosses for gca and sca both. The performance of single crosses can be used to predict the performance of double cross hybrids.

Improvement of Inbred Lines

It is a well-established fact that the performance of hybrids depends upon the performance and quality of the inbreds. In direct isolation of inbreds from source populations, the frequency of outstanding inbreds in low. Improvement of inbred lines to be used for producing a hybrid, is done specifically for disease and insect resistance or other characteristics like productivity and combining ability so as to increase the yielding ability of their hybrids. Generally, back cross and convergent methods are adopted for improving the inbred lines. For the incorporation of resistant genes in the inbred lines to be used for the production of multiple-resistant hybrids, backcross method can be utilized. The convergent method is adopted where it is required to combine desirable characteristics of two or more inbreds in a single genotype which can further be utilized for hybrid development.

Production for Hybrid Seed

Commercial production of F_1 hybrid seed of khol rabi depends on the two parental inbred lines each homozygous for a different incompatibility S-allele. Each line should be both self- and sib-incompatible but the two lines must be cross-compatible. In khol rabi, single-, triple-, 3-way or 4-way/double-cross hybrids can be produced.

Single-Cross Hybrid. A single-cross hybrid is developed by using two homozygous self-incompatible but cross-compatible lines in a planting design of either alternate or 2 to 4 rows of each (Figure 1). Pollination is done by insects. F_1 population of a single-cross hybrid is homogeneous and thus, preferred over triple- and double-cross hybrids. The only drawback is that it is expensive and uneconomical due to less quantity of hybrid seed.

Triple-Cross Hybrid. This method was advocated in kale (*Brassica oleracea* var. *acephala*) by Thompson (1964) in which two 3-way hybrids are involved in the production of hybrid (Figure 2). Although the hybrid seed yield is more and is less expensive than the single-cross hybrid seed but the uniformity is reduced. The only advantage of this method is the low cost of production of parental inbred lines due to less quantity of seed required for producing single-cross and 3-way hybrid seed.

3-Way Hybrid. The basic requirements for producing a 3-way hybrid is a single cross F_1 hybrid that is highly self- and sib-incompatible and similarity

FIGURE 1. Scheme for production of hybrid seed: single cross

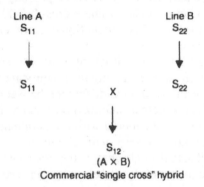

FIGURE 2. Scheme for production of hybrid seed: triple cross

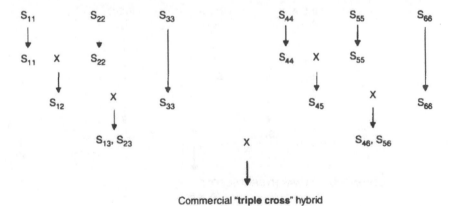

of two parental inbreds of each single-cross hybrid for horticultural characters so as to maintain uniformity in a 3-way hybrid approaching that of single-cross hybrid and both these requirements are fulfilled by identifying the S-allele genotypes and sexual-organ × S-allele interaction types in the I_1 generation (Wallace, 1979 b). A 3-way hybrid involves a cross between a single cross and an inbred line (Figure 3).

Double-Cross or 4-Way Hybrid. Odland and Knoll (1950) used this method for hybrid seed production in cabbage (Figure 4). Two inbred lines that are highly self- and sib-incompatible but reciprocally compatible are developed from a variety. Similarly, another two highly self- and sib-incompatible but reciprocally compatible inbred lines are developed from other variety. The basic requirements of a double-cross or 4-way hybrid are the same as for 3-way hybrid. Therefore, the single crosses of each pair of inbred lines must be highly self- and sib-incompatible but must cross reciprocally with each other. The single crosses are produced similar to single-cross hybrid. These single crosses are planted in 1:1 ratio for the final production of commercial F_1 seed resulting from their crossing with each other.

All the four inbred lines have different S-alleles. The knowledge gained during genotype identification regarding the reciprocal compatibility, reciprocal incompatibility or the reciprocal difference between the S-allele heterozygote and each of its counterpart homozygotes makes it possible to produce the single-cross seed using the inbred with the dominant allele as the female parent because any unintended selfs in the single cross will all be reciprocally incompatible with all actual single crosses thereby causing near total elimination of inbreds in the 3-way or double-cross hybrid seed.

FIGURE 3. Scheme for production of hybrid seeds: 3-way cross

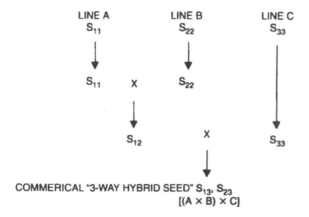

LINE A   LINE B   LINE C
S_{11} S_{22} S_{33}

S_{11} X S_{22}

S_{12} X S_{33}

COMMERICAL "3-WAY HYBRID SEED" S_{13}, S_{23}
[(A × B) × C]

FIGURE 4. Scheme for production of hybrid seeds: Double cross or 4-way cross

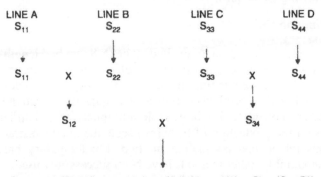

Commercial "double cross or 4-way" hybrid seed [(A × B) × (C × D)]

On the other hand, if an inbred homozygous for recessive S-allele is used as the female parent of the single cross which in turn is used as female parent of a 3- or 4-way hybrid, any unintended selfs within the single cross will be reciprocally compatible with the single cross (Type I) or compatible in one direction but incompatible in the other (Types II and III) depending on the sexual-organ × S-allele interaction types derived from the two S-allele interactions (dominance and co dominance) in factorial combination with the two sexual tissues namely, pollen and stigma. Type IV, in which the S-allele interaction for the heterozygotes is co dominance in both stigma and pollen, is the most desirable for use as female parent of the single-cross used in producing a 3- or 4-way hybrid because any unintended selfs in the single cross will be debarred from acting since the single cross (heterozygote) is reciprocally incompatible with both of its homozygous parental inbred lines.

For hybrid seed production, synchronization of flowering of self-incompatible inbred lines is an obstacle and this can be made possible by adjusting the time of transplanting of different lines and also by pinching the main stalk or picking of lower buds to synchronize flowering. In F_1 hybrid seed production, use of bees for pollination will definitely enhance the seed yield.

MAINTENANCE OF INBRED LINES

In the production of inbreds to be used as parents in hybrid seed production it is essential that self-compatibility is achieved in such a manner that self-incompatibility is again attained, i.e., fully functional in the selfed progeny. Thus self-compatibility is also known as pseudo-compatibility and this can be achieved by temporarily suppressing the self-incompatibility reaction by any of the following techniques.

Bud Pollination (BP)

In this method, the stigma, in the bud stage (5-8 mm bud length) about two to three days before opening of flowers, is self-pollinated by the mature pollen which causes good seed-setting due to the inactivity of S-alleles in the bud stage. This method is useful where only a small quantity of seed of inbred lines is required as in breeding. For large scale multiplication and utilization of inbred lines for the production of F_1 hybrid seed, the main constraint has been the tedious job in bud pollination method. For temporary breakdown of self-incompatibility, other methods have been successfully used.

Treatment with Carbon Dioxide

If the flowering plants of a self-incompatible line are placed in 2 to 5 percent carbon dioxide gas in an air tight greenhouse within 2 to 6 hours of pollination, self-incompatibility is temporarily eliminated. In practice, free pollination is allowed with the help of bees during the day followed by treatment of the greenhouse with the required concentration of carbon dioxide preferably in the evening so that the carbon dioxide gas is diffused during night and the conditions inside the greenhouse become normal by the following morning. Fertilization takes place due to temporary suspension of self-incompatibility. This treatment is repeated several times during the full flowering period.

Since the different self-incompatible lines respond differently to carbon dioxide treatment, therefore, the concentration and duration of treatment would have to be standardized for each line. If self-incompatibility cannot be broken by this method, the bud pollination method can be successfully used.

Electric Aided Pollination

Electric aided pollination can be used to self the open flowers by applying a direct electric potential difference of 100 V between pollen and stigma during pollination (Roggen et al., 1972; Roggen and Van Dijk, 1973). Although the number of seeds per fruit (siliqua) was lesser in electric aided pollination than with bud pollination method, but is compensated for by a higher percentage of fruit setting.

Thermally Aided Pollination

Application of heat during self-pollination of open flowers by means of an electric mini soldering iron caused seed-set in self-incompatible *Brassica oleracea* varieties and gave a considerably higher seed yield at 70°C and 80°C temperatures.

Pseudo-compatibility may also occur naturally due to (a) genetic factors (weak S-alleles, modifier genes, and effect of non-allelic genes), (b) environmental factors (high temperature, high humidity and high concentration of carbon dioxide gas), and (c) physiological factors (flower age and stage of flowering).

REFERENCES

Bauch, W. (1983). Use of modified sibling hybrids in breeding heading cabbage. *Archiv. Zuechtungsforsch*, 13:285.

Brar, J.S. (1975). Summer Institute on vegetable production techniques and methods of vegetable breeding at PAU. Ludhiana.

Dorsman, C. (1976). F $_1$ hybrids in some out-pollinating crops, in *heterosis in Plant Breeding*. Proc. 7th Congr., Eucarpia, Budapest, Janossy, A. and Lupton, F.G.H. (Eds.). Elsevier, Amsterdam, 197.

Fabig, F. (1973). Studies in line maintenance in hybrid breeding in *Brassica oleracea Pl.. Breed. Abstr.*, 45:695.

Fink, M. and Krug, H. (1998). Effects of short-term temperature changes on photosynthesis, respiration and tuber growth of kohlrabi (*Brassica oleracea* convar. *acephala* var. *gongylodes* L.). *Gartenbauwissenschaft*, 63 (6):254-262.

Gates, R.R. (1953). Wild cabbages and their effect on cultivation. *J. Genet.* 51: 363-372.

Gross, D. (1995). Validation and application of a simulation model for flower formation of kohl rabi depending on vernalization and devernalization. *Gartenbauwissenschaft*, 60 (3):97-101.

Haruta. T. (1962). Studies on the genetics of self- and cross-incompatibility in cruciferous vegetables (English transl. available from Minneapolis, MN: Northrup, King & Co. Takii) Plant Breed. and Exp. Sta., Kyoto, Japan Res. Bul. 2.

Niewhof, M. (1969). *Cole Crops*. Leonard Hill, London.

Odland, M.L. and Noll, C.J. (1950). The utilization of cross-compatibility and self-incompatibility in the production of F$_1$ hybrid cabbage. *Proc. Am. Soc. Hortic. Sci.*, 55:391.

Pearson, O.H. (1929). Observations on the type of sterility in *Brassica oleracea* var. *capitata*. *Proc. Am. Soc. Hortic. Sci.*, 26:34.

Roggen, H.P.J.R. and Van Dijk, A.J. (1973). Electric aided and bud pollination which method to use for self seed production in cole crops ? *Euphytica*, 22:260.

Roggen, H.P.J.R. and Van Dijk, A.J. (1976). Thermally aided pollination: A new method of breaking self-incompatibility in *Brassica oleracea* L. *Euphytica*, 25:643.

Roggen, H.P.J.R., Van Dijk, A.J. and Dorsman, C. (1972). Electric aided pollination: A method of breaking incompatibility in *Brassica oleracea* L. *Euphytica*, 21:181-184.

Rubatzky, V.E. and Hang, Chen (1996). Investigation and utilization of heterosis in kohlrabi (*Brassica oleracea* var. *gongylodes*). *Acta Horticulturae*, 467:127-132.

Schwerdtfegor, G. (1976). An overview of hybrid khol rabi breeding. *Land Wirtschaftskammer*, 54 (1):73-90.

Singh, H.B., Thakur, M.R. and Bhagchandani, P.M. (1959). Vegetable seed production in Kullu Valley-I. *Indian J. Hort.*, 16 (2):92-101.

Thompson, K.F. (1964). Triple-cross hybrid kale. *Euphytica*, 13:173.

Thompson, K.F. and Howard, H.W. (1959). Self-incompatibility in marrow stem kale, *Brassica oleracea* var. *acephala*. II. Methods for the recognition in inbred lines of plants homozygous for S alleles. *J. Genet*, 56: 324-340.

Thompson, K.F. and Taylor, J.P. (1966). Non linear dominance relationships between S alleles. *Heredity*, 21:345.

Tunwar, N.S. and Singh. S.V. (1988). *Indian Minimum Seed Certification Standards*. Published by the Central Seed Certification Board, Department of Agriculture and Cooperation, Ministry of Agriculture, Government of India, New Delhi.

Verma, T.S. and Sharma, S.C. (1999). *Post Vegetative Phase Characteristics of Biennial Vegetable Crops*. (Bulletin). I.A.R.I. Regional Station, Katrain (Kullu Valley), HP-175129 (India) pp. 26-27.

Verma, T.S., and Sharma, S.C. (2000). *Producing Seeds of Biennial Vegetables in Temperate Regions*. Directorate of Information and Publications of Agriculture, Indian Council of Agricultural Research, New Delhi.

Wallace, D.H. (1979a). Interaction of S-alleles in sporophytically controlled self-incompatibility of *Brassica*. *Theor. Appl. Genet.* 54:193-201.

Wallace, D.H. (1979b). Procedures for identifying S-allele genotypes of *Brassica*. *Theor. Appl. Genet.*, 54:249-265.

Watts, L.E. (1968). Natural cross pollination and the identification of hybrids between botanical varieties of *Brassica oleracea* L. *Euphytica*, 17:74-80.

A Review
of Hybrid Cauliflower Development

S. R. Sharma
Praveen K. Singh
Veronique Chable
S. K. Tripathi

SUMMARY. Cauliflower is an important vegetable crop which belongs to the family Brassicaceae and is grown in many countries like India, China, Italy, Europe, America, etc. It is grown for its highly suppressed 'prefloral fleshy apical meristem' branches called "curd." It is a cross-pollinated crop. There are different groups based on their characteristics. Multiple pollination mechanisms, e.g., self-incompatibility and male sterility, not only encourages cross-pollination but also found useful in the commercial hybrid seed production of the crop. As the main hindrance to the popularization of F_1 hybrids is unavailability and high cost of hybrid seed. In cauliflower, F_1 hybrids have been found advantageous for earliness, high early and total yield, better curd quality with respect to compactness and color, uniform maturity, resistance to insect pest, diseases and unfavorable weather conditions. Many studies has been

S. R. Sharma is affiliated with the Division of Vegetable Science, IARI. Pusa, New Delhi 110012.

Praveen K. Singh and S. K. Tripathi are affiliated with the Sungro Seeds Ltd., New Delhi 110088.

Veronique Chable is affiliated with the UMR Amelioration des Plantes et Biotechnologies Vegetales, 35653 Le Rheu, France.

[Haworth co-indexing entry note]: "A Review of Hybrid Cauliflower Development." Sharma, S. R. et al. Co-published simultaneously in *Journal of New Seeds* (Food Products Press, an imprint of The Haworth Press, Inc.) Vol. 6, No. 2/3, 2004, pp. 151-193; and: *Hybrid Vegetable Development* (ed: P. K. Singh, S. K. Dasgupta, and S. K. Tripathi) Food Products Press, an imprint of The Haworth Press, Inc., 2004, pp. 151-193. Single or multiple copies of this article are available for a fee from The Haworth Document Delivery Service [1-800-HAWORTH, 9:00 a.m. - 5:00 p.m. (EST). E-mail address: docdelivery@haworthpress.com].

http://www.haworthpress.com/web/JNS
Digital Object Identifier: 10.1300/J153v06n02_08

done on the aspects of genetic improvement, resistance for biotic, abiotic stresses and on bio-technological aspects. *[Article copies available for a fee from The Haworth Document Delivery Service: 1-800-HAWORTH. E-mail address: <docdelivery@haworthpress. com> Website: <http://www.HaworthPress. com> © 2004 by The Haworth Press, Inc. All rights reserved.]*

KEYWORDS. Cauliflower, breeding, improvement, seed production, hybrid seed production, resistance breeding, male sterility, self-incompatibility, biotechnology

INTRODUCTION

Cauliflower (*Brassica oleracea* L. var. *botrytis* L. 2n = 2x = 18) is one of the most popular *Brassica* vegetable after cabbage. This is cultivated worldwide in different climatic conditions, ranging from temperate to tropics during most of cropping seasons and is available round the year in the market. The word cauliflower comes from Latin term *caulis* and *floris*, meaning stem or stalk and flower, respectively. Its world wide total area and production is 8085 lakh ha and 157015 lakh tones, respectively (FAO production year book, 2002). It is grown for its white tender head or curd. The curd of cauliflower has been described as a prefloral structure, which has the characteristics of both the vegetative and reproductive apices (Sadik, 1962; Margara and David, 1978). The vegetative shoots follows the 5 to 8 phyllotaxy of leaves, but the leaf development reduced so that only bracts are formed. The lateral buds of the shoot meristem elongate and are much branched whose apices form the surface of the curd. The whole shoot system are much shortened and thicker and can give rise to the future inflorescence. So the present understanding is that, cauliflower curd is a prefloral fleshy apical meristem, which invariably precedes floral initiation compared to the closely related another *Brassica* vegetable broccoli (*Brassica oleracea* L. var. *italica* Plenck.), whose head are composed of flower buds.

Cauliflower is generally used as cooked vegetable either singly or mixed with potato, carrot, and peas. In raw form, it is also mixed with green salad or its pieces are dipped into sauces. It is also used in the preparation of pickle or mixed pickle with other vegetables.

Cauliflower is low in calories, but is a good source of ascorbic acid and contains substantial amount of protein, and nutrients like phosphorus, calcium, and iron.

Apart from India and China, the other major producers of cauliflower are France, Italy, United Kingdom, United State of America, Spain, Poland, Germany, and Pakistan.

ORIGIN AND EVOLUTIONARY HISTORY

Brassica oleracea L. grows wild in primitive form in Atlantic coasts of Europe. It was eventually brought to east Mediterranean region where it became fully domesticated and started giving rise to wide range of cultivated forms. So like other cultivated forms of cabbage group, the cauliflower is also believed to be descendent of wild cabbage (*Brassica oleracea.* var. *sylvestris.*), which is still found growing wild, in the coastal area of Mediterranean sea and western Europe. But recently, a polyphyllectic origin by incorporation of genes into the *Brassica oleracea* genome from different wild mediterranean species was suggested (Gustafsson, 1979; Snogerup, 1980). This resulted in giving rise to wide range of *Brassica oleracea* forms and their adaptation. But again possibility of introgression found is very less.

Schulz (1919) suggested *B. cretica.* as the probable progenitor of cauliflower. The taxonomical studies suggested that the progenitor of *Brassica oleracea* exists in the 9 chromosomes wild *B. oleracea* kale (Snogerup, 1980). RFLP studies also revealed that a primitive cultivated *Brassica oleracea* might have evolved from wild *B. oleracea* (Song et al., 1988 a & b; Figdore et al., 1988). The accessions of *B. oleracea* could be divided in to three groups. Thousand head kale and chinese kale; cabbage group (cabbage, collard, savoy cabbage, kohl rabi, Portugese cabbage) and broccoli (Marrow stem kale, broccoli, brussels sprouts, Jersey kale). Earlier studies by Crisp (1982) and Gray (1982) suggested that cauliflower originated from broccoli but above RFLP investigation showed that two cauliflower accessions phyllogenetically were more close to cabbage rather than broccoli group. A large divergence of cauliflower from other accessions suggested that it does not belong to either group. It is also possible that cauliflower have morphotype origin in cabbage group or may have independent origin from wild species such as *Brassica cretica* as suggested by Snogerup (1980).

According to Boswell (1949), it originated in the island of Cyprus, from where it moved to other areas like, Syria, Turkey, Egypt, Italy, Spain, and north western Europe. It was unknown in its present form before early mid age. Cauliflower might have originated gradually from the wild cabbage through mutation, human selection and adaptation and suppose to have been domesticated in the eastern Mediterranean region (Helm, 1963). According to Hyams (1971), cauliflower was first noticed, selected, and propagated in Syria.

The Herbalist, Dodens (1578) presented the first description and illustration of cauliflower. It was in cultivation in France around 16th century and was available in the markets in England as early as 1690. In the USA, it was first mentioned in 1806 but attained commercial status after 1920. It was introduced to India by Dr. Jemson, a botanist, at Kew garden London in 1822 and the Royal Agri-Horticultral Society, South Africa in 1824 (Swarup and Chatterjee,

1972). Seeds were imported from South Africa and England was also given to the growers in north India. In about 100 years (1822-1929), these growers through selections, though unconsciously evolved the present day Indian cauliflower with ability to grow under high temperature and humid conditions, with the ability to produce seeds in north Indian plains. It is likely that Cornish types has contributed most of the genes like long stalk, open growth habit, and yellowish, uneven, and strong flavored curds. Some of the leaf and curd characteristics were also contributed by 'Roscoff' 'Italian' and 'Northern' types. So the present day Indian (tropical) cauliflower is the result of introgression of all the above types. The Indian cauliflower has been recognised as a different type not only at national but at international level also by the earlier workers like Nieuwhof, 1969 (Netherland); Swarup and Chatterjee, 1972; Chatterjee and Swarup, 1972; Crisp, 1982 (England). The Indian cauliflower has been further divided on the basis of temperature requirement for curd development and maturity as, (a) Early (20-27°C), (b) Mid (16-20°C), and (c) Mid-late (12-16°C) and under north Indian plains, the respective period for production are August end-mid November, Late November-mid December, and late December-mid January. Mid late maturity group is followed by annual temperate types, which includes Snowball, Erfurt or Alpha strains maturing in January-February at a mean temperature range of 10°C-16°C (Chatterjee 1993; Singh and Sharma, 2000). So in India, only annual types are grown.

In addition to these two annual types, there are other developments in this crop, which took place independently in different regions of the world. They remained genetically isolated for a long period (except for the Italians or originals) and thus maintained their characteristic features. It would, therefore, be worthwhile to classify these broad groups so that a proper understanding and relationship of the present-day cultivars is possible. Swarup and Chatterjee (1972) classified these groups as shown in Table 1.

Crisp (1982) has classified the cauliflowers according to their phylogeny in Table 2.

However, Chatterjee (1993) recommended that further studies are required for separate grouping of the North European annual and Australian types.

THERAPEUTIC PROPERTIES

Cauliflower curd extract is used as a traditional medicine in the treatment of scurvy; as a blood purifier and as an antacid (Liebstein, 1927). Its seeds also have contraceptive properties (El-Dean, 1972). Cauliflower extract contains ascorbic acid, carotenoids, tocopherols, isothiocynotes, indoles, and flavonoids, which act as potential chemoperventive agent to check the initiation and promotion of carcinogenesis. Anti-tumor promoting activity of cauliflower ex-

TABLE 1

Cauliflower types	Country of origin	Probable period of first cultivation	Characters
Italians or Original	Mediterranean	16th Century	Plants short; leaves erect broad with rounded tips, bluish green; curds good not protected by leaves
Cornish	England	Early 19th Century	Plants vigorous; long stalked; leaves loosely arranged, broadly wavy; curds flat, irregular, loose, not protected, yellow, highly flavored
Northerns	England	19th Century	Leaves petiolate, broad, very wavy, serrated; curds good, well protected
Roscoff	France	19th Century	Plants short; leaves long erect, slightly wavy with pointed tip, midrib prominent, bluish green; curds white or creamy, hemispherical, well protected
Angers	France	19th Century	Leaves very wavy, serrated, greyish green; curds solid, white, well protected
Erfurt and Snowball	Germany and Netherlands	18th Century	Plants dwarf; leaves short, erect, glaucous green; curds solid, well protected
Indian cauliflower	India	Late 19th Century	Plants short, long stalked; leaves loosely arranged, broadly wavy; curds flat, somewhat loose, yellow to creamy, not protected and highly flavored

tracts *in vitro* short term assay system of inhibition of Epstein-Bar virus was reported by Koshimizue et al. (1988). Dried cauliflower powder contains indol-3 acetonitrile, 3,3-diindolylmethane, and 1-methoxyindol-3 carbaldehyde inducers of cytochrome P-450 dependent monoxygenase activity in mice (Bradfield et al., 1985; Bradfield and Bjeldanes, 1987). Cauliflower rich diet was also found to inhibit hepatic toxicity caused by food contaminants, polybrominated biphenyls and aflatoxin B_1 in rats (Stoewsand et al., 1978).

GENETIC RESOURCES

With the adoption of improved modern cultivars, the genetic variability in most of the cole crops including cauliflower present in the form of land races, and the primitive types are disappearing. Moreover, inadequate taxonomic knowledge of the wider variability found in this crop and efforts made to im-

TABLE 2

Group	Chief characteristics	Common types
Italian	Very diverse, include both annuals and biennials and curds with peculiar conformations and colors	Jezi, Naples (Autumn Giant), Romanesco, Flora Blanca
North-West European biennials	Derived within the last 300 years from Italian material	Old English, Walcheran, Roscoff, Angers, St. Malo
North European annuals	Developed in northern Europe for atleast 400 years. Origin unknown, perhaps Italian or Eastern Mediterranean	Lecerf, Alpha, Mechelse, Erfurt, Danish
Asian	Recombinants of European annuals and biennials developed within 250 years, adapted to tropics	Four maturity groups are recognized by Swarup and Chatterjee (1972)
Australian	Recombinants of European annuals and biennials and perhaps Italian stock, developed during the last 200 years	Not yet been categorized

prove it by transferring desirable traits from distant botanical varieties (cabbage or kale) instead of close relative (broccoli) has also resulted in loosing the variability. Replacement of open-pollinated cultivars with narrow genetic based F_1 hybrids has resulted in the genetic erosion of cauliflower and other cole crops. So, sincere efforts are needed to preserve the germplasm. Germplasm conservation is encouraged by an international organization: IPGRI (International Plant Genetic Resources Institute). At the European level, ECP-GR (European Cooperative Program for Crop Genetic and Resources networks) have grouped 35 countries. For *Brassica*, information is centralized in the Wageningen University (The Netherlands) in a Center for Genetic Resources (CGR). The European *Brassica* collection is spread in several countries. A comprehensive base collection of cultivated *Brassica oleracea* including cauliflower has been established at Vegetable Gene Bank of Horticulture Research International (HRI), Wellesbourne, Warkwick, UK; The Institute voor de veredeling van Tuinbouwgewassen Wageningen, The Netherlands; and the Instituto del Germoplasm, Bari, Italy (Van der Meer et al., 1984), in the INRA of Le Rheu, in France. In India, National Bureau of Plant Genetic Resources (NBPGR), New Delhi, has been assigned the duty of conserving the germplasm of all crops, including vegetables. Germplasm of tropical cauliflower types and that of temperate (Snowball) types are being maintained at Indian Agricultural Research Institute, New Delhi and Dr. Y. S. Parmar University of Horticulture and Forestry, Solan (Himachal Pradesh), respectively. Large collection of cauliflower is also available with the United State Department of Agriculture, Plant Introduction Service.

CROP BIOLOGY

Unlike other cole crops, most forms of cauliflower do not require a specific vernalization for the transformation of vegetative to the reproductive stage. However, winter cauliflowers and other biennial types require cold treatment for both curding and later to reproductive phase, while for the other cauliflowers, such cold treatment is not necessary. If the heads are removed in cabbage, the axillary buds develop but in cauliflower only curd gives rise to multiple reproductive shoots.

Root

The seedlings of Indian cauliflower are generally characterized by green colored hypocotyl and in snowball and other types by reddish colored hypocotyl. The reddish type is dominant and can be used as marker. The plant has branched tap root system. On transplanting, tap root becomes slightly thick and give rise to a number of lateral roots. The thickness of roots is between 0.5 mm to 1 cm. The root system is shallow, which spreads around the plant with in a depth of 20-30 cm.

Leaves

Leaves are large, generally oblong, sessile; longer and narrower than those of cabbage and have no axillary buds. The color may be bluish green, waxy green to glossy green. The glossiness is a recessive character and can be used as marker. The number of leaves may vary from 18-50. Generally, Indian cauliflower and winter cauliflower have higher number of leaves. In some cases leaf veins are thicker and prominent. The inner leaves in some types fold toward inside and cover the curd, which is known as self-blanching habit.

Plant Type

Four distinct plant types have been observed in cauliflower; No. 1 has completely flat leaves with exposed curd, No. 4 has completely erect leaves with small curd; No. 2 and No. 3 are intermediate types (Chatterjee & Swarup 1972). The No. 2 has semi-erect and close to No. 1 and No. 3 has erect leaves. No. 3 is most desirable type. Plant type 2 is most common in Indian cauliflower and type 3 in snowball types.

Stem

Cauliflower has short (< 15 cm), medium (16-20 cm), and long (> 21 cm) stem length. Early Indian cauliflower generally have long stem. Long-stemmed varieties are more suitable for cultivation during rainy season and for mechanical harvesting.

Inflorescence

The peduncles in the axils of the bracts formed by the main growing point branch repeatedly even up to fifth order. In the beginning, the peduncles do not grow length wise but becomes thick. So, roughly spherical, colourless, terminal head is formed which is composed of large number of naked apical meristems, called "curd." The curd formation invariably precedes floral initiation. Its inflorescence differs from that of cabbage in being more dwarf, cymose type and has umbrella shape. Such formation results from the absence of central main stem above the point where branching starts. However, in case of early Indian cauliflower, branching system is common.

Floral Biology

The flowers are typical of cruciferous family, having 4 sepals, 4 petals, 6 stamens, of which, 2 are short and 2 carpels have superior ovary. The color of corolla may be light yellow, yellow, dark yellow or white. The androecium is tetradynamous, i.e., having four long and two short stamens. The carpels have superior ovary with a 'false' septum and two rows of ovules. The flowers are attached with the stalk with a short pedicel. The anthesis of flowers depends on temperature and the flowers open mainly during hotter part of the day. The two functional nectarines are present, which are situated between the bases of ovary and short stamens and the other two inactive nectaries at the bases of pairs of long stamens. Honeybees are the usual pollinating agents, though bumble bees and other syrphid flies may also be responsible for pollination. Myers and Fisher (1944) reported that as a pollinating agent, wind could be practically disregarded. It has been found that stigma of *Brassica* spp. is receptive even 5 days before and 4 days after anthesis. The period from pollination to fertilization generally takes 24-48 hours, depending on temperature. The ideal temperature has been found to be 12-18°C. Continuous foggy weather with lower temperature affects both fertilization and seed setting. Similarly, higher day temperature causes pollen sterility, resulting in poor seed development. The pod maturity or harvest of pods may require 50-90 days, from the date of flowering, depending on the climatic conditions. The fruit is a siliqua and often called a pod. The seeds are small, globular, smooth, and dark brown in color. They are mostly embryo as the endosperm is generally absorbed in the formation of cotyledons. There are normally 12-20 seeds per pod and nearly 350 seeds make one gram.

Breeding Behavior

In cauliflower, the cross-pollination varies according to the types for example in summer cauliflower (Snowball or Erfurt types), there may be self-pollination as high as 70 percent (Nieuwhof, 1963), while in winter cauliflower and in early Indian cauliflower, the cross-pollination is quite high due to its self-incompatible nature.

The cole group, in general, is characterized by both self-incompatibility and protogyny. The self-incompatibility is of sporophytic type and has been studied by many workers. So far, 70 self-incompatible alleles have been isolated from the cole crops (Hodgkin et al., 1988). The long receptivity of the flowers, as reported earlier, is due to protogynous nature of the flower. Considering these two conditions, the breeding techniques and methods have been developed for its improvement.

POLLINATION CONTROL MECHANISMS

The flower structure of cauliflower is complete in nature but the crop is basically a cross-pollinated one. There are some varieties, which set seeds freely even in self-pollination conditions. There are two naturally occurring mechanisms for ensuring cross-pollination in a hermaphrodite species like cauliflower are self-incompatibility and male sterility. The use of both these systems is very useful in the commercial hybrid seed production (Singh, 2000).

Self-Incompatibility (SI)

Self-incompatibility is genetically controlled, physiological hindrance to self-fruitfulness or self-fertilization, and is probably the most important way to enforce out crossing. Selfing could be avoided by other factors like, embryo abortion, but self-incompatibility is prezygotic and prevents embryo formation. Self-incompatibility is, therefore, the prevention of fusion of fertile male and female gametes after self-pollination. Two types of incompatibility has been reported, gametophytic, and sporophytic. In the former system, the pollen-pistil interaction is genetically governed by the haploid genome of each pollen grain and the diploid genome of the pistil tissue, and in the latter system, by the genome of the somatic tissue (of the sporophyte) in which the pollen grains are formed. Members of *Brassica* family including cole crops posses homomorphic sporophytic self-incompatibility associated with trinucleate pollen and inhibition of pollen germination at stigma surface (Bateman,1955). In monofactorial sporophytic self-incompatibility, pollen from a compatible pollination adheres to the papillae of the stigma, imbibes, and then germinates.

It was reported that cutinase enzyme digests the cuticle by its action (Linskens and Heinen, 1962) and then pollen tube grows into the papillae. The growth of pollen tube continues down the style to effect fertilization. With light microscopy, it appeared that the tube grow down between the cuticle and the cellulose-pectin layer of the papilla, but electron microscopy has shown that the tube actually grows down inside the cellulose-pectin layer (Kroh, 1964; Elleman et al., 1988).

Inhibition of self-incompatible pollen takes place on the surface of the papilla (Christ, 1959) and it is accompanied by deposition of callose inside the papillae. There are two main reasons for the expression of self-incompatibility, (i) due to lack of adhesion, hydration and germination of the pollen grain, and (ii) the failure to penetrate the papillae.

Homomorphic incompatibility in both the systems is generally controlled by a series of alleles at a single locus S (Bateman, 1952). The structure of this locus is complex with at least 3 important genes described (Nasrallah 2000; Dickinson 2000): S locus glycoprotein (SLG), S locus receptor kinase (SRK) and S locus cysteine-rich (SCR). The 2 first ones are determinants of the SI specificity in stigma. SLG is a soluble cell-wall localized protein and SRK is a plasma-membrane anchored signaling receptor, the extra cellular domain of which shares similarity with SLG. SCR is the male determinant recently discovered by Nasrallah laboratory (Schopfer et al., 1999). It is a small highly charged and polymorphic cysteine rich protein, exclusively expressed in anthers during pollens development. The allele forms of the locus are designated "haplotypes." They shows a tremendous variability which reflects a molecular divergence in the organization and sequence of the S-locus genes. More than 90 haplotypes had been described for the species Brassica oleracea (Hodgkins et al. 1988; Ruffio-Chable et al., 2001). Modifier genes are reported to influence self-incompatibility (Nasrallah and Wallace, 1968). Nasrallah et al. (1985) also reported the presence of glycoprotein, which inhibits germination of pollen tube of homozygous self-incompatible lines.

Thompson and Taylor (1966) reported that ancestral Brassica oleracea was highly self-incompatible. On this basis, it is but natural that the cole crops would be self-incompatible. In cauliflower, systematic studies on self-incompatibility was initiated by Watts (1963,1965b). He found higher level of self-incompatibility in biennial winter and autumn types and low in European summer types (snowball, alpha and erfurt). This was also confirmed by Nieuwhof (1974), Hoser-Krauze (1979) and many other workers. Chatterjee and Mukherjee (1965) found that medium duration strains of cauliflower were more compatible than long duration strains and further reported that fully self-compatible to self-incompatible forms occur. The distribution of S-haplotypes were performed by Ruffio-Chable et al. (1997), ten S-haplotypes were detected by immunochemical analysis. Half of the plants analyzed (126 belonged to 82 open pollinated populations, representing the variability of the group) possessed the same haplotype designated S15 in the Ockendon nomenclature. The self-compatibility of summer and autumn types were caused by the presence of the SC-haplotype which would have lost the kinase activity of SRK (Ruffio-Chable, personal communication). Annual Indian cauliflower and biennial winter cauliflower have stronger self-incompatibility mechanism. A detailed investigation in Indian cauliflower of self-incompatibility revealed that inbreds/lines of maturity group I have strongest self-incompatibility followed by maturity group II, group III showed weak self-incompatibility (Murugiah,1978; Vidyasagar, 1981; Chatterjee and Swarup,1984; Sharma et al. 2001). Murugiah et al. (1983) reported that among the identified self-incompatible alleles with varying degree of dominance, high ranked Sd Sd and Sv Sv alleles retained their incompatibility throughout flowering period whereas mid-ranked SmSm and low ranked SaSa lost it gradually in Indian cauliflower having homozygous alleles. But loss in low ranked was slow than mid-ranked lines.

The dominance/independence relationship of the S-alleles in the pollen and pistil may differ. In a cross involving $S_1S_3 \times S_1S_2$ (male) as demonstrated by Frankel and Galun (1977) showed complex reaction of self-incompatible system as shown in Table 3.

Being a natural method, self-incompatibility has no adverse side effects, such as those often found with cytoplasmic or chemically induced sterility. However, it is often less than perfect. Although the possibility of using self-incompatibility to produce hybrids was suggested over 70 years ago (Pearson 1932), it was not until 1950 that they first appeared in Japan, and in 1954 in the USA (Wallace, 1979). However, in the recent years, use of self-incompatible

TABLE 3. Sporophytic self-incompatible system; $S_1S_3 \times S_1S_2$ (male)

Pollen reaction	Pistil reaction	Compatibility
Independent	Independent	Incompatible
S_1 dominant to S_2	Independent	Incompatible
S_2 dominant to S_1	Independent	Compatible
Independent	S_1 dominant to S_3	Incompatible
Independent	S_3 dominant to S_1	Compatible
S_1 dominant to S_2	S_1 dominant to S_3	Incompatible
S_1 dominant to S_2	S_3 dominant to S_1	Compatible
S_2 dominant to S_1	S_1 dominant to S_3	Compatible
S_2 dominant to S_1	S_3 dominant to S_1	Compatible

lines has become a standard practice for the production of commercial hybrid seed in several *Brassica* vegetable crops.

One of the most important aspects of self-incompatibility is that in this mechanism, pollen and nectar production are unaltered. This may not matter much with wind-pollinated plants, but with insect-pollination, it is very important. Some insects, especially honeybees are highly discriminatory when foraging amongst flowers. Faulkner (1971) suggested that only slight differences in flower color or UV reflectance may cause such behavior. Butler (1971) showed that bees were first attracted by color, but were unlikely to investigate a flower further if unable to sense any perfume. It is not likely, therefore, that bees will pay much attention to male-sterile flowers without nectaries. This is probably the reason why self-incompatibility has become more important in the production of hybrids in insect-pollinated crops, and this is particularly so in the case of cauliflower.

Male Sterility

Male sterility has also been reported in cauliflower. It is of two types, i.e., genic and cytoplasmic.

Genic Male Sterility

Male sterility in cole crops are mainly recessive character. A single recessive *ms* gene mutated from male fertile *Ms* gene has been reported in cauliflower by Nieuwhof (1961), Borchers (1966), Nieuwhof (1968), and Ahluwalia et al. (1977), and was designated as *ms-4* and *ms-C*. Van der Meer (1985) reported, male sterility under the control of duplicate dominant genes with cumulative effect. Dominant male sterility have been described in cauliflower (Ruffio-Chable, 1997). This has some possible practical value in hybrid seed production programs, because of inadequate and unreliable nature of self-incompatibility system in some of the cauliflowers. This sterility can be responsive to temperature and humidity.

Cytoplasmic Male Sterility (CMS)

Cytoplasmic male sterility is not apparently found in cauliflower or other cole crops but has been introduced from several other sources. Cytoplasmic male sterility has been reported in an identified cultivar of Japanese radish by Ogura (1968) and was introduced by transferring to *Brassica oleracea* genomes through repeated back cross with broccoli (Bannerot et al., 1974 and Mc Collum,1981). Later Dickson (1975) and Hoser-Krauze (1987) transferred it from broccoli to cauliflower. The Ogura type cytoplasmic male sterility was transferred into heat tolerant Indian cauliflower from kale and broccoli through repeated back crosses, four lines, MS-91, MS-51, MS-11, and MS-110 from the former and five lines, MS-01, MS-04, MS-05, MS-09, and MS-10 from the later were developed, which are now being used in heterosis studies by Sharma (2003). Pearson (1972) crossed *Brassica oleracea* (cabbage) with *Brassica nigra* and developed male sterile lines. Chiang and Crete (1987) introduced male sterility from *Brassica napus* into cabbage and later from cabbage to cauliflower (Crisp and Tapsell, 1993).

Both genic as well as cytoplasmic male sterility have been associated with physiological problems. Some forms of genetic male sterility are temperature sensitive and result in to self-pollination when used for F_1 seed production Nieuwhof (1968). Pearson (1972) type of cytoplasmic male sterility functional nectaries are not developed making them unsuitable for commercial hybrid seed production (Pelletier et al., 1983). Both the Ogura and McCollum type of cytoplasmic male sterile plants or their hybrids when grown at low temperature less than 12°C, show chlorosis and loss of vigor at their early stage of growth (Dickson, 1985; Hoser Krauze, 1989). High regeneration capacity from cultured mesophyll cells of a cauliflower line having Ogura system was reported by Jourdan et al. (1985). This was useful step in the possible production of cytoplasmic mutants, transgenics or recombinants superior male sterile genotypes. Non-chlorotic male sterile lines, using cybrids followed by protoplast fusion between sterile and normal genotypes have also been developed in cauliflower and other cole crops. Male sterile cybrids with normal photosynthesis and improved nectar secretion were obtained through chloroplasts exchange and mitochondrial recombination. The Ogura CMS system was first improved in *B. napus* in 1983, and then in *B. oleracea* in 1989 (Delourme and Budar, 1999).

By genetic engineering, it has become possible to develop female parents having barnase genes, which inhibit the activity of pollen producing tapetum cells and make them male sterile. The introduction of barstar genes in the male parent, which restore fertility in hybrid seed by inactivating the functioning of barnase genes responsible for disruption of pollen development in female parent (Reynaerts et al., 1993).

CROP IMPROVEMENT

Before the start of any crop improvement programme, the objectives should be clear so the strategic methodology can be adopted to achieve the targeted results. In cauliflower, besides yield, special emphasis is needed to improve its quality characters, including nutritiveness and insect-pest and disease resistance. The improvement in yield can be achieved through its component characters, which have direct or indirect effects. The important components are, curd size, weight, depth, compactness and color of the curd besides their uniformity in size and maturity. The plant type including frame, stem length, harvest index and resistance to common biotic and abiotic stresses also need attention of the breeder. Recently some intra-crosses were attempted between tropical and temperate types to transfer desirable traits in cauliflower. Honma and Cash (1986) reported three varieties, viz., Supreme, Beta-White, and One-up are developed by intra-crosses. While Supreme and Beta-White originated from the Pua-kea (tropical) and with Self-Blanche Snowball (temperate summer), the cultivar One-up was derived from a series of crosses involving Snowball M1, Pua-kea, vans Osena, February L, Early Fuji, and Self-Blanche. However, commercial possibilities of such varieties have to be seen. Gill et al. (1987) isolated a pure line, called Pusa Himjyoti, having retentive white curd from MGS 2-4 for July to October cultivation in hilly areas of Himachal Pradesh, India. Sel.12, a black rot resistant line has been developed, using a black rot resistant tropical type line SN-445 and Pusa Snowball-1 at IARI, Katrain, India (Gill et al., 1983).

A lot of work has been reported on the genetics of qualitative and quantitative traits, genetic advance, heritability and combining ability in cauliflower as shown in Table 4. The inheritance of qualitative characters was studied in detail in Indian cauliflower by Ahluwalia et al. (1977) and gene symbols for different traits were assigned as: stalk length-long St, short st; leaf apex-round Ro, pointed ro; habit-errect E, branching e; curd color-yellow Y, white y; flower stalk length-long F, short f; flower stalk color-variegated V, green v; siliqua length-long SL, short sl (Ahluwalia et al., 1977).

Combining ability was studied to select the parents for hybridization and good cross-combinations for production of hybrids. The inbred 103 exhibited the best gca for all characters and cross-combination 105 × 108 showed the maximum sca for yield potential in early Indian cauliflower (Lal et al., 1977). Further, Lal et al. (1978) found that parent 308, 303, 302 had high gca for curd weight and curd size index in mid-Indian cauliflower. In Snowball group, Lawyana was the best general combiner for curd weight, curd size, and leaf size. Sel.12 and Pyramis were best general combiners for early maturity and gross weight, respectively (Sharma et al., 1988). Line IHR3, IHR4, IHR9 and IHR36 were good combiners for most of the characters. The selection of parents on the basis of *per se* performance and general combining ability was effective (Pandey and Naik, 1986), days to curd initiation, curd weight, number of leaves and plant height and diameter, and curd weight highly influenced the plant weight (Pandey and Naik, 1985). According to Gangopadhayay et al. (1997), in early cauliflower, self-incompatible lines cc-13 and vv-(351) were found to be the best general combiners for earliness, curd color, compactness, and yield contributing characters, respectively.

TABLE 4. The genetics of quantitative characters of cauliflower*

Character	Nature of gene action	References
1. Curd weight	Dominance and epistasis	(Swarup and Pal, 1966)
	Pronounced over dominance and epistasis	(Singh et al., 1975)
	Additive and dominance gene action	(Singh et al., 1976a; Jyoti and Vashistha, 1986; Gangopadhayay et al. 1997; Sharma et al., 1988)
2. Curd to plant ratio	Partial dominance	(Kale et al., 1979)
3. Curd diameter	Predominance of dominance gene action	(Lal et al., 1979)
4. Curd size index	Pronounced over dominance and epistasis.	(Singh et al., 1975)
	Dominance and epistasis.	(Swarup and Pal, 1966; Lal et al., 1979)
	Additive dominant gene action.	(Singh et al.,1976; Sharma et al., 1988)
	Partial dominance.	(Kale et al., 1979)
5. Curd angle	Pronounced additive gene action.	(Lal et al., 1979)
	Additive arld dominant gene action.	(Dadlani, 1977; Chand, 1980)
6. Curd compactness	Polygenic.	(Nieuwhof and Garretson, 1961)
	Dominance and additive gene action.	(Lal et al., 1979)
	Additive	(Vashistha et al., 1985)
7. Maturity (i) Earliness	Partially dominant gene action.	(Watts, 1964)
	Dominance and epistasis.	(Swarup and Pal, 1966)
	Predominance of additive gene action.	(Singh et al., 1975; 1976b; Lal et al., 1979; Mahajan et al., 1996; Gangopadhayay et al., 1997)
	Additive gene action.	(Kale et al., 1979)
	Additive and dominant gene action.	(Sandhu and Singh, 1977; Sharma et al., 1988)
(ii) Lateness	Recessive polygenes	(Watts, 1963)

* Adapted from Chatterjee (1993)

BREEDING/SELECTION METHODS

Crop breeding is a breeder's activity, picking up useful characteristics and putting up them together to develop a variety having desirable traits. Population improvement method has been commonly followed for the improvement of cole crops. In India, mass selection has been widely used for the improvement of cauliflower (Sharma and Singh, 2003). Though this method is useful for the improvement of simply inherited traits, but is not much effective in case of polygenic characters. Moreover, this method is time consuming and hence not found as an ideal option. Based on progeny evaluation, modifications like, mass pedigree method and family selection method have proved better than mass selection (Nieuwhof, 1959). The choice between these methods depends also on the populations, according to their level of homogeneity due to the system of self-incompatibility. Recurrent selection method in cole crops have been found as a better option for the improvement of quantitative characters especially those, which are under the control of additive gene action. This method has been found effective for the improvement of curd compactness, yield and other economic characters in cauliflower. Significant improvement after one generation of recurrent selection in the yield (18-47%) and diameter, depth and weight of curd over the original material was reported by Tapsell (1989). Inbreds, thus developed have been used in the breeding of hybrids, synthetics and open pollinated varieties or in intervarietial hybridization program. In the recent past, a variety, 'Pusa Sharad' in mid-maturity group of Indian cauliflower has been developed using recurrent selection method (Sharma et al. 1999). The cross combinations involving inbreds, with low inbreeding depression and high heterotic residual effects due to additive gene action, which shows better response to selection, are used in the breeding of composite varieties. Back cross method has been commonly followed to transfer resistance from donor to recurrent parent. This method has been followed to develop 'Pusa shubhra' a cauliflower variety resistant to black rot, curd blight and riceyness (Singh et al., 1993). However, Kalloo (1988) has reported that family breeding method was found to be more important for the improvement of cauliflower and other cole crops. In this method, seeds of the selected plants on the basis of progeny testing are used to develop synthetics. Similarly disruptive selection method to break tight linkage has been recommended for the improvement of *Brassica* vegetables (Kalloo, 1996). In this method, only extreme type of population is selected and intermediate one is discarded.

Pusa Early Synthetic and Pusa Synthetic cauliflower varieties were developed by synthesizing 6 and 7 parents, respectively, and were released in India for commercial cultivation (Singh et al., 1997 and Gill, 1993). Development of synthetic variety is based on the exploitation of additive genetic variance. The desirable inbred lines to be synthesized are selected after testing their combining ability. To test general combining ability, diallel cross or top cross or poly cross method may be used. Two to seven or more such lines are selected for developing synthetic varieties. The advantage of synthetic varieties are many fold: (i) its seeds can be produced by the farmer himself from his own crop like any other open pollinated variety, (ii) it is particularly useful in the places where commercial seed industry is not well developed and mechanism (floral biology) is not available for commercial seed production, (iii) it serves as a reservoir of germplasm, and (iv) it adapts better to varying growing conditions unlike F_1 hybrids.

HETEROSIS BREEDING

In vegetable crops, Tamassy (1973) described heterosis in to three types, i.e., somatic, reproductive and additive, manifested in terms of greater vegetative growth, seed production, and tolerance or resistance to adversities, respectively. In cauliflower heterosis was first reported by Jones (1932). However, Haigh (1962) and Nieuwhof and Garretson (1961) fail to found appreciable amount of heterosis in European summer cauliflower (Snowball, Erfurt or Alpha type) which may be due to their narrow genetic base. Later, Watts (1965a) observed sufficient heterosis for earliness and curd size. Swarup and Pal (1966) and Pal and Swarup (1966) found appreciable heterosis in snowball types for earliness (5-7 days), curd weight (24.5-28.2%), curd size index (22.54-34.85%) over better parent. In different maturity groups of Indian cauliflower, appreciable heterosis for economical characters were reported by many workers. Kumaran (1971) made a study of three maturity groups of Indian cauliflower and recorded appreciable amount of heterosis for curd weight, plot yield, curd size index, maturity of curd, stalk length and leaf size index. Maximum percentage of heterosis observed by him was 22.85, 12.07, 24.64, 21.53, 49.76, and 32.46 percent, respectively. Swarup and Chatterjee (1972, 1974) investigated heterosis in Indian cauliflower and reported better manifestation of it in first maturity group, compared to other groups. Similarly, maturity group second was superior over maturity group third. In this respect, they recorded heterosis of 41.23, 16.00, and 23.41 percent for yield, curd size and curd weight, respectively in maturity group second, while in maturity group third, it was 22.22 and 27.00 percent for curd size and curd weight, respectively. Appreciable amount of heterosis was reported in Indian cauliflower for different characters in different groups (Deshpande, 1975; Singh et al. 1975; and Sandhu et al., 1977).

Verma (1979) using male sterile lines as female parent, observed sufficient heterosis over better parent. The observed heterosis was 93.3, 91.1, 55.6, 44.4, 20.0, and 2.2 for curd weight, plant height, curd size index, curd to plant ratio, early maturity and number of leaves, respectively. Hoser-Krauze et al. (1982) reported heterosis for earliness, curd diameter, curd weight and quality in 2 and 3 way reciprocal F1 hybrids, which were made using three self-incompatible Indian cauliflower and three temperate self-incompatible lines. Pandey and Naik (1985) studied heterosis in hybrids of different sub-species of *Brassica*. Brazilian broccoli line 137 was crossed with two cauliflower lines 138 and 149 from the USA and 10 Indian cauliflower cultivars. Heterosis for tallness, number of leaves and leaf area index was positive, except for the cross, Superfine Maghi × 137 which showed negative heterosis for leaf number. They also concluded that lines with high heterosis can be selected at the seedling stage itself.

Gangopadhayay et al. (1997) using four self-incompatible lines in maturity group first as female parent with 11 male parents in the early and 7 in mid group reported 31.2, 25.3, 34.5, 25.0, 16.6, 25.7, 53.5, 87.0, 49.5, 71.8 and 82.3% heterosis in early and 19.4, 17.7, 39.3, 42.9, 13.5, 20.8, 38.3, 20.6, 49.6, 51.9, and 63.2% heterosis in mid-group for days to maturity, days to 50% maturity, curd compactness, color, number of leaves, curd diameter, curd depth, curd size index, gross weight, marketable weight, and curd weight, respectively.

Breeding for Biotic Resistance

Cauliflower being a delicate crop is more prone to insect-pests and diseases. All the cole crops have common insect-pests and diseases problems. With the evolution of large number of cultivars for different seasons/climates, cauliflower and other cole crops are being grown round the year. All this has resulted in the continuous built-up of disease inoculum and insect population. The common diseases of cauliflower are, black rot, bacterial soft rot, sclerotinia rot, downy mildew, dark black spot (*Alternaria* spps.), cabbage yellows, club root, and wire stem. The important insect pest are diamond back moth, cabbage butterfly caterpillar, aphids, cabbage head borer, cutworm, and Bihar hairy caterpillar (*Spilosoma oblique*). Management of these diseases and pests using chemical pesticides is not only cumbersome and costly but also health hazardous. So it is imperative to have resistant varieties for which identified resistant sources and knowledge of genetics of resistance for a particular disease and pest is a pre-requisite. The related information for economically important biotic stress of cauliflower is discussed here.

Diseases

Downy Mildew: It is caused by an obligate parasite, *Peronospora parasitica* which can infect the crop at any stage of growth. It is systemic in nature and infection observed at seedling stage can reappear at curd and marketing stage (Crute and Gordon, 1987). In cauliflower, Igloo, snowball Y, Dok Elgon, and RS355 (Kontaxis et al., 1979); BR-2,CC and 3-5-1-1; EC177283, Ec191150, EC191157, Kibigiant, Merogiant, EC191140, EC191190, EC191179, and Noveimbrina (Singh et al., 1987; Mahajan et al., 1991); MGS2-3,1-6-1-4, 1-6-1-2 and 12C (Chatterjee, 1993); KT-9 (Sharma et al., 1991); Early Winter

Adam's White Head (Sharma et al., 1995); CC-13, KT-8, XX, 3-5-1-1, CC (Trivedi et al., 2000); Perfection, K1079, K102, 9311 F1 and 9306 F1 (Jensen et al., 1999); Kunwari-7, Kunwari-8, Kunwari-4 and First Early Luxmi (Pandey et al., 2001) were reported resistant to moderately resistant. Pusa Hybrid-2 (Singh et al., 1994) of Indian cauliflower and Pusa snowball K-25 (unpublished) of snowball type having resistant to downy mildew were released for commercial cultivation in India. Resistance to downy mildew has been ascribed to a single gene with dominant effect (Sharma et al., 1991; Mahajan et al., 1995; Jensen et al., 1999), single gene with recessive effects (Hoser-Krauze et al., 1984; Mahajan et al., 1995) or several genes (Hoser-Krauze et al., 1995).

Sclerotinia Rot. The disease is caused by *Sclerotinia sclerotiorum.* It has a very wide host range and can infect most of dicot crops, but is more severe in the seed crop of cauliflower, though, it may attack the crop at an early stage of its growth also. Moderately resistance to this pathogen was reported in EC131592, Janavon, EC103576, Kn-81, Early Winter Adam's White Head, EC162587, EC177283 (Kapoor, 1986; Baswana et al., 1991; Singh and Kalda, 1995; Sharma et al., 1995 and Sharma et al., 1997). Resistance is polygenically controlled and recessive in nature (Baswana et al., 1993; Sharma et al., 1997). Pusa Snowball K-25 developed by using EC103576 as resistant source with Pusa Snowball-1, has recently been released for commercial cultivation, which possess field resistance against this disease.

Black Rot. The disease is caused by a bacterium *Xanthomonas campestris* (Pam) Dawson. Yellowing of leaves starts from leaf margin and extend in the direction of the midrib, followed by blackening of veins (vascular bundles). Cauliflower lines reported resistant sources are Sn 445, Pua kea and MGS2-3 (Sharma et al., 1972); RBS-1, EC162587 and Lawyana (Sharma et al., 1995); Sel-12 (Gill et al., 1983); Sel-6-1-2-1 and Sel-1-6-1-4 (Chatterjee, 1993); Sakata 6, Takki's February, Nazarki Early, Henderson's Y 76 and Henderson's Y 77 (Moffett et al., 1976); Avans and Igloory (Dua et al., 1978).

Some of the above sources have been used in the development of resistant varieties. Pusa Shubhra was developed, using Pua kea and MGS2-3 lines and recommended for commercial cultivation (Singh et al., 1993). The resistance was dominant and governed by polygenes and the dominance components of variation were more pronounced than additive (Sharma et al., 1972). But Jamwal and Sharma (1986) reported that dominant resistance is governed by a single gene.

Alternaria Black Spot. In cole crops, the black leaf spot disease is caused by *Alternaria brassicae* or *Alternaria brassicicola.* Brown to black, small to elongated spots appears on leaves, stems of older leaves. In younger plants, it may cause symptoms like *Rhizoctonia solani.* When the fungus infects the curd, specially in case of seed crop, the disease is called as inflorescence blight. Resistance was found in Indian cauliflower lines, MGS2-3, Pua kea and 246-4 (Sharma et al., 1975); 23-7, 466, MS98, 210-21, Sel-9, 443-7 (Trivedi et al., 2000); IIHR142 and IIHR217 (Pandey et al., 1995); and Snowball KT-9 (Sharma et al., 1991). Resistance to curd blight is dominant in nature, pollygenically inherited and in general additive effects were found more pronounced than dominant one (Sharma et al., 1975). Pusa Shubhra having resistance to curd blight has been released for commercial cultivation (Singh et al., 1993). Both additive and dominant gene action played a role in resistance but partial dominance is more important (King and Dickson, 1994). No linkage was found by them between leaf color (red or green) and leaf spot resistance.

Club Root. The disease is caused by *Plasmodiophora brassicae* which has as many as nine races. Gall formation takes place on lateral roots and gives the shape of spindle. Walker and Larson (1960) reported resistance is recessive in nature and polygenically inherited. Gallegly (1956) also supported polygenic theory. Crisp et al. (1989) found differences for disease severity between 845 varieties of cauliflower and broccoli, were mainly attributed to selection pressure within locality. More recently, two lines of a resistant kale were selected for their highly resistance against a large range of pathotypes of the pathogen. These lines presented a sufficient level of resistance to be directly useful in the breeding program in order to develop cauliflower and broccoli hybrids resistant to clubroot (Mazanares-Dauleux et al., 2000).

Yellows. It is caused by *Fusarium oxysporum* f. *conglutinans*. Vascular tissues become yellow to brown, causing wilting of plants. It has two races type A and type B, the resistance in former is inherited monogenically, and in later polygenically and is dominant in nature as reported in cabbage by Walker (1930). Early purple cauliflower was reported resistant to yellows (Natti and Atkin 1956).

Pests

Cauliflower is infested by a large number of insect-pests. Many workers have reported varietal differences in susceptibility to various pests in cauliflower under field conditions. Resistance to lepidopterous pests in cauliflower and cabbage is attributed to non-preference in conjunction with either tolerance or antibiosis (Shelton et al., 1988). A glossy leaved cauliflower PI234599 was reported to be resistant to lepidopterous pests, diamond back moth (*Plutella xylostella*), cabbage looper (*Trichoplusia*), imported cabbage worm (*Artogeiso rapae*) by Dickson and Eckenrode (1980). Genetic resistance to lepidopterous pests has been quantitative with additive dominance and relatively low heritability, ranging from 22% to 47%. Ellis et al. (1986) found the PI 234599 resistant to other leaf pests (*Mamestra brassicae* and *Evergestis forficorlis*) but more succeptible to flea beetle, and in general, red leaved varieties are found more resistant to these pests than normal green leaved ones. The glossiness character of leaves though as such is not undesirable character, but associated with small sized plants, having poor curd quality. Recombinants has also not yielded encouraging results in breeding program. Resistance to cabbage head borer (*Hellula undalis* L. *Fabricius*) has been reported in cauliflower, ES-97, ES-96, Katiki (J.B), KW-5, KW-8, KW-10, Kunwari (RB), Kathmandu Local, Early Patna, EMS-30 and PSK-16 (Lal et al., 1991). Lal et al. (1994) also found resistance under field conditions in Indian cauliflower F_1 hybrids like aa X ES102, aa X Katiki (JB), aa X First Early, aa X First Crop, aa X Sel.100, aa X Sel.41 and aa X 824 have resistance to Bihar hairy caterpillar (*Spilosoma oblique* Walker). Aphids causes major losses to cole crops. The aphid species responsible for economic losses in cauliflower and other cole crops are cabbage aphid (*Brevicoryne brassicae*), green peach aphid (*Myzus persicae*) and turnip aphid (*Lipaphis erysimi*). Resistance to cabbage aphid has been reported in NY 13816, NY101181, NYIr9602, and NYIR 9605 but work on cauliflower is very scanty. Natural occurring compounds like, glucosinolates, pipecolic acid and β-nitroprionic acid in the tissues of *Brassica* plants are responsible for resistance to cabbage looper and imported cabbage worm. Breeding resistant varieties in cauliflower and other cole crops for most of pests and some diseases still remains elusive because hardly any resistant source with desirable degree of resistance is available in the germplasm. So, biotechnological tool like genetic engineering offers a safe and long lasting solution. Cauliflower can be conveniently transformed by using *Agrobacterium* provided suitable genes are available. Introduction of genes coding for *Bacillus thuringiensis* (Bt) insecticidal crystal protein; cowpea and soybean trypsin inhibitors and cytokinin biosynthesis enzymes are known to provide substantial protection against the insects in the crop (Kumar and Sharma, 1997) (Table 5).

These Bt genes in genetically transferred plants have proved quite successful. Similarly, there are some plant enzymes, which solublize fungal cell walls and cause membrane damage. Some of the antifungal proteins listed in Table 6 for different fungi responsible to cause disease in cauliflower and other crops (Kumar and Sharma, 1997). Transgenic plants expressing genes coding for such proteins could be developed for developing resistant varieties.

Biotechnology and Its Application

Recent development in the area of plant biotechnology can be used as an effective tool in speeding up and providing precision to the process of conventional breeding, creating genetic variability through harvesting genes from wild and relative species, and evolving novel genotypes through recombinant DNA (genetic engineering) technology. Following biotechnological tools have been employed in cauliflower.

TABLE 5. Important insect-pest of vegetable brassicas and potential gene that confer insect resistance

Common name	Scientific name	Useful genes
Diamondback moth	*Plutela xylostella*	*Cry 1A class*
Cabbage butterfly	*Pieris brassicae*	*Cry 1A class*
Stem-borer	*Hellula undalis*	*Cry 1A class*
Hairy caterpillar	*Spilosoma oblique*	*Cry 1A class*
Cut-worm	*Agrotis ipsilon*	*Cry 1A class*
Aphids	*Brevicome brassicae* *Myzus persicae* *Lipaphis erysimi*	Snowdrop lectin Snowdrop lectin
Thrips	*Caliothrips indicus* *Thrips tabaci*	*Cry 2A*
Leaf-minor	*Phytomyza hartiocola*	

TABLE 6. Important fungal diseases of vegetable brassicas and useful genes that confer fungal resistance

Disease	Causative fungus	Useful genes
Damping off	*Pythium* spp.	Permantis
Downy mildew	*Peronospora parasitica*	Chitinase
Wire-stem	*Rhizoctonia solani*	Glucanase
Leaf-spot	*Alternaria brassicae/* *Alternaria brassicicola*	Thionin
Cabbage yellows	*Fusarium oxysporum*	RIP
White rust	*Albugo candida*	Osmotin
Black-rot	*Xanthomonas campestris*	Chitinase
Soft-rot	*Erwinia carotovora*	Chitinase

Tissue Culture/Micropropagation

Mass multiplication of plants, using tissue culture technique, is more commonly used in cauliflower and other cole crops, especially in case of self-incompatible and male sterile lines. Explants like leaf, peduncle, pedicel, anther, meristem, tip and segments of root, stump and stem can be used for *in vitro* multiplication. For seed production, industrial production of *in vitro* hybrid parent plants has been performed by curd explants. Thousands plants can be obtained by this way (Kieffer et al 1994). Production of disease resistant plants *in vitro* have been discussed in detail by Ross (1980).

Anther and Microspore Cultures and Production of Dihaploids

Anther and microspore cultures are more suitable to develop homozygous lines in cauliflower and other *Brassica oleracea* crops. The anther culture is quick, which decrease the incidence of aneuploidy. However, the number of embryoides obtained from anther culture depends upon the factors like, sucrose concentration, growth hormone's concentrations, pretreatment. with high temperature and developmental stage of donor anther (Lillo and Hansen, 1987). Considerable differences occur among cauliflower genotypes in their efficiency to produce embryoides (Ockendon, 1988).

Uninucleate stage has been found the best for microspore or anther culture in most of *Brassica* species (Simmonds et al., 1989). The plants obtained from anther culture may be haploid, diploid, tetraploid and rarely aneuploids. Counting the chromosome numbers of nuclear cells is most reliable way to confirm ploidy. Ockendon (1988) obtained very few haploids, and mostly diploid and tetraploid regenerants. These results were confirmed by Boucault et al. (1991). For haploid embroids, the apices of such plants are treated with 0.05 percent colchicine at an early stage of growth to double their chromosome. These dihaploid (DH) plants, after selfing, form homozygous uniform population. The natural diploid plants in anther cultured may have arised spontaneously as doubled haploid, which may be used in breeding homozygous lines, depending upon their behavior after selfing.

Somaclonal Variation

Decapitation of cauliflower does not follow development of axillary branches as axillary buds are not commonly produced/developed in this crop. However, sometimes branches may develop from leaf scars near the base of stem or exposed roots. These shoots shows abnormality for morphological as well as reproductive characters, which may be analogous to somaclonal variation as recovered from cell culture or callus culture in many crops (Crisp and Tapsell, 1993).

Somatic Hybridization

Protoplast isolated from root tips, cotyledons, leaves and hypocotyls have been induced to form callus in *Brassica* species. The success of callus to differentiate into somatic embryoides depends upon a number of environmental factors and medium components. In cauliflower, Jourdan et al. (1990) developed an improved procedure for protoplast culture. Jourdan and Earle (1989) found that genotype played a critical role in determining the success rate of leaf protoplast culture in five cruciferous species including *Brassica oleracea.*

Protoplast fusion is a potential tool to create cybrids through hybridizing divergent, texas, which otherwise are non-crossable. The cybrid contains organelles from both the parents to create genetic variability. This also helps in bringing together cytoplasmic traits, which is rather impossible through conventional breeding. Schenck and Robbeln (1982) synthesized *Brassica napus* through protoplast fusion of *Brassica campestris* with *B. oleracea* for the first time. However, later, it was reproduced by many workers. Jourdan et al. (1989) developed cybrid plants through protoplast fusion of cauliflower and a variety of *Brassica napus* which have cytoplasmic derived resistant to atrazine.

These cybrids showed no segregation in population on selfing, which confirmed the cytoplasmic inheritance transfer of this character in the offspring. Protoplast fusion has also been used for the effective transfer of Ogura type of cytoplasmic male sterility traits from radish to cauliflower and other cole crops to devoid them from associated physiological abnormalities (Kagami et al., 1990).

Embryo Rescue

The embryo culture technique *in vitro* has been used to rescue non-variable interspecific hybrids in *Brassica* species (Ayotte et al., 1987). Embryos of the cross of *Brassica oleracea* and *Brassica napus* were rescued by them between 11-17 days after pollination. Later, they transferred triazine-resistance characters from *Brassica napus* into cauliflower, cabbage, broccoli, and kale. Usually it is difficult to obtain F_1 hybrid seeds by crossing *Brassica oleracea* with *B. campestris* but by using embryo rescue techniques, Inomata (1977) got success.

Molecular Breeding

It is now possible to select desirable genotypes from a segregating population with very high degree of precision and predictability by using biotechnological tools.

Restriction Fragment Length Polymorphism (RFLP) and Random Amplified Polymorphic DNA (RAPD) are the most fundamental tools, which have raised high hopes among the breeders. A detailed analysis of *Brassica* species has been carried out by many researchers using RFLP markers. A series of monosomic *Brassica oleracea* chromosome addition lines were established in genetic background of *B. campestris,* using RFLP markers, plant morphology and isozymes and these gene markers were mapped on some *Brassica oleracea* synteny groups (McGrath et al., 1990).

A detailed RFLP based genetic maps constructed in *Brassica oleracea* (Slocum et al., 1990; Landry et al., 1992), have also been used to study the origin of *Brassica oleracea* and its relationship with other *Brassica* species (Song et al., 1988, 1990) and among *Brassica oleracea* varieties (Osborn et al., 1989). RAPD has been used to identify club root resistant plants in segregating population (Grandclément et al., 1996). Molecular Assisted Selection (MSA) is still in infancy stage in India in cole crops. However, some work is in progress to develop tropical cabbage lines having resistance/tolerance to a serious pest diamond back moth (Unpublished) using transformation technique.

PRODUCTION OF F_1 HYBRID SEED

The advantage of using F_1 hybrids for commercial production in cole crops including cauliflower is well documented. Hybrids have been found advantageous by having uniformity in maturity, size and shape of the curds in cauliflower. Their superiority for other yield characters as earliness, quality parameters, adaptability, resistance to some pests and diseases was mainly due to the considerable breeding effort by breeding companies. In cauliflower, development of commercial hybrids in contrast to other *Brassica* vegetable picked up late and it is only since about 1985 they attained commercial reality (Crisp and Tapsell, 1993). The reasons may be unavailability of strong self-incompatible inbreds. But now almost all the cauliflower varieties grown in the developed countries are F_1 hybrids, mainly because economical interest and not for agronomical reasons. Local populations varieties are more adapted to their environment. The introduction of F_1 hybrid, not selected in one area, could be the cause of new diseases or the development of pests. But in India still the area under hybrid cultivars is less than five percent of the total area under cauliflower and the picture of many other developing countries is no way better than this. Indian cauliflower especially those types belong to early maturity groups are grown under adverse climatic conditions need immediate attention of breeders to develop commercial F_1 hybrids. Lots of variability within the commercial varieties have been observed in this group in size, shape and maturity time. Recently in the year 2003, an F_1 hybrid DCH-541 developed by Indian Agricultural Research Institute have been identified for growing under north Indian plains through All India Coordinated Research Project on Vegetable Crops (Sharma, 2003) in early maturity group of cauliflower. Commercial acceptance of hybrids, in addition to their superiority for desirable traits also depends upon the reasonable cost of F_1 hybrid seed production. The cost of the seed should be within the reach of common growers.

Hand emasculation and pollination have not been found economically effective in case of cole crops, since the seed quantity produced per pollination is very low. Cauliflower is a naturally cross-pollinated crop and to substantiate this has inbuilt genetic mechanism like self-incompatibility and male sterility. The self-incompatibility and male sterility have been used in the production of F_1 hybrid seed in most of the cole crop to avoid hand emasculation and to reduce the cost of F_1 seeds.

Self-Incompatibility

Keeping in view the advantage of self-incompatibility as already discussed in this chapter, it is being used more frequently in cauliflower and other cole crops. Production of hybrid seed involves, (i) selection of parent material, (ii) development of homozygous self-incompatible but cross-compatible lines, (iii) testing of combining ability, (iv) evaluation of cross-combinations/F_1 hybrids, and (v) production of commercial F_1 hybrid seed.

Selection of Parental Material

The parental material to be used for the development of homozygous lines and development of F_1 hybrids has to be chosen carefully keeping in view the general as well as specific requirement of the place where F_1 hybrids are to be marketed. The success of the hybrids to a greater extent depends upon the basic material. The specific requirement such as resistance to the particular disease or pest, suitability for a particular temperature regime must be taken care of as per need. About 6-8 years are required to develop F_1 cultivars from the initial selection of parental lines until release and marketing. Therefore, the parental material should be selected very carefully keeping in view the breeding objectives at the beginning of the hybridization program.

Development of Homozygous Self-Incompatible Lines

Almost all the plants of a variety are heterozygous for S-locus. A few plants (I_0) are selected on the basis of their superiority for desired traits at curd stage and selfed by bud pollination. The selsfed seed is sown in rows (I_1) and few buds are self-pollinated. In the next year single plant progenies are grown (I_2) and about 10-12 plants are selected at random in each progeny. The homozygous line is identified by self pollinating each plant of a family with a mixture of pollen taken from each plant of the family. The line showing no seed set or poor seed set is homozygous for self-incompatibility. About 50 percent of the lines are homozygous or heterozygous. The involvement of alleles in different homozygous lines is tested by reciprocal differences. The homozygous lines having only different set of S-haplotypes will be cross-compatible and can be used in hybrid seed production. Watts (1965b) suggested fertility index estimation using following formulae for determining self-incompatible lines within and between the progenies.

$$\text{Fertility Index} = \frac{\text{Average no. of seeds per siliqua from natural/ compatible cross-pollination}}{\text{Average number of seeds per siliqua from self-pollination in freshly opened flowers}}$$

If the plants of a line having fertility index more than two, the line is classified as self-incompatible and those with less than one are grouped as compatible. Lines/progenies having fertility index between one and two are designated as pseudo self-incompatible. Fluorescence microscopy technique suggested by Crehu (1968) can also be used to identify homozygous lines in I_1 generation by identifying pollen tube growth in the style. If the growth of the pollen tube in the style is absent then the line is considered as homozygous for self-incompatibility alleles (Wallace, 1979).

Testing of Combining Ability

The inbreds/parents to produce hybrid seed must be selected on the basis of gca (general combining ability) and sca (specific combining ability) estimates. Generally sca value of the cross combinations gives better prediction than gca of the parents. Combining ability generally studied by adopting diallel cross, top cross, poly cross mating designs. In the crossing program use homozygous self-incompatible but cross-compatible lines. In addition to this, the parental lines have synchrony in flowering, so can be used in commercial seed production.

Evaluation of F₁ Hybrids

Based on combining ability estimates the promising hybrid combinations are assessed in replicated trials. This is followed by multilocational evaluation. The hybrids having wider adoptability in addition to superiority for desirable traits are recommended for commercial production.

Production of Hybrid Seed

The common methods to produce hybrid seed commercially based on self-incompatible lines are as follows.

Multiplication of Homozygous Self-Incompatible Lines. Homozygous self-incompatible lines are developed with the objective to devoid self-pollination but for commercial F_1 seed production sufficient seeds of these parental lines are needed. Many methods have been proposed to overcome self-compatibility in cauliflower, to multiply the seeds of inbreds to be used in commercial hybrid seed production. The self-incompatibility system becomes effective only one or two days before anthesis. So bud pollination found very effective to produce the seed of homozygous self-incompatibles lines. But bud pollination is labor intensive, requires more hand and complete isolation from pollen of other lines to avoid contamination. So it is not cost effective.

An alternative to bud pollination is CO_2 treatment. The plants are grown in green house and CO_2 treatment having concentration between 3-5% at 100 percent relative humidity is given for 8-24 hour immediately after pollination in airtight chamber (Hinata, 1975). This technique is as effective as bud pollination and equally reliable. It requires high capital investment but running cost is low. The high concentration of CO_2 may be lethal to honeybees and even in low concentration they may not be active hence bumble bees may be used as pollinator. High concentration of CO_2 may be injurious to human being hence proper precautions should be taken in releasing the CO_2 and try to avoid its leakage, etc.

Sharma (1999) suggested spraying of 5% sodium chloride (NaCl) solution at the time of full bloom to overcome self-incompatibility in Indian cauliflower. Carafa and Carratu (1997) treated *Brassica olerecea* stigma with drop of salt solution and normal pollen tube growth was observed after 24 hours on selfing. Other method suggested by different workers has little application in mass multiplication of self-incompatible lines and are not much reliable. These includes steel brush pollination (Roggen and Van Digk, 1972), double pollination (Ockendon and Currah, 1978), electric aid pollination (Roggen at al. 1972), end season pollination (Fraser, 1984 and Ockendon, 1978), heat treatment (Johnson, 1971 and Roggen and Van Dijk, 1976), and organic solvent treatment (Gonai and Hinata, 1989 and Ockendon, 1978) of stigma in *Brassica olerecea*.

Cloning the parent by *in vitro* culture of curd explant is very effective when a laboratory is available for the breeder. Moreover, there is no risk to observe an evolution of the self-incompatibility in the progeny of the parent.

Crossing Methods

Single Cross Hybrid: Two self-incompatible but cross-compatible inbred lines with different S-haplotypes are planted in alternate rows, i.e., 1:1 or 1:2 or 1:3. The seed produced by both the lines is identical except that the characters are maternally inherited. Single cross hybrids are more uniform than double or top cross hybrids.

Double Cross Hybrid: Four homozygous self-incompatible inbreds with different S-haplotypes are required. Here in the first year single cross hybrids having different alleles are produced. Then, these F_1 hybrids are crossed with each other in the following season to produce double cross as follows:

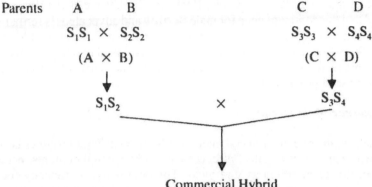

Commercial Hybrid

The double cross hybrids are less uniform but higher seed yield of commercial hybrid is obtained using F_1 combination in the production.

Top Cross: This method is more commonly used in the production of cabbage hybrid in United State of America. Here a single self-incompatible line is used as female parent and generally the most popular variety or superior inbred line as pollen parent. As soon as the pod setting is over, the pollen parent is removed. The planting ratio of female to pollen parent is generally kept as 2:1 or 3:1 or 4:1. The commercial F_1 seed of cauliflower Pusa Hybrid-2 is being produced by using top cross method at IARI, New Delhi. In India, most of private seed companies are also using top cross method in the commercial hybrid seed production.

In the commercial F_1 hybrid seed production of cauliflower and other *Brassica* vegetables problems of loss in vigor, continuously in breeding, pseudo-compatibility, environmental factors affecting self-incompatibility level and restriction of pollination within one parental line by bees are some important problems faced by producers. These above factors encourage increase in number of sibs thereby contaminating the hybrid seed quality. Frankel and Galun (1977) suggested that to get uniform visit of pollinator vector, morphological similarity in the parental lines may be increased and in addition to honey bees, flies may also be used. Synchronization in flowering of the parental lines is very important since most of the cauliflower types has cymose type of inflorescence and flowering period is very short.

Genic Male Sterility

In commercial hybrid seed production, the use of genic male sterility is limited because the male sterility phenotype is unstable and fertile plants must be discarded before flowering. However, with the standardized methods available for vegetative propagation both *in vitro* and *in vivo* of selected male sterile plant of cauliflower, the recessive male sterility can be used for commercial F_1 seed production (Ruffio-Chable, 1994). Ruffio-Chable et al. (1993) also reported that the dominant gene for male sterility and advocated its further use in hybrid production. Genetically engineered digenic male sterility of Mariani et al. (1990, 1992), an herbicide resistance gene is linked to *Ms* gene. By spraying herbicide the survived or resistant selected male sterile plants can be used in hybrid seed production.

Cytoplasmic Male Sterility

The cytoplasmically induced male sterility from different sources is associated with number of negative effects such as chlorophyll contents, nectar secretion, flower morphology and yield. The chlorophyll deficiency found in Ogura system is corrected by obtaining cybrid cytoplasm in *Brassica oleracea* L. (Pelletier et al., 1989) and now being used in the commercial production of F_1 hybrids in cauliflower. In different types of cauliflower number of lines having normal plant type with Ogura system are available for commercial F_1 seed production (Earle and Dickson, 1995 and Sharma 2003, unpublished).

Use of cytoplasmic lines is going to gain more commercial status, since there will be no danger of sibs. The male sterile lines in commercial F_1 production may be planted with the pollen parent in the ratio of 2:1 or 3:1 or 4:1 depending upon varietal characters. While selecting the pollen parent besides good combining ability, it also has similarity for most of morphological characters including plant height and synchrony in flowering with that of male sterile plant. Immediately after pod setting, the pollen parent is removed to avoid mixture and providing sufficient space to the female plant producing hybrid seed.

GENERAL SEED PRODUCTION METHOD

Cauliflower and broccoli except a few biennial types unlike most of other cole crops do not require vernalization for transforming from vegetative to reproductive phase. However, at the time of flowering and pollination mild temperature is required. In India, only annual types are grown. The seed production of tropical Indian cauliflower and temperate Snowball types is done in the plains and middle hills of north India, respectively. Sowing to raise seed crop in hills is done in September but in case of tropical type crop cultivation time is strictly followed. The cultural practices including sowing/transplanting time adopted to raise the commercial crop for vegetable production is followed to raise the seed crop. This helps in the production of true to type curds. The cauliflower curd is most important economical character besides others to be considered at the time of selections. This crop is highly thermo-sensitive for its growth and curd development hence sowing time should be followed strictly. After curd formation additional dose of fertilizer NPK is applied to raise the seed crop. Soft rot of curd caused by *Erwinia caratovora* and sclerotinia rot caused by *Sclerotinia sclerotiarum* are most serious diseases of seed crop besides black rot and Alternaria black leaf spot. A control schedule starting from curd initiation to pod development is recommended (Kapoor et al., 1988). Crop is spread with Bavistin (0.1%) followed by Dithane M-45 (0.2%) + Bavistin (0.1%) at 15-20 days interval. At least 4-5 sprays are needed. This also keeps the crop free from other foliar fungal diseases. The soft rot and black rot are managed by Streptocycline (0.01-0.02%) sprays. Aphids and Cater pillars of diamond back moth and *Pieris* spp. are more common. Since honey bees and other flies are the main pollinating agent, so spray the crop in late evening hours with safer insecticides or biopesticides.

Seed to seed method is followed since transplanting after curd development make the plant more prone to *Alternaria*, *Erwinia*, and other defects. Curd formation invariably precedes floral initiation. In Indian cauliflower, it takes about 25-40 days from curding to full bloom depending upon the variety. The optimum temperature for fertilization and seed development is 12-18°C. Cultivars of early group are more susceptible to diseases like *Erwinia* and *Alternaria*, distortion of inflorescence due to prevailing high temperature and humidity. Distortion of inflorescence is more common when the day temperature is more than 30°C. The commercial seed growers delay the transplanting of the early maturity group to get higher seed yield. This should not be encouraged since the selection of seed producing plants is in late sowing form curd at lower temperature. This affects the maturity time and is the main reason for longer harvesting period in early cultivars. Compact solid curd gives low seed yield due to lesser number of bolters. Some plants even do not bolt. Scooping of central portion or longitudinal half cutting or partial marginal cutting of curd is practiced to get more number of bolters. But scooped or half cut curds become more prone to diseases. Spraying of the crop with gibrellic acid 50-100 ppm at curd development and bolting stage helps in increasing the number of bolters and total seed yield by loosening the curd after proper development.

ROUGING

At least four inspections/rouging are required in cauliflower. First rouging is done at vegetative stage to remove off type plants and second at curd initiation. Third rouging is done after curd formation. Plants forming loose, ricey, fuzzy, and buttons are rejected. Blind, deformed, and diseased plants are also rejected. Fourth rouging is done after bolting but before flowering, plants with peripheral and uniform bolting are kept for seed production. Early and late bolters are also rejected.

ISOLATION

The cole crops are highly cross-pollinated. Percentage of cross-pollination varies from 72-95 percent depending upon the crop (Watts, 1968). In cauliflower cross-pollination varies from type to type. In European summer type, selfing may be as high as 70% (Nieuwhof, 1963). While in winter cauliflower and early Indian cauliflower cross-pollination is quite high as these types have high level of self-incompatibility. But this relationship may not be always true. The botanical varieties of *Brassica oleracea* group, i.e., cabbage, cauliflower, knol-khol, Brussels sprouts, kale, collard and other wild allies freely cross with each other. Hence, proper isolation among these crops is needed to keep purity of seeds. Similarly growing of different cultivars of each crop also require proper isolation. Isolation of 3000 m for breeder seed and 1500 m for certified seed is recommended. Honeybees and bumble bees are main pollinating agents. Arbitrary isolation distances are reported based on their flights.

HARVESTING AND SEED YIELD

When the pods attain yellow color, should be harvested, delayed harvesting results into seed shattering. Hence, 2-3 harvesting are required. About 50-60 days are needed for pod maturity after fertilization. Seeds of early types are ready for harvesting in February-March and mid-maturity in March-April months in north Indian plains. The harvest is cured under shed for 3-5 days and then gently thrashed after sun drying for 2-3 days. After cleaning, the seed is properly graded using grading machine and before storage should be dried to minimum moisture (about 5%). However, the seed of snowball types are ready for harvesting by June in India. Seed yield of Indian cauliflower may vary between 500-600 kg/ha and Snowball types from 300-500 kg/ha.

REFERENCES

Ahluwalia, K.S., Swarup, V. and Chatterjee, S.S. (1977). Inheritance of qualitative characters in Indian cauliflower. *Veg. Sci.* 4 : 67-80.
Anonymous (2002). FAO production year book. FAO, Rome, Italy.
Ayotte, R., Harney, P.M. and Souza-Machado, V. (1987). The transfer of triazine resistance from *Brassica napus* L. to *B. oleracea* L. 1. Production of F₁ hybrids through embryo rescue. *Euphytica*, 38: 615.

Bannerot, H.O., Boulidard, L., Cauderon, Y. and Tempe, T. (1974). Cytoplasmic male sterility transfer from *Raphanus* to *Brassica*. In Eucarpia Cruciferae Conference, Scotish Hort. Res. Institute, Mylnefield, Dundee, Scotland. p. 52.

Baswana, K.S., Rastogi, K.B. and Sharma, P.P. (1991). Inheritance of stalk rot resistance in cauliflower (*B. oleracea* var. *botrytis* L.). *Euphytica* 57:93-96.

Baswana, K.S., Rastogi, K.B. and Sharma, P.P. (1993). Inheritance studies in cauliflower. *Veg. Sci.* 20:56-59.

Bateman, A.J. (1955). Self-incompatibility system in angiosperms III Cruciferae. *Heredity* 9: 53.

Bateman, A.J. (1952). Self-incompatibility systems in angiosperms. I. Theory. *Heredity* 6: 285-310.

Borchers, E.A. (1966). Characteristics of male sterile mutant in cauliflower (*Brassica oleracea* L.). *Proc. Am. Soc. Hort. Sci.* 88: 406-410.

Bosewell, V.R. (1949). Our vegetable travelers. *Natural Geog*, 96: 145-217.

Boucault, L., Chauvin, J.E., Margale, E. and Hervé, Y. (1991). Etude de caractères morphologiques et iso-enzymatiques sur des plantes issues de culture d'anthères chez le chou-fleur (*Brassica oleracea* var. *botrytis*). *Agronomie* 11: 727-736.

Bradfield, C.A. and Bjeldanes, L. (1987). Dietary modification of xenobiotic metabolism: Contribution of indolylic compounds present in *Brassica oleracea*. *J. Agric. Food Chem.* 35: 896.

Bradfield, C.A., Chang, Y. and Bjeldanes, L. (1985). Effects of commonly consumed vegetables on hepatic xenobiotic-metabolizing enzymes in the mouse. *Food Chem. Toxicol.* 23: 899.

Butler, C.G. (1971). The World of Honey Bee. Collins, London, 185 pp.

Carafa, A.M. and Carratu, G. (1997). Stigma treatment with saline solution: A new method to overcome self-incompatibility in *Brassica oleracea*. L. *J. Hort. Sci.* 72: 531-535.

Chand, J. (1980). Studies on selection for yield and quality in cauliflower. Ph.D. Thesis, PG School, IARI, New Delhi.

Chatterjee, S.S. and Swarup, V. (1972). Indian cauliflower has a still greater future. *Indian Hort.* 17: 18-20.

Chatterjee, S.S. and Swarup, V. (1984). Self-incompatibility in Indian cauliflower. *Eucarpia Cruciferae Newl.* 9 : 25-27.

Chatterjee, S.S. (1993). Cole crops. In Vegetable Crops. Bose, T.K., Som, M.G., and Abir, J. (eds) Naya Prokash. India 125-223 pp.

Chatterjee, S.S. and Mukherjee, S.K. (1965). Selection and maintenance of cauliflower (*B. oleracea* var. *botrytis* L.). *Indian J. Hort.* 22: 60-68.

Chiang, M.S. and Crete, R. (1987). Cytoplasmic male sterility in *Brassica oleracea* induced by *Brassica napus* cytoplasm, female fertility and restoration of male fertility. *Can. J. Plant. Sci.* 67: 891-897.

Christ, B. (1959). Entwicklungsgeschichtliche und physiologische Untersuchungen uber die Selbssterilitat von *Cardamine pratensis* L. *Z. Bot.* 47: 88-112.

Cre hu, G. Du. (1968). Early testing of pollen stigma compatibility relationship in *Brassica oleracea* by flurescence. Proc. Brassica Meeting of *Eucarpia* Dixon, G. E. (ed): 34-36.

Crisp, P. (1982). The use of an evolutionary scheme for cauliflowers in the screening of genetic resources. *Euphytica* 31:725.

Crisp, P., Crute, I.R., Sootherland, R.A., Angell, S.M., Bloor, K., Burgess, H. and Gourdon. P.L. (1989). The exploitation of genetic resources of *Brassica oleracea* in breeding for resistance to club root (*Plasmodiophora brassicae*). *Euphytica* 42:215.

Crisp, P. and Tapsell, C.R. (1993). Cauliflower (*Brassica oleracea* L.). In: Genetic Improvement of Vegetable Crops. Eds. G. Kalloo and B.O. Bergh, Pergamon Press. Oxford, pp.157-178.

Crute, I.R. and Gordon, P.L. (1987). Downy mildew. *Rep. Natn. Veg. Res. Sta.* for 1986/7: 54.

Dadlani, N.K. (1977). Studies on selection for yield and quality in cauliflower. Ph.D. Thesis, PG School, IARI, New Delhi.

Delourme, R. and Budar, F. (1999). Male sterility. In Gómez-Campo (ed.), Biology of *Brassica* Coenospecies. Elsevier Science : 185-216.

Deshpande, A.A. (1975). Genetical studies in early Indian cauliflower. M.Sc. Thesis, PG School, IARI, New Delhi.

Dickinson, H.G. (2000). Pollen stigma interactions, so near yet so far. *Trends in Genetics* 16(9): 373-376.

Dickson, M.H. and Eckenrode, C.J. (1980). Breeding for resistance in cabbage and cauliflower to cabbage looper, imported cabbage worm and diamond back moth. *J. Amer. Soc. Hortic. Sci.* 105: 782-785.

Dickson, M.H. (1975). G1117A, G1102A and G1106 A cytosterile broccoli inbreds. *Hort. Sci.* 10: 535.

Dickson, M.H. (1985). Male sterile persistent white curd cauliflower NY7642A and its maintainer NY7642B. *Hort. Sci.* 20:397.

Dodoens, R. (1578). A nieuve herbal ....Antverpiae, transl. H. Lyte. London.

Dua, I.S., Suman, B.C. and Rao. A.V. (1978). Resistance of cauliflower (*Brassica oleracea* var. *botrytis*) to *Xanthomonas comestris* influenced by endogenous growth substances and relative growth rate. *Indian J. Experimental Biology* 16: 488-491.

Earle, E.D. and Dickson, M.H. (1995). *Brassica oleracea* cybrids for hybrid vegetable production. In Terzi, M., Cella, R. and Falavigna, A. (eds.). Current Issues in Plant Molecular and Cellular Biology, Kluwer Academic publishers, Dordrecht, pp. 171-176.

El-Dean Mahmoud, A.A.G. (1972). Study of Indigenous (Folkways) Birth Control Methods in Alexandria, MS thesis, University of Alexandria Higher Institute of Nursing, Alexandria, Egypt.

Elleman, C.J., Willson, C.E., Sarker, R.H. and Dickinson, H.G. (1988). Interaction between the pollen tube and the stigmatic cell wall following pollination in *Brassica oleracea. New Phytol.*109: 111-117.

Ellis, P.R., Hardman, J.A. and Crisp, P. (1986). Investigations of the resistance of cabbage cultivars and breeding lines to insect-pests at Wellesbourne. Rep. I.O.B.C. meeting on Int. Crop Protection in Vegetables. Rennes, France, p. 7.

Faulkner, G.J. (1971). The behaviour of honeybees (*Apis mellifera*) on flowering Brussels sprout inbreds in the production of F_1 hybrid seed. *Hortic. Res.* 11: 60-62.

Figdore, S.S., Kenard, W., Song, K.M., Slocum, M.K. and Osborn, T.C. (1988). Assessment of the degree of restriction fragment length polymorphism in *Brassica. Theor. Appl. Genet.* 75: 833.

Frankel, R. and Galun, E. (1977). Pollination mechanism, reproduction and plant breeding, Monographs on Theoretical and Applied Genetics. Springer Verlag, Berlin, Heidelberg, NY.

Fraser, R.S.S. (1984). Genetic engineering: Techniques, challenges and some possible applications in Horticulture. *Sci. Hortic.* 35: 34.

Gallegly, M.E. (1956). Progress in breeding for resistance to club root of broccoli and cauliflower. *Phytopathology.* 46: 467.

Gangopadhyay, K.K., Gill, H.S. and Sharma S.R. (1997). Heterosis and combining ability studies in early group of Indian cauliflower involving self-incompatible lines. *Veg. Sci.* 24: 26-28.

Gill, H.S. (1993). Improvement of cole crops. Advances in Horticulture–Vegetable Crops, vol. 5 eds. Chadha, K.L. and Kalloo, G., Malhotra Publishing House, New Delhi:287-303.

Gill, H.S., Lakhanpal, K.D., Sharma, S.R. and Bhagchandani, P.M. (1983). K-1 a valuable addition to snowball group of cauliflower. *Indian Hort.* 27:23-24.

Gill, H.S., Sharma, S.R. and Kapoor, K.S. (1987). Pusa Himjyoti cauliflower for the hills. *Indian Hort.* 32:13-16.

Gonai, H. and Hinata, K. (1989). Studies on self-incompatiblity in *Brassica*: Influence of organic solvent on seed fertility. *Jpn. J. Breed.* 19:153.

Grandclément, C. and Thomas, G. (1996). Detection and analysis of QTLs based on RAPD markers for polygenic reistance to *Plasmodlophora brassicae* Woron in *Brassica oleracea. Theor. Appl. Genet* 93: 86-90.

Gray, A.R. (1982). Taxonomy and evolution of broccoli (*B. oleracea* L. var. *italica* Plenck). *Econ. Bot.* 36: 397.

Gustafsson, M. (1979). Biosystematics of *Brassica oleracea* group in *Eucarpia Cruciferae* Conference. Wageningen. pp. 11-21.

Haigh, J.C. (1962). Cauliflower: *Ann. Rep. Glasshouse. Crop. Res. Instt.* Littlehampton, Sussex.

Helm, J. (1963). Die Lacimiaten sipp-en von *Brassica oleracea* L. *Kulturplanze*, 11: 92-210.

Hinata, K. (1975). Effect of CO_2 for overcoming self-incompatibility of different genotypes in *Brassica oleracea* L. *Incompatibility News Lett.* 5: 60.

Hodgkin, T., Lyon, G.D. and Dickinson, H.G. (1988). Recognition in flowering plants: A comparison of the *Brassica* self-incompatibility system and plant pathogen interactions. *New Phytol.* 110: 557-569.

Honma, S. and Cash, J.N. (1986). 'Supreme' 'Beta White' and 'One Up' cauliflower. *Hort. Sci.* 21: 1076.

Hoser-Krauze, J., Gabryl, J. and Antosik, J. (1982). Influence of the cytoplasm of Indian self-incompatible lines on the earliness and quality of F1 cauliflower curds. *Eucarpia Cruciferae Newl.* 7: 12.

Hoser-Krauze, J., Lakowaska-Ryk, E. and Antosik, J. (1995). The inheritance of resistance of some *Brassica oleracea* L. cultivars and lines to downy mildew (*P. parasitica*). *Ex. Fr. Jour. of Applied Genetics.* 36: 27-33.

Hoser-Krauze, J., Lakowska-Ryk, E. and Antosik, J. (1984). Resistance of cauliflower & broccoli (*Brassica oleracea* L. var. *botrytis* L.), seedlings to downy mildew (*Perenospora parasitica*). *Eucarpia Cruciferae Newl.* 9: 92.

Hoser-Krauze, J. (1989). Comparison of suitability of sources of cytoplasmic male sterility (cms) and self-incompatibility for breeding F_1 hybrids of cauliflower. *Biuletyn Warzyniczy* supplement, I: 29.

Hoser-Krauze, J. (1979). Inheritance of self-incompatibility and the use of it in the production of F_1 hybrids of cauliflower (*B. oleracea* var. *botrytis* sub var. *cauliflora* DC). *Genetica Polonica* 20: 341-367.

Hoser-Krauze, J. (1987). Influence of cytoplasmic male sterility source on some characters of cauliflower (*Brassica oleracea* var. *botrytis* L.). *Genet. Polonica*. 28: 101.

Hyams, E. (1971). Plants in the Service of Man, London: pp 33-81.

Inomata, N. (1977). Production of interspecific hybrids between *Brassica campestris* and *B. oleracea* by culture *in vitro* of excised ovaries. I. Effects of yeast extracts and casein hydrolysable on the development of excised ovaries. *Japanese J. Breed.* 27: 295.

Jamwal, R.S. and Sharma, P.P. (1986). Inheritance of resistance to black rot (*Xanthomonas campestris*) in cauliflower (*B. oleracea* var. *botrytis* L.). *Euphytica* 35: 941-943.

Jensen, B.D., Hockenhull, J. and Munk, L. (1999). Seedling and adult plant resistance to downy mildew in cauliflower (*Brassica oleracea* var. *botrytis*). *Pl. Pathology* 48: 604-612.

Johnson, A.G. (1971). Factors affecting the degree of Self-Incompatibility in inbred lines of Brussels sprout. *Euphytica*. 20: 561.

Jones, H.A. (1932). Vegetable Breeding at the Univ. of California. *Proc. Amer. Soc. Hort. Sci.* 29 (5): 572-581.

Jourdan, P.S. and Earle, E.D. (1989). Genotypic variability in frequency of plant regeneration from leaf protoplasts of four *Brassica* species and *Raphanus sativus*. *J. Am. Soc. Hort. Sci.*, 1124: 343.

Jourdan, P.S., Earle, E.D. and Murschler, M.A. (1989). Synthesis of male sterile, triazine resistant *Brassica napus* by somatic hybridization between cytoplasmic male sterile *B. oleracea* and atrazine resistant *B. campestris*. *Theor. Appl. Genet.* 78: 445.

Jourdan, P.S., Earle, E.D. and Mutschler, M.A. (1990). Improved protoplast culture and stability of cytoplasmic traits in plants regenerated from leaf protoplasts of cauliflower (*Brassica oleracea* species *botrytis*). *Plant Cell Tissue Org. Cult.* 21: 227.

Jourdan, P.S., Earle, E.D. and Mutschler, M.A. (1985). Efficient plant regeneration from mesophyll protoplast of fertile and cms cauliflower (*Brassica oleracea* cv. *botrytis*). *Eucarpia Cruciferae Newl.* 10: 94.

Jyoti, S. and Vashistha, R.N. (1966). Gene effects studies of curd weight in mid season cauliflower (*Brassica oleracea* var. *botrytis* L.). *Haryana J. Hort. Sci.* 15: 263-266.

Kagami, T., Abamatsu, T. and Shiga, T. (1990). Effective transfer of Ogura type ctyoplasmic male sterile traits from *Brassica campestris* or *Raphanus sativus* to *B. oleracea* by protoplast fusion. In: *Proc of the 6th Cruciferae Genetics Workshop;* McFerson, J.R., Kresovich, S. and Dwyer, S.G. (Eds.) Cornell University, NY : 26.

Kale, P.N., Chatterjee, S.S. and Swarup, V. (1979). Heterosis and combining ability in Indian cauliflower (*Brassica oleracea* L. var. *botrytis* L.). *Veg. Sci.* 6: 11-20.

Kalloo, G. (1988). Vegetable Breeding Vol. I. CRC Press Boca Raton, Florida. 105-135.

Kalloo, G. (1996). Vegetable Brassica. Oil seed and vegetable Brassicas: Indian perspective (ed.). Chopra, V.L. and Prakash. S. Oxford IBH publication, New Delhi pp:103-131.

Kapoor, K.S. (1986). Disease resistance in cauliflower (*Brassica oleracea* var. *botrytis*) against *Sclerotinia sclerotiarum* (lib) De Bari. *Veg. Sci.* 13: 285-288.

Kapoor, K.S., Gill, H.S. and Sharma, S.R. (1988). An observatory for predicting sclerotinia rot of late cauliflower. *Indian Hort.* 33: 23-24.

Kieffer, M., Fuller, M.P., Jellings, A.J. (1994). The rapid mass production of cauliflower propagules from fractionated and graded curd. *ISHS Symposium on Brassicas, Ninth Crucifer Genetics Workshop.* 15-19 November 1994. Lisbon-Portugal.

King, S.R. and Dickson, M.H. (1994). Identification of resistance to *Alternaria brasicicola* in *Brassica oleracea. Eucarpia Cruciferae Newl.* 16: 126-127.

Kontaxis, D.G., Mayberry, K.S. and Rubatzky, V.E. (1979). Reaction of cauliflower cultivars to downy mildew in Imperial Valley. *California Agriculture* 33: 19.

Koshimizu, K.H., Ohigashi, H., Tokuds, A., Kondo, and Yamaguchi, K. (1988). Screening of edible plants against possible anti tumor promoting activity. *Cancer Lett.* 39: 247.

Kroh, M. (1964). An electron microscopic study of the behaviour of *Cruciferae* pollen after pollination. In: Linskens H.F. (ed.) Pollen Physiology and Fertilization. North Holland, Amsterdam: pp. 221-224.

Kumar, P.A. and Sharma, S.R. (1997). Improving vegetable brassicas biotechnologically. *Ind. Hort.* 41: 55-56.

Kumaran, N.M. (1971). Studies on combining ability, gene effects and heterosis in cauliflower. Ph.D. Thesis, PG School, IARI, New Delhi.

Lal, G., Chatterjee, S.S. and Swarup, V. (1978). Studies on combining ability in Indian cauliflower. *Genet Agr.* 32: 85-97.

Lal, G., Chatterjee, S.S. and Swarup, V. (1979). Genetics of important characters in Indian cauliflower. *Genet Agr.* 33: 181-190.

Lal, G., Swarup, V. and Chatterjee, S.S. (1977). Combining ability studies in early Indian cauliflower. *J. Agri. Sci.* 89: 169-175.

Lal, O.P., Gill, H.S., Sharma, S.R. and Singh, R. (1991). Field evaluation in different cultivars of cauliflower against the head borer (*Hellula undalis* Fabricius) (Lepidoptera: pyralidae). *J. Ent. Res.* 15: 277-281.

Lal, O.P., Sharma S.R. and Singh, R. (1994). Field resistance in different hybrids of cauliflower against Bihar hairy caterpillar (*Spilosoma obliqua* Walker) (Lepidoptera: arctidae). *J. Ent. Res.* 18: 45-48.

Landry, B.S., Hubert, N., Lincoln, S.E., Etoh, T., Crete, R. and Chiang, M.S. (1992). A genetic map for *Brassica oleracea* based on RFLP markers detected with expressed DNA sequences and mapping of resistance genes to race 2 of *Plasmodiophora brassica (Woronin). Genome* 35: 409.

Liebstein, A.M. (1927). Therapeutic effects of various food articles. *Am. Med.* 33: 33.

Lillo, C. and Hansen, M. (1987). An anther culture of cabbage, Influence of growth temperature of donor plants and media composition on embryo yield and plant regeneration. *Norwegian J. Agri. Sci.* 1: 105.

Linskens, H.F. and Heinen, W. (1962). Cutinase-Nachweis in pollen. *Z. Bot.* 50: 338-347.

Mahajan, V., Gill, H.S., Sharma, S.R. and Singh, R. (1996). Combining ability studies in Indian cauliflower (*Brassica oleracea* var. *botrytis* L.) Group III. *Veg. Sci.* 23: 166-170.

Mahajan. V., Gill, H.S., Singh, R. (1991). Screening of cauliflower germplasm lines against downy mildew. *Cruciferae Newl.* 14/15, 148-149.

Mahajan, V., Gill, H.S. and More, T.A. (1995). Inheritance of downy mildew resistance in Indian cauliflower(group III). *Euphytica* 86:1-3.

Manzanares Dauleux, M.J., Divaret, I., Baron, F. and Thomas, G. (2000) Evaluation of French *Brassica oleracea* landraces for resistance to *Plasmodiophora brassicae*. *Euphityca* 113 : 211-218.

Margara, J. and David, C. (1978). Les etapes morphologiques du developpement du meristem de chou-fleur (*Brassica oleracea* L var botrytis) comptes Rendus, I Academic de Science Paris, Series D 278. 1369-1373.

Mariani, C., DeBeucheleer, M., Truttner, S., Leemans, J. and Goldberg, R.B. (1990). Induction of male sterility in plants by a chimaeric ribonuclease gene. *Nature* 347: 737-741.

Mariani, C., Gossele, V., DeBeucheleer, M., DeBlock, M., Goldberg, R.B., DeGreef, W. and Leemans, J. (1992). A chimaeric ribonuclease inhibitior gene restores fertility to male sterile plants. *Nature* 357:384-387.

Mc Collum, G.D. (1981). Induction of an alloplasmic male sterile *Brassica oleracea* by substituting cytoplasm from "Early Scarlet globe" radish (*Raphanus sativus*). *Euphytica* 30:855.

McGrath, J.M., Quiros, C.F., Harada, J.J. and Landry, B.S. (1990). Identification of *Brassica oleracea* monosomic alien chromosome addition line with molecular markers reveals extensive gene duplication. *Mol. Gen.* 233: 198.

Moffett, M.L., Trimboli, D. and Bonner, I.A. (1976). A bacterial leaf spot disease of several *Brassica* varieties. *Apps. News Letter* 593: 30-32.

Murugiah, S. (1978). Studies on Self-incompability in Indian cauliflower. M.Sc. Thesis, PG School, IARI, New Delhi.

Murugiah, S., Vidyasagar, Chatterjee, S.S. and Swarup, V. (1983). Inter allelic relationship and ability of S-alleles in Indian cauliflower. Abstract No. 44, XV International Congress of Genetics, New Delhi.

Myers, C.H. and Fisher, W.J. (1944). Experimental methods in cabbage breeding and seed production. *Men. Cornell Agric. Univ. Exp. Sta.* 259: 29.

Nasrallah, J.B. (2000). Cell-cell signaling in the self-incompatibility response. *Current Opinion in Plant Biology* 3 : 368-373.

Nasrallah, J.B., Doney, R.C. and Nasrallah, M.E. (1985). Biosynthesis of glycoprotiens involved in the pollen-stigma interaction of incompatibility in developing flowers of *Brassica oleracea* L. *Planta* 165:100-107.

Nasrallah, M.E. and Wallace, D.H. (1968). The influence of modifier genes on the intensity and stability of self-incompatibility in cabbage. *Euphytica* 17 : 495.

Natti, J.J. (1958). Resistance of Broccoli and other crucifers to downy mildew of crucifers. *Phytopath.* 50 : 93-97.

Nieuwhof, M. (1959). Experiences in breeding pointed head cabbage at the Institute of Hort. Plant Breeding III, IV and V (Conclusions). *Zaadbelangen* 12 : 204-205. 218-219, 236-238 and 250-251.

Nieuwhof, M. (1969). Cole crops: Botany, Cultivation and Utilization World Crops Books, Leonard Hill (Books) Ltd., London.

Nieuwhof, M. and Garretsen, F. (1961). The solidity of the cauliflower curd. *Euphytica* 10: 301-306.

Nieuwhof, M. (1961). Male sterility in some cole crops. *Euphytica* 10: 351-356.

Nieuwhof, M. (1963). Hybrid breeding in early spring cabbage. *Euphytica* 12:189-197.

Nieuwhof, M. (1968). Effect of temperature on the expression of male sterility in Brussels sprout (*B. oleracea* L.var. *gemmifera* DC.). *Euphytica* 17: 265-273.

Nieuwhof, M. (1974). The occurrence of self-incompatibility in cauliflower (*B. oleracea* var. *Botrytis* L. subvar. *Cauliflora* D.C.) and possibilities to produce uniform varieties. *Euphytica* 23: 473-478.

Ockendon, D.J. (1975). The S-alleles collection of *Brassica oleracea*. *Incompatibility Newsl* 5: 82-84.

Ockendon, D.J. (1988). The ploidy of plants obtained from anther culture of cauliflowers (*Brassica oleracea* var. *botrytis*). *Ann. Appl. Biol.* 113: 319-325.

Ockendon, D.J. and Currah, L. (1978). Time of cross and self-pollinated effect the amount of selfed seed set by partially self-incompatible plants of *Brassica oleracea*. *Theor. Appl. Genet.* 52: 233.

Ockendon, D.J. (1978). Effect of hexane and humidity on self-incompatibility in *Brassica oleracea*. *Theor. Appl. Genet.* 52: 113.

Ogura, H. (1968). Study on the new male sterility in Japanese radish with special reference to the utilization of this sterility towards the practical raising of hybrid seeds. *Mem. Fac. Agric.* Kagoshima Univ. 6. p. 39.

Osborn, T.C., Song, K.S., Kennard, W.C., William, P.H., Slocum, M.K., Figdore, S.S. and Suzuki, J. (1989). Genome analysis of *Brassica* using RFLP markers. In: Proc. of the 5th Cruciferae Genetic Workshop, Rep. No. 4. Genetic Resource Conservation Programme, Quiros, C.F. and McGuire, P.E. (Eds.): 10 University of California.

Pal, A.B. and Swarup, V. (1966). Gene effects and heterosis in cauliflower II. *Indian J. Genet* 26 (3): 282-294.

Pandey, K.K., Pandey, P.K., Singh, B., Kalloo, G. and Kapoor, K.S. (2001). Sources of resistance to downy mildew disease in Asiatic group of cauliflower. *Veg. Sci.* 28: 55-57.

Pandey, S.C., Naik, G., Ramkishan, and Sridhar, T.S. (1995). Breeding resistant varieties in cauliflower and cabbage. In: Agroecosystem Management (Mandal et al., eds.), IIHR, Bangalore.

Pandey, S.C. and Naik, G. (1985). Factor analyzing of yield components in cauliflower (*Brassica oleracea* var. *botrytis*). *Eucarpia Cruciferae Newl.* 10: 68-70.

Pandey, S.C. and Naik, G. (1986). Genetic and combining ability studies in cauliflower. *Eucarpia Cruciferae Newl.* 11: 36-37.

Pearson, O.H. (1932). Breeding plants of the cabbage group. *Calif. Agric. Exp. Sta. Bull.* 532: 3-22.

Pearson, O.H. (1972). Cytoplasmically inherited male sterility characters and flavor component from the species cross (*B. nigra* L. Koch) into *Brassica oleracea*. L. *J. Amer. Soc. Hortic. Sci.* 97: 397.

Pelletier, G., Primard, C., Vedel, F., Chetrit, P., Remy, R., Rousselle, P. and Renard, M. (1983). Intergeneric cytoplasmic hybridization in Cruciferae by protoplast fusion. *Mol. Gen. Genet.* 191:244.

Pelletier, G.A., Ferault, M., Lancelin, D. and Boulidard, L. (1989). CMS *Brassica oleracea* cybrids and their potential for hybrid seed production. XII Eucarpia Congress, Gottingen, Germany (Poster Abstract).

Reynaerts, A., Vandewiele, H., Desutter, G. and Janssens, J. (1993). Engineered genes for fertility and their application in hybrid seed production. *Scientia Hort.* 55 : 125-139.

Roggen, H.P.J.R. and Van Dijk, A.J. (1976). Thermally aided pollination: A new method of breaking self-incompatibility in *Brassica oleracea* L. *Euphytica* 25: 643.

Roggen, H.P.J.R. and Van Dijk. A.J. (1979). Breaking self-incompatibility of *Brassica oleracea* L. by steel brush pollination. *Euphytica* 21: 424.

Roggen, H.P.J.R., Van Dijk, A.J. and Dorsman, C. (1972). Electric aided pollination: A method of breaking self-incompatibility in *Brassica oleracea* L. *Euphytica* 21:181.

Ross, C.L. (1980). Embryo culture in production of disease resistant *Brassica*. In: Tissue Culture Methods for Plant Pathologist, Ingram, D.S. and Helgeson, J.P. Eds., Blackwell Scientific Publication, Oxford, UK. 225.

Ruffio-Chable, V. 1998. Complexity of self-incompatibility phenotype in *Brassica*: Its measure and some thoughts about its genetic control. *Acta Horticulturae* 459: 281-288.

Ruffio-Chable, V. and Gaude, T. (2001). S-haplotype polymorphism in *Brassica oleracea. Acta Horticulturae* 546: 257-261.

Ruffio-Chable, V., Hervé, Y., Dumas, C. and Gaude, T. (1997). Distribution of S-haplotypes and its relationship with self-incompatibility in *Brassica oleracea*. Part 1. In inbred lines of cauliflower. *Theor. Appl. Genet.* 94: 338-346.

Ruffio-Chable, V., Bellis, H. and Herve, Y. (1993). A dominant gene for male sterility in cauliflower (*Brassica oleracea* var. *botrytis*). Phenotype expression, inheritance and use in F_1 hybrid production. *Euphytica* 67: 9-17.

Ruffio-Chable, V. (1994). *Les systemes d'hybridations chez le chou-fleur (Brassica oleracea var. botrytis). Application a l'amelioration genetique*, thesis, ENSA de Rennes.

Sadik, S. (1962). Morphology of the curd of cauliflower. *Am. J. Bot.* 49:290.

Sandhu, J.S. and Singh A.K. (1977). Inheritance of maturity period and curd weight in cauliflower (*Brassica oleracea* var. *botrytis* L.). *Haryana J. Hort. Sci.* 6: 161-165.

Sandhu J.S., Thakur, J.C. and Nandpuri, K.S. (1977). Investigations on hybrid vigour in cauliflower (*Brassica oleracea* L. var. *botrytis* L.). *Indian J. Hort.* 34 : 430-434.

Schenck, H.R. and Robbelen, G. (1982). Somatic hybrids by fusion of protoplasts from *Brassica oleracea* and *Brassica campestris. Z. Pflanzenzucht,* 89: 278.

Schopfer, C.R., Nasrallah, M.E. and Nasrallah, J.B. (1999). The male determinant of self-incompatibility in *Brassica. Science* 286: 1697-1700.

Schulz, O.C. (1919). Das pflanzenreich Ed. A. Engler Leipzig 4. 1-290.

Sharma, S.R., Gill, H.S. and Kapoor, K.S. (1988). Heterosis and combining ability studies in late cauliflower (*Brassica oleracea* var. *botrytis* L.). *Veg. Sci.* 15: 55-63.

Sharma, S.R., Singh, R. and Gill, H.S. (1999). Cauliflower Pusa Sharad. *Ind. Hort.* 44: 7-8.

Sharma, B.R., Dhiman, J.S., Thakur, J.C., Singh, A. and Bajaj, K.L. (1991). Multiple disease resistance in cauliflower. *Advances in Horticultural Science.* 5: 30-34.

Sharma, B.R., Swarup, V. and Chatterjee, S.S. (1972). Inheritance of resistance to black rot in cauliflower. *Can. J. Gen. Cytol.* 14: 363-370.

Sharma, B.R., Swarup, V. and Chatterjee, S.S. (1975). Inheritance of resistance to drying at bolting and seed formation stage in cauliflower. *Indian. J. Genet.* 35(1):44-48.

Sharma, S.R., Gill, H.S. and Kapoor, K.S. (1997). Inheritance of resistance to white rot in cauliflower. *Indian J. Hort.* 54: 86-90.

Sharma, S.R., Kapoor, K.S. and Gill, H.S. (1995). Screening against Sclerotinia rot (*Sclerotinia sclerotiarum*), Downy mildew (*Peronospora parasitica*) and black rot (*Xanthomonas compestris*) in cauliflower (*Brassica oleracea* var. *botrytis* sub var. *cauliflora* DC). *Indian J. Agric. Sci.* 65: 916-918.

Sharma, S.R., Singh, R., Vinod, Singh, D.K. and Edison, S.J. (2001). Self-incompatibility level in different maturity groups in Indian cauliflower. Diamond Jubilee Symp.–Hundred Years of Post Mendelian Genetics and Plant Breeding, Nov. 6-9, New Delhi, pp. 155-156.

Shelton, A.M., Hoy, C.W., North, R.C., Dickson, M.H. and Bernard, J. (1988). Analysis of resistance in cabbage varieties to damage by Lepidoptera and Thysanoptera. *J. Econ. Entomol.* 81: 636-640.

Simmonds, D., Arnison, P. and Keller, W. (1989). Haploid and cell biology of microspore embryogenesis. In: Proc. of the 5th Cruciferae Genetics Workshop, Rep No. 4, Ouiros, C.F. and McGuire, P.E. (Eds.). University of California, Davis, 11.

Singh, D.P., Swarup, V. and Chatterjee, S.S. (1975). Genetical studies in Indian cauliflowers (*Brassica oleracea* L. var. *botrytis* L.). I. Heterosis and combining ability in maturity group I. *Veg. Sci.* 2: 1-7.

Singh, D.P., Swarup, V. and Chatterjee, S.S. (1976b). Genetical studies in Indian cauliflower III (*Brassica oleracea* L. var. *botrytis* L.). Heterosis and combining ability in maturity group-III. *Veg. Sci.* 3: 47-53.

Singh, R. and Sharma S.R. (2001). Cole crops. In: Thamburaj, S. and Singh, N. (Eds.) Textbook of Vegetables, Tubercrops and Spices. Directorate of Information and Publications of Agriculture, Indian Council of Agricultural Research, New Delhi. pp. 76-146.

Singh, P.K. (2000). Utilization and seed production of vegetable varieties in India. *Journal of New Seeds* 2: 37-42.

Singh, R., Gill, H.S. and Sharma, S.R. (1993). Breeding cauliflower for downy mildew resistance. Golden Jublee symposium–Hort. Res.–changing scenario–Bangalore pp. 67.

Singh, D.P., Swarup, V. and Chatterjee, S.S. (1976a). Genetic studies in Indian cauliflower(*Brassica oleracea* L. var. *botrytis*). heterosis and combining ability in maturity group II. *Veg. Sci.* 3: 41-46.

Singh, H. and Kalda, T.S. (1995). Screening of cauliflower germplasm against sclerotinia rot. *Indian J. Genet.* 55: 98-102.

Singh, R., Chatterjee, S.S., Swarup, V., Trivedi, B.M. and Sen, B. (1993). Evolution of Pusa Shubhra–A resistant cauliflower to black rot and curd blight diseases. *Indian J. Hort.* 50: 370-372.

Singh, R., Gill, H.S. and Chatterjee, S.S. (1997). Breeding Pusa Early Synthetic cauliflower. *Veg. Sci.* 24: 23-25.

Singh, R., Gill, H.S. and Sharma S.R. (1994). Breeding of Pusa Hybrid–2 cauliflower. *Veg. Sci.* 21:129-131.

Singh, R., Trivedi, B.M., Gill, H.S. and Sen, B. (1987). Breeding for resistance to black rot, downy mildew and curd blight in Indian cauliflower. *Eucarpia Cruciferae Newsl.* 12: 96-97.

Slocum, M.K., Figdore, S.S., Kennard, W.C., Suzuki, J.Y. and Osborn, T.C. (1990). Linkage arrangement of restriction fragment length polymorphism in *Brassica oleracea. Theor. Appl. Genet.* 80: 75.

Snogerup, S. (1980). *Brassica.* Crops and Wild Allies. Eds. Tsunoda, S., Himda, K. and Gomezcampo, C. Japan Scientific Societies Press. Tokyo.

Song, K.M., Osborn, T.C. and William, P.H. (1988a). *Brassica* taxonomy based on nuclear restriction fragment length polymorphism (RFLPs). Genome evolution of diploid and amphiploid species. *Theor. Appl. Genet.* 75: 784-794.

Song, K.M., Osborn, T.C. and William, P.H. (1988b). *Brassica* taxonomy based on nuclear restriction fragment length polymorphism (RFLPs). Premilinary analysis of subspecies with *B. rapa* (Sync *Campestris*) and *B. oleracea. Theor. Appl. Genet.* 76: 593-600.

Song, K.M., Osborn, T.C. and William, P.H. (1990). *Brassica* taxonomy based on nuclear restriction fragment length polymorphisms (RFLPs). 3–Genome relationship in *Brassica* and related genera and the origin of *B. oleracea* and *Brassica rapa* (Syn. *Campestris*). *Theor. Appl. Genet.*, 79: 497.

Stoewsand, G.S., Babish, J.B. and Wimberly, H.C. (1978). Inhibition of hepatic toxicities from polybrominated biphenyls and aflatoxin B_1 in rats fed cauliflower, *J. Environ. Pathol. Toxicol.* 2: 399.

Svensson, V. (1954). Choose the right variety of cauliflower. *Weibulls Allehanda*: 34-38.

Swarup, V. and Chatterjee, S.S. (1974). Heterosis in Indian cauliflowers. Proc. XIX Ind. Hort. Cong. I Section Vegetables p. 670.

Swarup, V. and Pal, A.B. (1966). Gene effects and heterosis in cauliflower-I. *Indian J. Genet.* 26: 269.

Swarup, V. and Chatterjee, S.S. (1972). Origin and Genetic improvement in Indian cauliflower. *Economic Botany* 26: 381-394.

Tamassy, I. (1973). Some problems of breeding for heterosis in vegetable crops. *Kerteszeti Kozi* 64: 113.

Tapsell, C.R. (1989).Breeding winter maturing cauliflower in south-west England. *J. Hortic. Sci.* 64: 27-34.

Thompson, K.F. and Taylor, J.P. (1966). Non-linear dominance relationships between S-alleles. *Heredity* 21: 345-362.

Trivedi, B.M., Sen, B., Singh, R., Sharma, S.R., and Verma, J.P. (2000). Breeding multiple disease resistance in mid season cauliflower. Proc. Indian Phytopath. Soc. Golden Jubilee International Conference on Integrated Plant Disease Management for Sustainable Agriculture. Vol. 2: 699-700.

Van der Meer, O.P. (1985). Male sterility in cole crop–a serial story. *Eucarpia Cruciferae Newsl.* 10: 58.

Vander Meer, O.P., Toxopeus, H., Crisp, P., Roelofsen, H. and Astley, D. (1984). The collection of land races of cruciferous crops in EC countries. Final report of EC Research Programme 0890. Instituut voor de veredeling van Tuinbouwgewassen, Wageningen, the Netherlands, Rapport 198: 210.

Vashistha, R.N., Neog, S.J. and Pandita, M.L. (1985). Gene effects studies of curd compactness in cauliflower. *Haryana. Agric. Univ. J. Res.* 15: 406-409.

Verma, S.S. (1979). Studies on heterosis breeding in cauliflower. Ph.D. Thesis, PG School, IARI, New Delhi.

Vidyasagar. (1981). Self-incompatibility and S-alleles in Indian cauliflower. Ph.D. Thesis, PG School, IARI, New Delhi.

Walker, J.C. and Larson, R.H. (1960). Development of first club root resistant cabbage variety. Bul. No. 5, Wisconsin Agr. Exp. Stn. 12 pp. 547.

Walker, J.C. (1930). Inheritance of *Fusarium* resistance in cabbage. *J. Res.* 40: 721.

Wallace, D.H. (1979). Procedures for identifying S-alleles genotypes in *Brassica. Theor. Appl. Genet.* 54: 249-265.

Watts, L.E. (1963). Investigations breeding. I. Studies on self-incompatibility. *Euphytica* 12: 330-340.

Watts, L.E. (1964). Studies on maturity in F1 and F2 generation of cauliflower from crosses between summer, autumn and winter types. *J. Hort. Sci.* 39: 84-91.

Watts, L.E. (1965a). The inheritance of curding periods in early summer and autumn cauliflower. *Euphytica* 14: 83-90.

Watts, L.E. (1965b). Investigations into the breeding system of cauliflower. II. Adaptation of the system to inbreeding. *Euphytica* 14: 67-77.

Watts, L.E. (1968). Natural cross-pollination and the identification of hybrids between botanical varieties of *Brassica oleracea.* L. *Euphytica* 17: 74-80.

Objectives of Okra Breeding

B. S. Dhankhar

J. P. Mishra

SUMMARY. Okra is a seed propagated hot weather crop sensitive to frost, low temperature, and waterlogging as well as drought conditions. It is a multipurpose crop due its various uses. It is grown in many countries and cultivars from different countries have certain adapted distinguishing characteristics specific to the country to which they belong. In home consumption India tops the world. Genetic diversity exists for number of characters. Genetic resistance involving inter-specific crosses have been exploited commercially for Yellow Vein Mosaic virus. Hybrids are very much popular in this crop and the hybrid seed production is based on hand emasculation and hand pollination. Proper isolation is necessary between two fields of two varieties for maintaining the genetic purity of the stock. *[Article copies available for a fee from The Haworth Document Delivery Service: 1-800-HAWORTH. E-mail address: <docdelivery@ haworthpress. com> Website: <http://www.HaworthPress.com> © 2004 by The Haworth Press, Inc. All rights reserved.]*

KEYWORDS. Okra, genetic diversity, crop biology, inheritance, heterosis, pollination control mechanisms, seed production, hybrid seed production

---

B. S. Dhankhar is affiliated with the ICAR, New Delhi, India.

J. P. Mishra is affiliated with the Division of Vegetable Science, Indian Agriculture Research Institute, Pusa, New Delhi 110012, India.

[Haworth co-indexing entry note]: "Objectives of Okra Breeding." Dhankhar. B. S., and J. P. Mishra. Co-published simultaneously in *Journal of New Seeds* (Food Products Press, an imprint of The Haworth Press. Inc.) Vol. 6, No. 2/3. 2004. pp. 195-209; and: *Hybrid Vegetable Development* (ed: P. K. Singh, S. K. Dasgupta, and S. K. Tripathi) Food Products Press, an imprint of The Haworth Press, Inc., 2004, pp. 195-209. Single or multiple copies of this article are available for a fee from The Haworth Document Delivery Service [1-800-HAWORTH, 9:00 a.m. - 5:00 p.m. (EST). E-mail address: docdelivery@haworthpress.com].

INTRODUCTION

Okra, also known as lady's finger, *bhindi*, gumbo, etc., is a seed propagated hot weather crop sensitive to frost, low temperature (below 15°C), water-logging as well as drought conditions. It is grown in tropical and sub-tropical regions and also in warmer parts of temperate regions. The crop thrives well under hot humid climate. Best plant growth and fruiting is observed at around 25°C average temperatures with high relative humidity (65-85%). Seed germination is fast at 30-35°C temperatures. Temperatures lower than 25°C slow down germination while above 42°C slow down plant and fruit growth and cause drying/desiccation and dropping of flower buds. The immature fruits are cooked as vegetable. The fruits are fairly good in nutritive value and 100 g consumable portion contains 10.4 g dry matter, 3100 calorie energy, 1.8 g protein, 90 mg calcium, 1.0 mg iron, 0.1 mg carotene, 0.07 mg thiamin, 0.08 mg riboflavin and niacin, and 18 mg vitamin C, with almost comparable constituents, barring few, in leaves (Grubben, 1977). In Far-East countries, a form of *Abelmoschus* ssp. is used as leafy vegetable. Dry fruit shell and stem are used in the manufacture of paper and cardboard. Dry seed kernels of okra contain 13-22% edible oil and 20-24% edible protein and may be used as a substitute for edible oil. Dehydrated and frozen fruits are also marketed for off-season consumption.

In home consumption India tops in the world with an availability of 3.419 million tonnes green fruits produced from over 0.349 million hectares (Anonymous, 2001). In addition to home consumption, the immature fresh fruits from India are exported to its neighboring countries like Singapore, Mauritius, Malaysia, Sri Lanka, Bangladesh and other countries while, Bahrain, Qatar, Kuwait, Saudi Arabia, Muscat, Iran and Abu Dhabi are among upcoming potential markets for such exports (Anonymous, 2001). Cultivars Pusa A-4, Pusa Sawani, Varsha Uphar, and Parbhani Kranti are preferred types for export from India.

Okra has been an important crop in Afghanistan, India, Iran, Pakistan, Turkey, and Yugoslavia, and cultivars from these countries have certain adapted distinguishing characteristics specific to the country to which they belong.

CROP BIOLOGY

The genus *Abelmoschus* is accepted to be of Asiatic origin, though opinions differ for the origin of *A. esculentus* as India (Masters, 1875), Ethiopia (de Candolle, 1883; Vavilov, 1951), West Africa (Chevalier, 1940; Murdock, 1959), Tropical Asia (Grubben, 1977) and Hindustani Centre of Origin chiefly India, Pakistan, Burma (Zeven and Zukovsky, 1975).

The cultivated species, *Abelmoschus esculentus* (L.) Moench. belongs to the family Malvaceae under the order Malvales. The genus *Abelmoschus* has been separated from the genus *Hibiscus* because in the earlier case calyx, corolla and staminal column are fused at the base and fall together after the anthesis. The cultivated okra has somatic chromosome number 2n = 130 and is considered to be an amphidiploid of *A. tuberculatus* (2n = 58) and an unknown species with 2n = 72. *A. esculentus* with 2n = 72, 108, 120, 132, and 144 indicates a series of polyploids of x = 12 (Datta and Naug, 1968). Later, polyploid nature of *A. esculentus* was also suggested by Charrier (1984).

The plant is mostly robust, erect (occasionally branched), annual herb with well developed tap root system; stem, petiole, leaf veins with or without red pigmentation; leaves alternate, palmately lobed, hirsute and serrate; flowers solitary, usually axillary with lobed epicalyx. Calyx generally non-lobed, tubular; corolla with 5 large yellow petals, purplish red pigmentation on one or both sides of claw; tubular staminal column with numerous stamens, attached to the corolla base; stigma 5-10 lobed, velvety red/purplish; fruit 5-10 chambered, beaked capsule, with or without sutures/ridges; seed dark green, grey or dark brown, rounded.

Flowering starts from 3rd to 7th node from the base and continues along with the growing stems/branches. Stigma becomes receptive 2 hours before anthesis (protogyny) and receptivity persists till 3-4 hours after anthesis with varying degree of receptivity. Anthers dehisce at the time of anthesis. It takes about 45-50 days from sowing to flowering and fruit set. The fruits develop to marketable (unripe harvest) maturity in 5-6 days and to drying and seed harvest in 30-32 days from anthesis.

INHERITANCE

An indicative account of inheritance of characters reported by various workers is presented in Table 1.

In addition, pigmentation of calyx, corolla, fruit color were observed to have monogenic control (Kolhe and D'Cruz, 1966). While in case of plant height and spininess, the heterozygotes were intermediate indicating polygenic control (Jasin, 1967). One to three groups of dominant genes controlled days to flower and number of fruits per plant, while 4 to 5 groups of dominant genes controlled plant height. Kulkarni (1976) observed that all the three yield component characters showed over-dominance, and further analysis of gene effects showed that duplicate and complimentary types of epistasis were more pronounced (Kulkarni et al., 1973, 1978).

In inter-specific crosses involving *A. manihot* ssp. *manihot,* the susceptibility to YVMV was controlled by two dominant genes (Thakur, 1976; Sharma

TABLE 1. Gene effects and inheritance of principal characteristics of okra.

Characteristic	Observation	References
Plant height	Dominance gene action Non-additive gene action Both additive and dominance high heritability (79%)	Kulkarni, 1976 Rao and Sathyavathi, 1977 Ramu, 1976; Lal et al., 1977
Internode length	High heritability (79%)	Lal et al., 1977
Dark red stem	Simple inheritance	Erickson and Couto, 1963
Pubescence of foliage	Incomplete dominance	Nath and Dutta, 1970
Light red petiole	Simple inheritance	Erickson and Couto, 1963
Degree of leaf lobing	Simple inheritance (cut/lobbed)	Kalia and Padda, 1962; Jasin, 1967
	Monogenic with incomplete dominance	Nath and Dutta, 1970
Resistance to Jassids	Dominant, polygenic control	Sharma and Gill, 1984
Resistance to YVM virus	High heritability (70%) Digenic control (recessive) Monogenic	Padda et al., 1970 Singh et al. 1962; Thakur 1976, Sharma and Dhillon, 1983; Jambhale and Nerker, 1981; Dutta, 1988.
Earliness	Simple inheritance	Erickson and Couto, 1963
Days to flowering	High heritability (65%)	Padda et al., 1970
	Non-additive gene action	Rao and Sathyavathi, 1977
Calyx color	Simple inheritance (Dominance)	Kalia and Padda, 1962
Petal blotch	-do-	Erickson and Couto, 1963
Base color	-do-	Kalia and Padda, 1962
Petal venation color	-do-	-do-
Pod color	Simple inheritance (multiple alleles) Two dominant genes One dominant gene (white/green)	Kalia and Padda, 1962 Nath and Dutta, 1970 Jasin, 1967
Pubescence of pods	Complex inheritance	Miller and Wilson, 1939
	Simple inheritance Incomplete dominance	Kalia and Padda, 1962 Nath and Dutta, 1970
Fruit shape	Two genes Digenic with epistasis (angular/round)	Kalia and Padda, 1962 Jasin, 1967
	Several genes	Miller and Wilson, 1939

Characteristic	Observation	References
Weight of fruit	Average heritability (48%)	Ngah and Graham, 1973
Pods per plant	Additive gene action and Average heritability	Rao and Sathyavathi, 1977
	High heritability	Rao and Kulkarni, 1977
Seeds per pod	Very high heritability (94%)	Padda et al., 1970
Seed weight	High heritability (77%)	Padda et al., 1970
Yield	High heritability (63%)	Padda et al., 1970

and Dhillon, 1983), while Jambhale and Nerker (1981) observed single dominant gene to control YVMV resistance in *A. manihot* and *A. tetraphyllus*. A
single dominant gene controlled YVMV resistance in *A. tetraphyllus* was also
observed by Dutta at IIHR, Bangalore, India (pers. comm.). Resistance to
jassids in *A. manihot* ssp. *manihot* var. *Ghana* was under polygenic control
(Sharma and Gill, 1984). Both additive and dominant gene effects were significant but additive gene effects and dominance × dominance interactions appeared relatively more important. Resistance against jassids has been reported
in IHR-21 (India) while against Shoot and Fruit Borer in *A. tuberculatus*, cvs.
Red-I, Red-II, Red Wonder, and against Root Knot nematode in the cv. Long
Green Smooth. Resistance to *Fusarium oxysporum* f. sp. *vasinfectum* has been
reported in *A. manihot* P.I. 379584 while resistance against powdery mildew
(*Erysiphe* sp.) has been reported in *A. tetraphyllus*, *A. manihot*, *A. manihot* ssp.
manihot, and *A. moschatus*. True resistance to YVM virus has been reported in
A. manihot (some forms), *A. pungens, A. crinitus, A. panduraeformis*, and *A.
vitifolius*. Symptom less carrier nature of YVM resistance is observed in *A.
manihot* ssp. *manihot, A.* ssp. 'Ghana', *A. tuberculatus*, and *A.* ssp. (West African okra). This group could be utilized for breeding YVM resistant hybrid varieties.

A. *manihot* ssp. *manihot* contributes dark green fruit color and more number of fruits per plant in F_1 while *A. manihot* sp. *tetraphyllus* contributes to increase in number of branches, fruits and ridges on fruits. *A. panduraeformis*
(2n = 24) could be utilized to incorporate jassid and drought tolerance in the
progeny. Natural allopolyploids like *A. esculentus* an amphidiploid of *A.
tuberculatus* (2n = 58) and *A.* ssp. (2n = 72) and West African okra an
amphidiploid of *A. esculentus* (2n = 130-140) and *A. manihot* (2n = 60-68) are
widely cultivated showing considerable out crossing in nature even at species
level. West African okra is characterized with red leaf veins, late flowering,

reduced number of sub-calyx bracts, pods mounted at right angles to stem and large number of seeds in fruits.

VARIABILITY AVAILABLE FOR UTILIZATION

In *A. esculentus* types, i.e., cultivated okra, genetic diversity exists for number of ridges and their presence/partial presence or absence on fruits, extent of hairiness and pigmentation on plant parts including fruits, ratio between length and width (diameter) of fruits, attachment of fruits on plant, lobing and lamina surface of leaves and also branching and height of the plant. Some of these characters are also influenced by environmental conditions.

In India, maximum variability in plant type and fruit type is available in Eastern part (Orissa, West Bengal, Bihar) and North-Eastern part (Assam, Meghalaya, Sikkim) where land races and primitive types can still be found with poor yields and varying degree of tolerance to stresses. In hilly areas (warmer parts), smooth fruited variety may develop hard hairs and sometimes non-pigmented ones develop partial pigmentation.

A. esculentus forms have adapted to agro-climates, requirement and preference of the consumers. As such the plant type, and fruit type from different countries may look different (due to adaptation) but these are all *A. esculentus*, barring few. In West Africa, Soudanian (*A. esculentus*) and Guinnean (*A. caillei*) forms have got established. In The Philippines, Sounde (extra early) and Gbogboligbo types (both *A. esculentus*) have got established. In Papua New Guinnea, Fiji and Solomon islands forms of *A. manihot*, both flowering types (Aidiba) as well as non-flowering types (Aibika) have got established as leafy vegetables suitable for rejuvenation after multiple cuttings and perpetuated through easily rooted stem cuttings. The presence of some form of *A. esculentus* having round non-lobed leaves, short fruits and probably with $2n = 72$ in Saurashtra and Dwarka region of India and some part of Japan (used by Teshima, 1933 and Ugale et al., 1976) indicate genetic variability within *A. esculentus* itself.

Magnitude of genetic variability was observed as high for number of fruits per plant, plant height, onset of flowering, number of nodes, stem diameter, leaf color and leaf shape (Kuwada 1964). Complex characteristics such as yield must be related to many individually distinguishable characteristics. Yield is related to height, branching, stem diameter, leaves per plant and other characteristics. High correlation coefficients ($r = 0.4$) were found in case of color characteristics of cotyledons–(stem and leaf), leaves (vein base, petiole of first leaf) and pods, and simply indicated that anthocyanin (red color) tends to be produced in various organs at the same time. Total seed yield is related principally to number of pods per plant and secondly to seeds per pod.

Seed oil content and Gossypol content were neither related *inter se* nor with other characters. Relatively low correlations among most of the listed characteristics suggested that selection can proceed independently for several characters with considerable success.

Siemonsma (1982) classed West African Guinean okra (2n = 185-198) as a separate species different from *A. esculentus* (2n = 130), as the former are tropical, short day types, late flowering with high seed content in the fruits. This group of West African okra has very low homogeneity within itself which can be used for improving tropical as well as temperate okra, particularly for imparting dark green fruit color and YVM resistance.

HETEROSIS

Inter-varietal hybrids in okra have, of late, been found useful. Some of the examples are given in Table 2.

Important characteristics in a hybrid combination should be, medium plant height, up-right branching, high number of fruit bearing nodes, lower position

TABLE 2. Some studies showing promising heterosis

Improved characters	F1 hybrid combinations	Source
Early flowering, early maturity, higher fruit weight and yield	H 398 × Pusa Sawaniand H-398 × Pusa Makhamali	Joshi et al., 1958; Raman and Ramu, 1963
Higher germination, early flowering, more plant height and higher yield	Local 5 (Malaysian) × Emerald (American), Local 7 (Malay.) × Gold Coast (Amer.), Local 5 (Malay.) × Gold Coast (Amer.), Local 7 (Malay.) × Emerald (Amer.)	Jalani and Graham, 1973
More fruits/plant and higher yield	Pusa Sawani × Smooth Long Green	Sharma and Mahajan, 1978
More plant height and higher yield	American Seven Ridged × Pusa Sawani	Sharma and Mahajan, 1978
Improved yield	New Selection × AE 91	Poshiya and Shukla, 1986
Higher plant height, fruits/plant and fruit weight	Balady × Gold Coast	El-Maksoud et al., 1986

of first fruiting node, deeply lobed leaves, short internodes, dark-green fruit color, earliness, tolerance to important pests and diseases. Potential for high nutrient and water up-take in order to improve production is one single important factor responsible for success of a genotype/hybrid.

So far, major breeding objective has been the yield. Earlier workers have reported yield improvement in okra from 49 to 60 percent through heterosis breeding (Joshi, et al., 1958; Partap and Dhankhar, 1980). However, there is still scope for yield improvement (Hoque and Hazarika, 1993). In okra, fruit number has been found to be the most effective factor contributing to yield, followed by plant height, fruit weight and size, and earliness. Improving yield and ensuring its sustainability under adverse conditions through resistant hybrids is the major objective of heterosis breeding.

Hybrids are also aimed to have improved mineral utilization capacity for different elements. Combining resistance to one or more pests (jassids, aphids, borers), diseases (wilt, yellow vein mosaic) through morphological or biochemical (sugars, phenols, alkaloids, phytoallexins) processes may be an important object. For processing purposes fruits of the hybrids should have less fibre, less mucilage, high protein, vitamins, dry matter, and minerals and high dehydration/rehydration ratio for dehydration while for canning and freezing the fruits should be small, tender, dark green with low dry matter and rich in nutrients.

POLLINATION CONTROL MECHANISM

The flowers in okra are bisexual, large, yellow, showy and attract insects both beneficial as well as harmful. It takes about 20-22 days from sowing to flower bud initiation and a similar period from bud initiation to the opening of flower. Several environmental as well as plant factors decide the flower size, number of anthers per flower or pollen per anther, size and number of stigmatic lobes, first flowering node, time of anthesis, receptivity of stigma, etc. Among factors within the plant are plant health, nutritive status and turgidity of cells and among environmental temperature and relative humidity, wind velocity, rain during pollination and type and population of insects in the vicinity. While an unhealthy plant may produce small flowers with low number of anthers/pollen and low stigma receptivity, dry air may cause small flower, low pollen number and fertility, and drying and desiccation of pollen causing failure in pollination. These factors will affect the post pollination and pre-fertilization process also which needs 2-6 hours to complete. These will also interfere in the perpetuation process by way of affecting number and quality of seed.

Okra is an often cross-pollinated crop. However, pollination is mainly through self-pollination. While growing, the style pushes the stigma through staminal column and the stigmatic lobes come in contact with the bursting anthers and pollen get stuck to the sticky surface. This necessarily causes self-pollination. Presence of insects in the field help in sib/self-pollination.

Proper isolation distance (400 m) kept between the fields of two varieties safe guards the risk of cross pollination by insects. A 6-8 feet tall barrier crop will be a better guard against high flying insects and in that case isolation distance may be reduced.

For strictly controlled pollination, individual flower buds when fully mature, are bagged with perforated butter paper bags to restrict the entry of any foreign pollen. Enclosures made with fine mosquito net also restrict cross pollination in covered plants. Single plants could be covered using 3-ring muslin cloth bags of 50 cm diameter and appropriate height, tying the top and bottom of the bag with some support. Larger enclosures could be made to suit the requirement.

Anthesis takes place from 6.0 to 8.0 a.m. The anthers burst open at the time of anthesis though these are matured an hour earlier to anthesis. The flowers droop down after few hours of pollination and drop down next day. The present day cultivated okra has complex genotype due to involvement of diverse parents in the process of resistance breeding. As such, fresh studies on the flower biology are required.

DEVELOPMENT OF HYBRIDS

The process of development of hybrids in okra is not different from other vegetable crops, viz., development of inbreds, testing of combining ability and production of F_1 hybrid seeds. Here, breeding lines/cultigens/varieties stable for phenotypic characters can be taken as parents to study combining ability and develop hybrids. It should be kept in mind that due to high chromosome number and polygenic control of major economic characters, 100% homozygosity in the parents is difficult to achieve.

Both use of heterozygosity *per se* and pooling of favorable genes in F_1 is important in okra. It is not only the yield and its contributing characters that should be taken care of but there may be many additional objectives, which are not related with yield directly. For example, upright unbranched plant will give higher plant density and higher yield. Deeply cut, smooth leaves will harbor less pathogens and pests. Short internodes will give higher proportion of plant dry matter in the form of economic part as fruit. Besides attractive color, fruit shape and smoothness of fruits may be desirable to combine in the hybrids. Several biotic and abiotic stresses limit production in certain areas.

These may be YVM virus, fruit and shoot borer, wilt, powdery mildew or may be soil salinity conditions. Incorporation of resistance or tolerance to one or more of these in the F_1 hybrid is possible and will help in realizing sustainable yield even under problematic environment. Bringing thermo- and photo-insensitivity in the F_1 hybrid and breeding short day types will enable us to take the crop over extended period. Most of the varieties (except Pusa Sawani) are medium to late and photosensitive. Moisture stress causes considerable yield reduction in most of our improved varieties. Our earlier variety Pusa Sawani has shown tolerance to salts, for germination and vegetative growth, up to 6 mmhos/cm NaCl and 8 mmhos/cm Na_2SO_4.

PRODUCTION OF HYBRID SEED

Hybrid seed production has three major components;

1. Selection of suitable parents,
2. Crop production and controlled pollination, and
3. Seed Extraction.

Selection of parents is done on the basis of their combining ability. For proper identification of the variety (parent), clear cut/distinct characters of identification are known/noted so that any undesired type of plant could be rouged out. Such characters may be plant height, number of branches, color, size and lobing of leaves, pigmentation on leaf, stem, petiole, internodal length, length of pedicel, shape of thalamus, ridges and beak on the fruits, etc. The cross-pollination is effected through hand pollination, and enough (400 m) spacing between the two parents and similar distance from fields of other okra varieties is maintained. Cross-pollination by hand gives hybrid seed while left over flowers in the two parents help in their maintenance as proper isolation does not permit cross-pollination. However, crossing blocks should be kept separate so that any negligence in labeling the crossed fruit may not give hybrid seed mixture in otherwise harvest of sib/self seed in parental blocks.

The hand emasculation is practiced on fully developed and mature buds on female parent in the evening for pollination in the next morning. Such buds are fully swollen and pale green/light yellow. Buds after emasculation are bagged to avoid contamination from foreign pollen. The mature buds on pollen parent are similarly covered with perforated butter paper bags for collecting uncontaminated pollen next morning for pollination. Pollen could be collected in petri dishes (anthers/staminal column/pollen) and pollination could be done with brush. Reciprocal differences in crossing direction have not shown any marked variable effect in the hybrid. As such both parents could be used as male as well as female. The fruits developing from such cross-pollination will produce 100 percent hybrid seed. The practice so far is hand emasculation followed by cross-pollination which gives 100% hybrid seed. Since sufficient pollen is produced by flowers, good seed production could be expected if the stigmatic surface is fully covered with pollen. First one or two flowers as well as rest of flowers beyond 15th-20th node, depending upon agro-climate, should not be taken for seed production as they produce low number of pollen and also low seed content, sometimes physiologically immature. Normally 50-55 healthy seeds could be expected from one capsule and 10-12 quintals hybrid seed per hectare.

Emasculation is a tedious and time consuming process and hand pollination together increases cost of F_1 seed. One easier method could be to remove staminal column along with petals from mature buds one day advance of anthesis on male fertile seed parent and pollinate such emasculated flowers with brush next morning using pollen collected from bagged (covered) flowers of the pollen parent keeping pollinated flowers un-bagged. The pollen parent also may not require bagging of flowers if isolated suitably from other crossable types. This way male parent will also get multiplied in isolation if properly cared. For this type of forced cross-pollination planting design having 2 rows of seed parent alternated with 2 rows of pollen parent, i.e., in 1:1 ratio could be time saving and more economic. Economics of these methods need to be worked out.

Induction of male sterility though use of Gamma rays @ 50-60 kR (Dutta, 1971) and gameticides FW 450 and MH @ 0.4-0.5% (Dubey & Singh, 1968) have neither been perfect nor stable and as such could not be used in hybrid seed production. The cost of hybrid seed could be brought down once some type of stable male sterility is developed. Hybrids within *A. esculentus* are popular and commercial. However, hybrids between *A. esculentus* and West African okra could give longer production life and higher yield even under humid regions.

Recently genic male sterility in okra controlled by two recessive genes has been reported from IIHR, Bangalore, India (Dutta, 1996), which when planted in 2 rows MF:2 rows MS design in isolation, took 3 hours to pollinate 375 flowers and produce 1 kg seed as against 9 1/2 hrs needed to emasculate and pollinate 375 flowers and harvest 1 kg hybrid seed in male fertile seed parent, i.e., 70% saving of time. Fifteen laborers per day for 50 days could help harvesting 8-10 quintals F_1 hybrid seed. At cultivation cost of Rs. 20,000 and procurement price of Rs. 100 per kg seed, the expected net return to the farmer comes around Rs. 60,000. The fruits are harvested at bursting stage and left for further drying for 4-5 days. Seed is taken out from dry fruits by crushing or beating the pods gently. After cleaning, the seed is further dried to 8% moisture level before storage or packing.

The crop should be planted in a plot in sunny situation. Seed crop needs rich nutrition as maximum number of fruits is retained on the plant till they dry up and seed development continues till drying of the fruit. Calcium @ 60 kg/ha should be applied in the form of slaked lime in addition to the NPK requirements (120:50:60 kg/ha in normal soils). Spacing among the rows and plants should be kept at 60 cm × 30-45 cm. Early sowing in the season is always beneficial as it allows longer fruiting span. Sowing dates should be adjusted such that the harvesting for seed is done in a rain-free dry weather. Cultural practices are to be adjusted such that the plants attain vigor and maintain it in the form of vegetative health and fruiting. Drying/dieing/diseased leaves should be invariably removed from the plots. Soil should never be allowed to dry as it will affect the development and quality of the seed. Adequate plant protection measures should be taken as the seed crop attracts such insect pests also which feed upon the seed. Undesirable plants, including those of okra, in the parental blocks need removal and regular inspections are advised to maintain highest purity of parents. Though very low cross-pollination has been observed to occur yet proper isolation should be kept. Fruits for seed extraction should not be harvested before bursting stage as the seeds may not be physiologically mature and it may affect germination adversely. Care should be taken so that seed does not get damaged during extraction from fruits. Shellers should be preferred for seed extraction.

MAINTENANCE OF INBREDS/PURE LINES

Pure lines/Inbreds could be multiplied in the fields isolated at 400 m from crossable forms. Such maintenance should be done in a plot in which okra had not been grown for previous three consecutive years in order to avoid contamination with unwanted germinating plants. Full details of distinct characters of such lines should be kept handy. Rogueing of undesirable plant types is done minimum at three stages–before flowering on the basis of vegetative characters, at the time of flowering based on floral characters, and at fruit ripening stages. Several more inspections for rogueing are suggested. Inter-crossing naturally or through mixed pollen within population could be allowed to keep together all the genes. Plots and surroundings should be kept free from hosts of YVM virus and YVM affected plants should be continuously removed at least till 60% of fruiting is over. Such multiplication should be done in a season best suited for plant growth and fruiting and long enough to allow maximum number of fruits on plant.

CONCLUSION

Okra is gaining importance with regard to its nutritional supplement in the human diet, particularly for protein. It is gaining importance also as a crop for export and newer destinations are added day by day. Seed contains ample high quality edible oil and is looked as an alternative source oil for future. As such improvement in quality as well as production has to be prioritized. For this development of hybrids with high yielding potential coupled with stress resistance for harvesting sustainable yields even under prone environments and seasons has to be focussed with greater emphasis.

REFERENCES

Anonymous, (2001). Indian Horticulture Database 2001. National Horticulture Board. Govt. of India. 197p.

Anonymous, (2001a). Export Statistics for agro and food products. India 2000-2001. Agricultural and Processed Food Products Export Development Authority. New Delhi. p. 451.

Charrier, A. (1984). Genetic resources of genus *Abelmoschus* (Med.) (okra). IBPGR, Rome. 61p.

Chevalier, A. (1940). L'origine la culture et les usages de cing *Hibiscus* de la section *Abelmoschus*. *Rev. Bot. Appl.* 20:319-328, 402-419.

Datta, PC and Naug, A. (1968). A few strains of *Abelmoschus esculentus* (L.) Moench., their karyological study in relation to phylogeny and organ development. *Beitr. Biol. Pflanzen.* 45:113-126.

De Candolle, AP. (1883). *Origine des plantes cultivees*. Noble Offer Printing, New York, p. 150-151.

Dubey, RS and Singh, SP. (1968). Longevity of the pollen of okra and comparison of four qualitative traits for pollen viability. *Indian J. Hort.* 25(3-4): 210-213.

Dutta, OP. (1971). Effect of gamma irradiation on seed germination, plant growth, floral biology and fruit production in *Abelmoschus esculentus*. Proc. III Intl. Symp. Sub-Trop. and Trop. Hort. IIHR, Bangalore, India p. 21.

Dutta, OP. (1984). Breeding okra for resistance to yellow vein mosaic virus and enation leaf curl virus. Annu. Rep. IIHR, Bangalore, India p. 21.

Dutta, OP. (1996). Techniques of hybrid seed production in cucurbits and okra. (In) *Vegetable Hybrids and Their Seed Production*. Divn. of vegetable Crops, IARI, New Delhi, India. pp. 162-167.

El-Maksoud, MA, Hetal, RM and Mohamed, MH. (1986). Studies on an inter-varietal cross and hybrid vigour in okra. *Annals Agric. Sci.* 29(1): 431-478.

Ericksson, NHT and Couto, FAA. (1963). Inheritance of four plant floral characters in okra (*Hibiscus esculentus*). *Proc. Amer. Soc. Hort. Sci.* 83: 605-608.

Grubben, GJH. (1977). Okra. *Tropical Vegetables and Their Genetic Resources*. IBPGR, Rome. pp. 111-114.

Hoque, M. and Hazarika, GN. (1993). Hybrid vigour for some economic traits in okra. In *Heterosis Breeding in Crop Plants–Theory and Applications*. MM Verma, Irk, DS and Chadha, GS (Eds.) *Symp. Proc.*, Ludhiana. pp. 22-23.

Jalani, BS and Graham, KM. (1973). A study of heterosis in crosses among local and American varieties of okra. *Malaysian Agric. Research* 2(1): 7-14.

Jambhale, ND and Nerker, YS. (1981). Inheritance of resistance to okra yellow vein mosaic disease in inter-specific crosses of *Abelmoschus. Theor. Appl. Genet.* 60: 313-316.

Jasin, Abdul Zabbar. (1967). Inheritance of certain characters in okra (*H. esculentus* L. *Diss. Abstr.* 28(1): 3-13.

Joshi, AB, Singh, HB and Gupta, PS. (1958). Studies in hybrid vigour. III. Bhindi. *Indian J. Genet.* 18: 57-68.

Kalia, HR and Padda, DS. (1962). Inheritance of leaf and flower characters in okra. *Indian J. Genet.* 22(3): 252-254.

Kolhe, AK and D'Cruz, R. (1966). Inheritance of pigmentation in okra. *Indian J. Genet.* 26: 112-117.

Kulkarni, RS. (1976). Gene action in Bhindi. *Mysore J. Agric. Sci.* 10: 332.

Kulkarni, RS, Swami Rao, T and Virupakshappa, K. (1976). Gene action in Bhindi. *Research J. Kerala Agri. Univ.* 14:13-20.

Kulkarni, RS, Swami Rao, T and Virupakshappa, K. (1978). Genetics of important yield components in Bhindi. *Indian J. Genet. Pl. Breed.* 38: 160-162.

Kuwada, H. (1964). Studies on inter-specific crossing pattern between *Abelmoschus esculentus* (L.) Moench. and *A. manihot* (L.) Medicus and the various hybrids and polyploids derived from the above two species. *Kagawa Agric. Coll. Tech. Bull.* 15:79-88 (Jap.).

Master, MT. (1875). Malvaceae. In Hooker, JD (Ed.). Flora of British India. Ashford, Kent. 1: 317-348.

Miller, JC and Wilson, WF. (1939). A preliminary report on okra breeding in Louisiana. *Tech. Bull.* 35: 551-553.

Murdock, GP. (1959). *Africa, Its People and Their Culture History*. New York. McGraw Hill 556 p.

Nath, P and Dutta, OP. (1970). Inheritance of fruit hairiness, fruit skin colour and leaf lobing in okra. *Can. J. Genet. Cytol.* 12: 589-593.

Ngah, AW and Graham, KM. (1973). Heritability of four economic characters in okra (*Hibiscus esculentus*). *Malaysian Agric. Research* 2(1): 15-21.

Padda, DS, Saimbhi, MS and Singh, D. (1970). Genetic evaluation and correlation studies in okra. *Indian J. Hort.* 27:39-41.

Partap, PS and Dhankhar, BS (1980). Heterosis studies in okra, *Abelmoschus esculentus* (L.) Moench. *Haryana Agric. Univ. J. Res.* X (3): 336-341.

Poshia, VK and Shukla, PT. (1986). Combining ability analysis in okra. *GAU Research Jour.* 12(1): 25-28.

Raman, KR and Ramu, N. (1963). Studies on inter-varietal crosses and hybrid vigour in Bhindi. *Madras Agric. J.* 50: 90-91.

Ramu, PM. (1976). Breeding investigation in Bhindi, *Abelmoschus esculentus* (L.) Moench. *Mysore J. Agric. Sci.* 10: 146.

Rao, TS and Kulkarni, RS. (1977). Genetic variation in Bhindi. *HAU Jour. Research.* 7(1-2): 58-59.

Rao, TS and Sathyavathi, GP. (1977). Influence of environment on combining ability and genetic components in Bhindi, *Abelmoschus esculentus* (L.) Moench. *Genet. Polon.* 18:141-147.

Sharma, BR and Mahajan, YP. (1978). Line x tester analysis of combining ability and heterosis for some economic characters in okra. *Scientia Horticulturae.* 9(2): 111-118.

Sharma, BR and Dhillon, TS. (1983). Genetics of resistance to yellow vein mosaic virus in inter-specific crosses of okra. *Genet. Agrar.* 37: 267-276.

Sharma, BR and Gill, BS. (1984). Genetics of resistance to cotton jassid, *Amarasca biguttula* (Ishida) in okra. *Euphytica* 33: 215-220.

Siemonsma, JS. (1982). West African okra–morphological and cytogenetical indications for the existence of a natural amphi-diploid of *Abelmoschus esculentus* (L.) Moench. and *A. manihot* (L.) Medikus. *Euphytica* 31:(1): 241-252.

Singh, HB, Joshi, BS, Khanna, PO and Gupta, PS. (1962). Breeding for field resistance to YVMV in Bhindi. *Indian J. Genet.* 22: 137-144.

Teshima, TK. (1933). Genetical and cyto-genetical studies in an inter-specific hybrid of *Hibiscus esculentus* and *H. manihot. J. Fac. Agri. Hokkaido Univ.* 34: 1-155.

Thakur, MR. (1976). Inheritance of yellow vein mosaic in a cross of okra species, *Abelmoschus esculentus* and *A. manihot* ssp. *Manihot. SABRAO J.* 8: 69-73.

Ugale, SD, Patil, RC and Khuspe, SS. (1976). Cyto-genetic studies in the cross between *A. esculentus* and *A. tetraphyllus. J. Maharashtra Agric. Univ.* 1(2-6): 106-110.

Vavilov, NI. (1951). The origin, variation, immunity and breeding of cultivated plants. *Chron. Bot.* 13: 1949-1950.

Zeven, AC and Zhukovsky, PM. (1975). *Dictionary of Cultivated Plants and Their Centres of Diversity.* Centre of Agricultural Publishing and Documentation. Wageningen, The Netherlands. 219 p.

Hybrid Loofah

P. K. Singh

S. K. Dasgupta

SUMMARY. It is an old world genus, consisting of two cultivated and two wild species besides a wild new world species. Significant heterosis has also been reported in this genus and is being commercially exploited. It is an insect pollinated crop. Isolation is needed to prevent out crossing by insects when hybrid seed is produced by open pollination. Hybrid seed is produced commercially by the hand pollination or by defloration, the manual removal of male flowers from the female parent so that all of its open pollinated seed will be hybrid. *[Article copies available for a fee from The Haworth Document Delivery Service: 1-800-HAWORTH. E-mail address: <docdelivery@haworthpress.com> Website: <http://www.HaworthPress.com> © 2004 by The Haworth Press, Inc. All rights reserved.]*

KEYWORDS. Loofah, floral biology, crop biology, heterosis, hybrid seed production

INTRODUCTION

It is an old world genus, consisting of two cultivated and two wild species, besides a wild new world species. *Luffa cylindrica* (spongegourd) and *Luffa acutangula* (ridgegourd or angled luffa) are commercially grown in several

---

P. K. Singh and S. K. Dasgupta are affiliated with the Sungro Seeds Ltd. B. N. Block, Shalimar Bagh, New Delhi 110088, India.

[Haworth co-indexing entry note]: "Hybrid Loofah." Singh, P. K., and S. K. Dasgupta. Co-published simultaneously in *Journal of New Seeds* (Food Products Press, an imprint of The Haworth Press, Inc.) Vol. 6, No. 2/3, 2004, pp. 211-215; and: *Hybrid Vegetable Development* (ed: P. K. Singh. S. K. Dasgupta, and S. K. Tripathi) Food Products Press, an imprint of The Haworth Press, Inc., 2004, pp. 211-215. Single or multiple copies of this article are available for a fee from The Haworth Document Delivery Service [1-800-HAWORTH. 9:00 a.m. - 5:00 p.m. (EST). E-mail address: docdelivery@haworthpress.com].

Digital Object Identifier: 10.1300/J153v06n02_10

countries and are of economic importance, while *L. echinata* and *L. graveolens* are other species.

CROP BIOLOGY

The plants are fast growing, trailing or climbing vines with stems and trendils. The genus *Luffa* is monoecious with annual vines. Tendrils are branched; leaves are 5-7 lobed, nearly glabrous. Inflorescence is clustered racemose flowers mostly unisexual, yellow colored, large and showy, mostly monoecious (staminate and pistillate flowers separately in the same plant), have campanulate showy corolla, calyx forming a perianth tube, calyx lobes alternating with corolla lobes, filament free, two stamens having two locules and one unilocular. Pstillate flowers are borne singly in short peduncles. Usually staminate and pistillate flowers are borne in different axils but sometimes may be on the same node. Hermaphrodite form, which bears only bisexual flowers, like "satputia" cultivar of *Luffa acutangula*, is rare. Fruit is essentially a (inferior) berry, even though called pepo.

Thakur and Choudhury (1967) reported that *L. cylindrica* as female was easily crossable with *L. acutangula* as male. *Luffa cylindrica* flowers in the early morning (4-8 a.m.) and remain suitable for selfing/crossing almost throughout the day, while *L. acutangula* flowers in the afternoon and evening (5-8 p.m.) and remain open throughout the night and are ready for pollination and selfing in the early morning/forenoon. Anther dehiscence coincides with flower anthesis. The stigma has a longer duration of receptivity and pollen grains are highly fertile. Autotetraploidy was induced in *L. acutangula* (4x = 52); however, such autotetraploids have no economic value (Roy and Dutt 1972). Pathak and Singh (1949) were successful in making reciprocal crosses between these two species. The F1 plants were generally intermediate between the parents. The F1 showed various irregularities, like univalents, rings, chains of four chromosomes, chromatin bridges and fragments at metaphase. The percentage of good pollen ranged from 18-40%. Thus, the species are not crossable in nature and the F1 appears to be not of much practical value.

For the improvement of this crop, knowledge of desired plant and fruit characteristics is essential. Information on some important traits are given below:

1. The lowest node number at which first pistillate (or hermaphrodite) flower appears, an indicator of earliness.
2. High female to male ratio as it results in higher fruit yield.
3. Uniformly thick cylindrical fruits free from bitterness.

4. Types having fruits, which remain tender and non-fibrous, optimum size fruits at edible maturity.
5. High fruit yield (high fruit number and weight at edible maturity).
6. Fruits without blossom end.
7. Fruit color (dark green/green/white) and seed color (black/white) as per the regional preference.
8. Resistance to powdery mildew, downy mildew, and insects.

The genetics of flowering habit are controlled by two independent suppressor gene A and G. Gene A suppressing maleness and G femaleness. The genotypes of various flowering habits, i.e., monoecious and trimonoecious, andromonoecious, gynoecious and hermaphrodite have been studied by Choudhury and Thakur (1965). The corolla color, type of androecium of *L. cylindrica*, and the fruit surface and seed surface of *L. acutangula* are monogenically controlled. The bitterness is controlled by *Bi* gene (Thakur and Choudhury, 1966).

HETEROSIS

Tozu (1981) observed that self-pollination resulted in inbreeding depression, while F1 hybrids showed some heterosis. Sahni et al. (1987) reported that fruit number per stem, female flower number per stem, first female flowering node and branch number per stem controlled by non-additive gene effects. They also advocated heterosis breeding for improvement in fruit diameter and fruit length. Qinghua et al. (1996) and Jianning (2000) developed early and vigorous FengKang and Yalu No. 1 hybrids, respectively. Seed companies are marketing many F1 hybrids across the world. Yield has a positive correlation with the number of fruits per plant. Fruit length and days to flowering exhibit high heritability and genetic advance.

POLLINATION CONTROL MECHANISM

The *Luffa* sp. is monoecious, having both male and female flowers in the same plant. A hermaphrodite form, which bears only bisexual flowers is also found. *Luffa* flowers 6-10 weeks after sowing, depending on cultivar, ecology, and cultural practices. The proportion of female flowers is increased naturally by long days and high temperature (Huyskens et al. 1993) and chemically by spraying with phytohormones. Hand pollination is best done in the morning if flowers are open (Larkcom 1991), but in ridgegourd, flowers have been reported to be open at 4-4.30 p.m. (Deshpande et al. 1980). Hand pollination can be avoided by introducing beehives.

HYBRID SEED PRODUCTION

Hybrid seed production steps in *Luffa* sp. is similar to bottlegourd.

1. Development of inbred lines.
2. Testing of combining ability.
3. Production of F1 hybrid seed.

Production of hybrid seeds involves the bagging of male flowers of male parent and female flower of female parent one day prior to anthesis and pollination. After pollination, cover the female flower with the butter paper bag. A simple pollination produces a large number of seeds in these crops. Therefore, the cost of hybrid seed production is not high. Mostly line X line hybrids are produced and marketed, as no report of male sterility is available in these crops.

MAINTENANCE OF INBRED LINES

Uniform inbreds of *Luffa* could be developed with comparable vigor of open pollinated varieties. They can be maintained through open pollination by growing them at proper isolation distance. Isolation distance of 1000 m is quite appropriate. Inbreds can also be maintained by hand pollination. Care must be taken for rogueing out undesirable off types at right stages of plant growth, e.g., (i) before flowering, (ii) at the time of flowering, and (iii) at the time of fruit set and maturity.

SEED PRODUCTION

Luffa is a highly cross-pollinated species. Therefore, maintenance of proper isolation distance between two varieties/inbred lines/crossable species is necessary. Bumblebees are the chief insect pollinators for this species. For better fruit set and higher seed yield, pollinators should be in abundance in the field at the time of flowering. Another important requirement in production of high quality seed is rogueing of diseased and off type plants from the seed crop before they start flowering.

REFERENCES

Choudhury, B. and Thakur, M.R. (1965). Inheritance of sex forms in Luffa. Indian J. Genet. and Plant Breeding 25: 188.

Deshpande, A.A., Bankapur, V.M. and Venkatasubbaiah, K. (1980). Floral biology of ridgegourd. Mysore J. Agric. Sci. 14 (1): 527.

Huyskens, S., Mendlinger, S., Benzioni, A. and Ventura, M. (1993). Optimization of agrotechniques in the cultivation of *L. acutangula*. J. Hort. Sci. 68 (6): 989-994.

Jianning, S.L., Hgo, G., Luo, J.N., Luo, S.B., and Gary, H. (2000). Breeding of new F1 hybrid 'Yalu No.1' of *Luffa acutangula*. Roxb. China-Vegetables. 3: 26-28.

Larckom, J. (1991). Oriental vegetables: The complete guide for garden and Kitchen. London John Murray: 232.

Pathak, G.N. and Singh, S.N. (1949). Studies in the genus *Luffa*. I. Cytogenetic investigations in the interspecific hybrid *L. cylindrica* × *L. acutangula*. Indian J. Genetics. 9: 18-26.

Roy, R.P. and Dutt, B. (1972). Cytomorphological studies in individual polyploids of *Luffa acutangula* Roxb. Nucleus 15: 17.

Sahni, G.P., Singh, R.K., and Saha, B.C. (1987). Genotypic and phenotypic variability in ridge gourd. Indian Journal of Agril. Sciences. 57(9): 666-668.

Thakur, M.R. and Choudhury, B. (1966). Inheritance of some qualilatative characters in *Luffa* sps. Indian J. Gen. Plant Breed. 26: 76.

Thakur, M.R. and Choudhury, B. (1967). Interspecific hybridization in genus *Luffa*. Indian J. Hort. 24:87.

Tozu, T. (1981). Heterosis in spongegourd (*Luffa* sps.). Fourth international SABRAO congress, 4-8 May 1981, at University Kebangsaa Malaysia and Federal Hotel, Kulalampur.

Qinghua, C., Tao, H., Qiyoung, Z., Xinzhou, H., Yue, L., Qh, C., Huang, T., Qy, Z., Xz, H., and Ye, L. (1996). Breeding of new hybrid FengKang of *Luffa acutangula* Roxb. China-Vegetables. 2: 7-8.

Heterosis in Bittergourd

T. K. Behera

SUMMARY. Bittergourd is an important cucurbit fruit vegetable grown in the tropics. It has rich nutritional and medicinal value. Hybrids are becoming popular in this crop. Gynoecy is also reported in this crop which could be a useful tool to exploit heterosis on commercial scale with more cheaper rates. But at present hybrid seed is produced by hand pollination without emasculation. *[Article copies available for a fee from The Haworth Document Delivery Service: 1-800-HAWORTH. E-mail address: <docdelivery@ haworthpress.com> Website: <http://www.HaworthPress.com> © 2004 by The Haworth Press, Inc. All rights reserved.]*

KEYWORDS. Bittergourd, crop biology, floral biology, heterosis, inbred lines, hybrid seed production

INTRODUCTION

Bittergourd, *Momordica charantia*, is an important curcurbits fruit vegetable grown in tropics. The crop is of Asiatic origin and the probable place of origin is either China or India. It is cultivated in Malaysia, China, The Philippines, tropical Africa, and North and South America. Immature fruit is a good source of vitamin C, and also contains vitamin A, phosphorus, and iron. The tender vine tips are an excellent source of vitamin A, and a fair source of protein, thia-

---

T. K. Behera is affiliated with the Division of Vegetable Science, Indian Agriculture Research Institute, Pusa Campus, New Delhi 110012, India.

[Haworth co-indexing entry note]: "Heterosis in Bittergourd." Behera, T. K. Co-published simultaneously in *Journal of New Seeds* (Food Products Press, an imprint of The Haworth Press, Inc.) Vol. 6, No. 2/3, 2004, pp. 217-221; and: *Hybrid Vegetable Development* (ed: P. K. Singh, S. K. Dasgupta, and S. K. Tripathi) Food Products Press, an imprint of The Haworth Press, Inc., 2004, pp. 217-221. Single or multiple copies of this article are available for a fee from The Haworth Document Delivery Service [1-800-HAWORTH, 9:00 a.m. - 5:00 p.m. (EST). E-mail address: docdelivery@haworthpress.com].

min, and vitamin C. The leaves and fruits are excellent sources of vitamin B, iron, calcium, and phosphorus. It has twice the amount of beta-carotene than in broccoli and twice the calcium content of spinach. Leaf juice is used for curing cough, as a purgative and antihelminthic to expel intestinal parasites and for healing wounds. Studies have suggested that bittergourd contains a hypoglycemic polypeptide, plant insulin responsible for blood sugar lowering effect.

CROP BIOLOGY

The plant is a fast-growing, trailing or climbing vine with thin stems and tendrils. The leaves are heart-shaped, 5-10 cm in diameter, cut into 5-7 lobes. Male and female flowers are borne separately on the same plant, and require insects for pollination. Flowers are borne singly in the leaf axils. Male flowers appear first and usually exceed the number of female flowers by about 20:1. The flower opens at sunrise and remains open for only one day. The fruits are characterized by continuous or discontinuous ridges through out fruit length. Immature fruits are light green, oblong, pointed at the blossom end and have white flesh. As the fruits begin to mature, the surface gradually turns yellow or orange. At maturity, it tends to split open, revealing orange flesh and bright red placenta to which the seeds are attached. Seeds are tan and oval, with a rough etched surface; there are about 5 to 7 seeds per gram.

HETEROSIS

Heterosis in bittergourd for yield/vine ranges from 27.3 to 86.1% over better parent. Several studies on heterosis in relation to yield and quality traits of bittergourd have been made at Indian agricultural research Institute. One bittergourd hybrid, Pusa Hybrid 1 is developed and resealed for commercial cultivation under north Indian plains. It gives 42% heterosis over better parent and its fruits are suitable for making pickles and dehydration (Sirohi, 2000).

POLLINATION CONTROL MECHANISM

Bittergourd is monoecious crop having both male and female flowers are in same plant. Gynoecism was reported by Ram et al. (2002). The first male flower appears at 45-55 days after sowing if environmental conditions are optimal. The anthesis and anther dehiscence occurs early in the morning. The stigma is receptive 24 h before and 24 h after anthesis. Hand pollination can be avoided either by introducing beehives. Pollen loses viability as the day advances and may be fully inviable by midday (Desai and Musmade, 1998). Flowering behavior varies among cultivars and climatic conditions (Deshpande et al., 1979).

Long days cause the male flowers to bloom up to 2 weeks before the female flowers, while short days have the reverse effect (Huyskens et al., 1992). Spraying with gibberellic acid @ 25-100 mg/L at the 2-4 leaf stage increases female flower numbers and can last for up to 80 days (Wang and Zeng, 1996). Treatment with 2,4-D, maleic hydrazide and Cycocel (chlormequat) has also increased female flowers and yields, but reduced vine length and leaf area (Kabir et al., 1989). Pruning lower laterals increases the total number of flowers per plant by increasing the number of flowers on higher laterals (Rasco and Castillo, 1990).

HYBRID SEED PRODUCTION

Since bittergourd is a monoecious plant, its seed is produced by hand pollination without emasculation. The female flowers of the female parent and male flowers of the male parent are tied with butter paper bags (size 7.5 cm × 12 cm) about 24 h before they open. It is less time consuming and also profitable if corolla of the buds of male flowers of male parent is covered with non-absorbent cotton instead of covering them with butter paper bags. The following day when the flowers open, the male flowers are collected after removing the cotton or butter paper bags and their pollen grains are dusted directly on the stigma of the female flowers after removing the butter paper bags (Sirohi, 2000).

After completing the hand pollination, the female flowers are covered again by butter paper bags and tagged with small labels on which the details of the crossing should be written. It is better to remove the butter paper bags from the crossed female flowers 4-5 days after pollination for better fruit set. Approximately 10% of the plants of male parent in a population are enough for the supply of pollen grains (1 male parent plant is required for pollinating 9 plants of female parent). In bittergourd, anthesis takes place early in the morning and hence pollination is very effective if it is done between 7.30 and 9.30 a.m. The seeds are extracted from the ripe fruits collected from the female parents.

MAINTENANCE OF INBRED LINES

For achieving uniformity in the hybrids, it is essential to inbreed the parents. The degree of inbreeding required in the parents will be determined by the extent of uniformity desired. In bittergourd, the parents/inbreds are maintained in pure form by selfing without loss of vigor. However, the selfing precedes selection and the first one or two female flowers on the plants must be selfed, otherwise selfing in later stages often fails to set fruits. It is not necessary to produce inbred lines and homozygous varieties can be used directly as inbreds as in self-pollinated crops like tomato, eggplant, and sweet pepper (Vishnu Swarup, 1991). When inbreds are to be produced, selection is done between the inbred lines to obtain promising inbreds. These inbred lines are tested for their combining ability through single and diallel crosses. Based on their combining ability, the most promising inbred lines or parents are used in the production of F_1 hybrids.

SEED PRODUCTION

Since bittergourd is highly cross-pollinated, maintaining the proper isolation distance between the two varieties/inbred lines is necessary. Honeybees are chief insect pollinators for this crop. For better fruit set and higher seed yield, honeybees should be in abundance in the field at the time of flowering. An isolation distance of 0.5-1.0 km should be given between two varieties/inbred lines (Sirohi, 1997).

Another important requirement in production of high quality seed of bitter gourd is rogueing of diseased and off-type plants (whose size, shape, color or other characteristics are not typical of the variety) from the seed crop before they start flowering. Proper rogueing of off-type plants require a lot of experience and knowledge, as there are several varieties available within this crop.

REFERENCES

Desai, U. T. and Musmade, A. M. (1998). Pumpkins, squashes and gourds. In: Handbook of vegetable science and technology: Production, composition, storage and processing (Eds., Salunkhe. D. K. and Kadam, S. S.). New York, Marcel Dekker, 273-298.

Deshpande, A. A., Venkatasubbaiah K., Bankapur V. M. and Nalawadi U. G. (1979). Studies on floral biology of bitter gourd (*Momordica charantia* L.). Mysore Journal of Agricultural Sciences 13(2): 156-159.

Huyskens, S., Mendlinger, S., Benzioni, A. and Ventura. M. (1992). Optimization of agrotechniques for cultivating *Momordica charantia* (karela). Journal of Horticultural Science 67(2): 259-264.

Kabir. J., Chatterjee, R., Biswas, B. and Mitra. S. K. (1989). Chemical alteration of sex-expression in *Momordica charantia*. Progressive Horticulture 21: 1-2.

Ram, D., Kumar Sanjeet, Banerjee, M. K. and Kalloo, G. (2002). Occurrence, identification and preliminary characterization of gynoecism in bittergourd. Indian J. Agric. Sci. 72(6): 348-349.

Rasco, A. O. and Castillo, P. S. (1990). Flowering patterns and vine pruning effects in bittergourd (*Momordica charantia* L.) varieties 'Sta. Rita' and 'Makiling'. Philippine Agriculturist 73: 3-4.

Sirohi, P. S. (1997). Improvement in cucurbit vegetables. Indian Hort. 42(2): 64-67.

Sirohi, P. S. (2000). Pusa Hybrid 1: New bittergourd hybrid. Indian Hort. 44(4): 30-31.

Vishnu Swarup (1991). Breeding procedures for cross-pollinated vegetable crops. ICAR, New Delhi, 29-31.

Wang, Q. M. and Zeng, G. W. (1996). Effects of gibberellic acid and Cycocel on sex expression of *Momordica charantia*. [Chinese]. Journal of Zhejiang Agricultural University 22(5): 541-546.

Current Trends in Onion Breeding

A. S. Sidhu
S. S. Bal
Mamta Rani

SUMMARY. Onion is an important vegetable crop worldwide and has been used in various forms as food. There are different types of onion available depending on shape, size and color. To a lesser extent the processing industry uses it in the form of dehydrated onion flakes and powder, one of the remarkable features of onion is its excellent transportability. The normal flower in onion is perfect. The flowering structure is called an umbel. It is an aggregate of many small flowers. Photoperiod is an important factor in bulb development. Normally onion is a long day plant. Heterosis has been commercially exploited in onion as male sterile lines are available in this crop. The male sterility was reported in the crop long back in 1936. Honeybees are used for necessary transfer of pollen from male parent to female parent. More research is needed on factors affecting attractiveness of onions to honeybees. The paper discusses in detailed information regarding research in heterosis, pollination control mechanisms, breeding for hybrids, and techniques related to hybrid seed production. *[Article copies available for a fee from The Haworth Document Delivery Service: 1-800-HAWORTH. E-mail address: <docdelivery@haworthpress.com> Website: <http://www.HaworthPress.com> © 2004 by The Haworth Press, Inc. All rights reserved.]*

A. S. Sidhu, S. S. Bal and Mamta Rani are affiliated with the Department of Vegetable Crops, Punjab Agricultural University, Ludhiana 141004, India.

[Haworth co-indexing entry note]: "Current Trends in Onion Breeding." Sidhu, A. S., S. S. Bal, and Mamta Rani. Co-published simultaneously in *Journal of New Seeds* (Food Products Press, an imprint of The Haworth Press, Inc.) Vol. 6, No. 2/3, 2004, pp. 223-245; and: *Hybrid Vegetable Development* (ed: P. K. Singh, S. K. Dasgupta, and S. K. Tripathi) Food Products Press, an imprint of The Haworth Press, Inc., 2004, pp. 223-245. Single or multiple copies of this article are available for a fee from The Haworth Document Delivery Service [1-800-HAWORTH, 9:00 a.m. - 5:00 p.m. (EST). E-mail address: docdelivery@haworthpress.com].

http://www.haworthpress.com/web/JNS
© 2004 by The Haworth Press, Inc. All rights reserved.
Digital Object Identifier: 10.1300/J153v06n02_12

KEYWORDS. Onion, heterosis, male sterility, breeding pollination control mechanisms, maintenance of inbred lines, hybrid seed production

INTRODUCTION

Onion is an important vegetable crop worldwide and has been used as food since time immemorial. It is used for flavoring or seasoning the food, both at mature and immature bulb stages, besides being used as salad and pickle. To a lesser extent, it is used by the processing industry for dehydration in the form of onion flakes and powder, which are in great demand in the world market. It is one of the few versatile vegetable crops that can be kept for a fairly long period and can safely withstand the hazards of rough handling including long distance transport.

India ranks first in area and second in production after China. Onion is grown in an area of 2.6 lakh hectares producing about 27 lakh tones of bulbs for local consumption as well as for export purpose. It is grown in all continents with world production of about 25 million tones and is commercially grown in a little over hundred countries of the world. However, about three-forth of global production is accounted for by 18 countries, important of which are China, India, USA, USSR, Japan, Spain, Turkey, Brazil, Iran, etc. India's share in the world production is about 11 percent (Currah and Proctor, 1990). Of the fifteen vegetable crops listed by FAO (Anon., 1973), onion falls second only to tomato in terms of annual world production. In overall cropping patterns, onion area represents about 0.1 percent of gross cultivated area under all crops in the country and about 7 percent of the total area under all vegetable crops.

Cytology: The *Allium cepa* species are diploid with basic chromosome number of x = 8 (2n = 16). Occasional tetraploids have also been reported. Of 8 pairs of chromosomes, 7 are metacentric or submetacentric and one is satellited subtelocentric chromosome with nucleolus organizer region. Onion chromosomes are relatively large and spread easily during slide preparation for cytogenetic studies. Heterochromatic DNA is distinguishable from euchromatic DNA by Geimsa C banding, G banding, Q banding and florescence. The heterochromatic segments lie near the ends of chromosomes and show low quinacrine florescence and appear as dark bands with geimsa staining. The satellites are also deeply stained. There are morphological and cytological similarities between the species of section *cepa*, but still strong crossing barriers exist between them. This prevents gene flow between the two even where sympatric distribution of two species occurs. Introgression of genetic material from wild to cultivated species is also difficult. Low success has been recorded in several interspecific hybridizations.

CROP BIOLOGY

The Onion Plant Family: Liliaceae (Lily) or Amaryllidaceae (Amaryllis)

Species: Allium cepa

This plant is a genus of strong smelling bulbous herb either of the Lily Family or Amaryllis Family. It includes more than 300 species of which about 70 are cultivated, some as ornaments, most as vegetables. Onions, garlic, leek, chive, and their relatives, are all members of this family.

The onion that we eat (Allium cepa) is a hardy biennial grown as an annual from seed or from sets (small bulbs). It is usually grown for its firm, ripe bulbs, but also grown for its immature stems (shallots and green onions). These two types are known as "bulbous" and "bunching." Onions are "tunicate" bulbs, which mean that the scales are covered by a thin skin known as tunic. These scales are also called "wrapping scales." The bulb is a modified shoot or flower bud that forms underground, though close to the surface of the soil. The thick scales that protect the bud are actually swollen leaf bases. The scales are anchored to a tough basal plate (the flat end of the bulb) from which the roots will grow. The layers of scales store food to nourish the bulbs when the plant's top growth dies back. The bulb actually contains nearly everything that the embryonic bud will need to grow and bloom, including a lot of water.

The sulfur is absorbed out of the soil and into the onion through its roots. When onion is cut into, the sulfur compounds are released into the air. When it reaches the saline solution that your tears are made up of, it combines and becomes a mild form of sulfuric acid. This is what makes people cry. Especially sweet onions contain very small amounts of the sulfur-containing compounds, only about 50% of the typical levels found in other varieties. We taste sugar instead of the sulfur and they are much sweeter. Ironically, sugar levels in sweet onions are only slightly greater than those of some storage onions.

Floral biology: According to Pike (1986), flowering of onion is initiated by environmental factors. The primary inductive factor is cool temperature. Temperature of 40°F or below for 1 week will generally induce flower formation in bulbs or in growing plants with four or more leaves. However, temperature prior to and following the 40°F, week can alter flower induction. Very small seedlings do not normally respond to cool temperatures. The larger the plant, generally the more easily it can be induced to initiate flower development. When the onion plant is induced to flower, the shoot apex ceases to produce leaf primordia and initiates the inflorescence. The inflorescence may consist of a few to more than 2000 flowers per umbel. The flower stalk (scape or seedstem), which bears the umbel consisting of spathe and the flowers, is actually a one-internode extension of the stem. The stalk is initially a solid structure, but with growth it becomes hollow as it develops. The number of seedstems produced per plant depends on the number of lateral buds contained on the stem, which is compact base plate on the bottom of the bulb (Figure 1).

Plants grown from seed usually produce only one seedstem if induced to flower. Plants grown from bulbs may produce 6 or more seedstems since several lateral buds may be present that formed during development of the bulb. It should be noted that it is common for plants to produce bulbs and seedstems when grown during winter and into the spring. This is due to the fact that one or more buds remain vegetative and produce leaves that form the bulb, while a lateral bud is initiated to form a seedstem. The plant then has both a bulb and a seedstem present at the same time.

The flowering structure is called an umbel (Figure 2). It is an aggregate of many small inflorescences (cymes) of 5-10 flowers, each of which opens in a definite order, causing flowering to be irregular and to last for 2 or more

FIGURE 1. The stem of an onion is very compact and generally not seen by the casual observer. The leaf base enlarges upon bulb initiation and form the bulb. Upon flower initiation, seed stems form in the apex of the leaf axis and elongate up through the bulb.

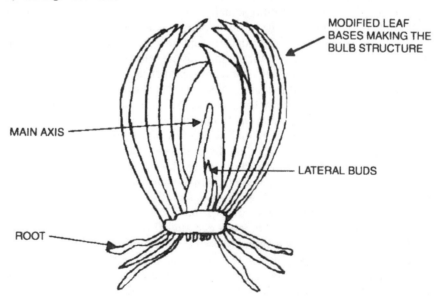

MODIFIED LEAF BASES MAKING THE BULB STRUCTURE

MAIN AXIS

LATERAL BUDS

ROOT

FIGURE 2

weeks. If the plant produces two or more seedstems, the flowering sequence may actually occur for over a month (Figure 3).

Each individual flower is made up of six stamens, three carpels united with one pistil, and six perianth segments (Figure 4). The pistil contains three locules, each of which contains two ovules. The flower also contains nectaries, which secrete nectar to attract insects. The stigma becomes receptive at this time; and as a result of delayed female maturity (protandry), cross-pollination is favored. After pollination, the seeds develop; as they are mature, the capsules dry and split from the apex and down the center of each locule, which allows the seeds to fall free upon maturity.

The normal flower in onions is perfect, but genetic and cytoplasmic sterility variations were discovered and reported by Jones and Emsweller (1933) in a single-plant segregate of the cultivar Italian Red. Male-sterile plants developed from this original plant produced normal flowers except that the pollen did not develop into a viable stage. The inheritance was determined by Jones and Clarke (1943) to be conditioned by a single recessive nuclear gene *ms/ms*, and a cytoplasmic factor, where one cytoplasm is considered normal (N) and the other sterile (S). To be made sterile, the onion plant must have the genetic and cytoplasmic condition *Sms/ms*. The discovery, propagation, and techniques of maintaining male sterility in the onion have provided an excellent method for producing hybrid seed.

FIGURE 3

FIGURE 4. Each individual onion flower within the umbel is complete, having six stamens, three carpels united with one pistil, and six perianth segments. The pistil contains three locules each of which contains two ovules.

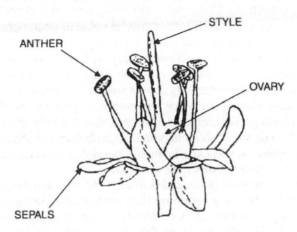

The flowering stalk of onion is an apical extension of the stem but without nodes and internodes. The growth of flower stalk ceases when umbels start to flower. The length of scape (flower stalk/seed stem) is controlled by genetic factors, long stalks being dominant over dwarf. Spherical umbel, which terminates the stalk, may have 50 to 2000 flowers, common range being 200-600. Inbreds produce significantly less flowers per head than heterozygous plants. Each flower is attached to a slender pedicel. Secondary umbels (those raising from branches that have already flowered) have approximately 30-50% of the number of the flowers in the primary heads. The flower stalk of onion reaches to a final length of 1-2 meters. The growth of flowering stalk is stopped when umbels start flowering.

Length of the day (Photoperiod): Photoperiod is an important factor in bulb development. This term describes the effect of day length on bulb formation. Some onion varieties have a short day length response, forming bulbs when the days are 12 hours or less. Other varieties have a long day length response, forming bulbs when there are 15 or more hours of daylight.

Onion is normally regarded as a long day plant and the bulb formation is promoted by long-day conditions. The cultivars differ greatly in day length requirement and there are some cultivars like "Cojumatlan" (Mexican cultivar) where bulb formation may start under short day condition. Correctly speaking it is not a short day cultivar and apparently it may appear so, since the bulb formation may become effective under relatively shorter photoperiodic condition. In fact, photoperiodic requirement in onion is a quantitative character and each cultivar needs a minimum day length for bulb formation, which is known as critical value. This critical value in combination with temperature influences bulb initiation. The so-called short day cultivars when pass the critical value limit, will initiate bulb formation and development and is continued under long day conditions. Conversely, a long day cultivar will not be able to initiate bulb formation unless its longer critical value of photoperiod is reached. Aoba (1967) has reported that under long day and high temperature, leaf growth was inhibited and ceased completely within 25-30 days of the start of long day treatment. When short day (8 hours light) treatment was given to the plants for 30 days, before the long day treatment, growth was accelerated and large bulbs were formed. Kato (1964) found that the effect produced by long photoperiod during bulb formation was neutralized if long day treatment was interrupted by a period of short days and the bulb-forming phase was reverted to vegetative phase. It was experimentally possible to reverse the bulb formation phase, even after the tops have fallen over.

Interesting findings of Heath and Hollies (1965) showed that if onion plants were kept under short day condition they continued growing for three years without bulbing. Kononkov et al. (1969) tried some cultivars, from Europe, Israel and U.S.A., in Cuba and found that the only cultivars from the countries approximately of the same day length of Cuba, produced bulbs. Those of far northern region did not produce any bulb. Austin (1972) showed that the response of different cultivars to day length remained in the same order, relative to each other over a wide range of photoperiodic treatments.

Growth requirements: Onions are usually grown on loam or silt loam soils having good water holding capacities. Good crop rotation is a necessity. Rotated crops may include small grains, potatoes, and sugar beets. The fields are prepared in the fall, fumigated, bedded, and the residue ploughed under. They are left to mellow over winter. Onions grow well in the soil with a pH pf 5.5 to 7.0 though the onion's sensitivity to acid soils lead them to grow best in soil with a pH between 6.2 and 6.8.

Bulbs can reproduce by means of underground offsets called bulblets (called "sets"). However, most growers use onion seed. Onions grown from seed will mature into bulbs in about five months. Seed-raised onions are most important commercially, because it is their ripe bulbs that can be shipped or used for storage. There are approximately 9,500 onion seeds per ounce. An average of one to three pounds are used per acre, depending on the desired size. Most onions are direct seeded, though sometimes over wintered sets are transplanted in spring. Onion seeds are usually planted at a depth of 1/2 to 1 inch. Growers usually use precision seeding techniques, placing individual seeds at a predetermined spacing within a row. The result is highly uniform crop with high yield of the desired size.

Depth of seeding has an effect on bulb shape since the onion stem plate (the base of the onion bulb) forms at the point where the seed germinates. Shallow planting results in flatter bulbs, while the deeper seed placement results in taller, and sometimes top-shaped bulbs. The most common planting practice is two rows planted per raised bed.

Growers usually prefer to not have to irrigate until the onion plant has emerged from the soil. Then the fields are irrigated regularly until the plant has matured. Soil conditioning is necessary. Onion plants are shallow-rooted and plant growth is good when nitrogen is carefully applied with irrigation water. It is important that excellent plant development is achieved before bulbing starts. The best foliage and root development occurs when the temperatures are cool (55 to 75°F). After bulbing begins, high temperature and low humidity continuing through out the harvest process are important. Through the entire growing process, adequate soil moisture must remain constant.

Maturation and harvest: A plant is considered to be mature when it stops growing. When this happens, the grower stops irrigating. The bulb will continue to grow even after the water is withheld. When 25-50% of the onion leaf tops have fallen over, the plant is ready to be lifted from the ground. The lifters move carefully through the fields so as not to create clods. The plants are carefully lifted from the ground and are left to wilt. They must dry rapidly (cure) before they are topped. If an onion has not cured properly, it is very susceptible to neck rot. Ten or more days later, when leaves are dry, a mechanical onion topper cuts the tops off the onion bulbs. Clods are removed or covered with a drag so that the bulbs can drop softly to the ground. Any breaks in the skin can leave the bulb susceptible to rot and other diseases. Again, the onions are left to lie in the fields. They will cure for at least three or more weeks before they are lifted again and bulked for storage.

HETEROSIS

Onion is one of the pioneer crops in which heterosis has been commercially exploited since about four decades. Although India is one of the leading onion producers, not much emphasis was given to heterosis breeding in the past. One of the main components for exploiting heterosis in onion is the availability of male sterility. In India, progress in the development of suitable male sterile and fertile inbred lines remained very slow in the past few decades. Sen and Srivastava (1957) attempted to develop F_1 hybrids in onion as early as in 1948 using exotic male sterile lines and Indian local male stocks. The exotic male sterile lines were found unsuitable in the photoperiodically different environment in India. Later, very few workers attempted to test different hybrid combinations for heterosis and combining ability studies using male sterile lines (Pathak et al. 1987).

Male sterility has been isolated from indigenous germplasm by several workers in India–Patil et al. (1973) in cv. Niphad 2-4-1, Pathak et al. (1980) in cv. Nasik White Globe (IIHR 20). Further studies indicated strong cytoplasmic factor responsible for male sterility in cv. Bombay White Globe (Pathak et al. 1986). This male sterility has been transferred to several breeding lines by backcross breeding method. In the hybrids of 3 male sterile testers and 20 inbred lines, various economic traits have been studied. Positive heterosis was observed in 9 hybrids over better parent for total bulb yield and it ranged from 47.9-89.5%, while heterosis over best parent for marketable bulb yield was over 35% in three hybrids, i.e., $MS_1 \times NER-1$, $MS_1 \times IIHR21-1$, and $MS_8 \times$ IIHR 52-1 (Pathak et al., 1987). Popandron (1998) obtained three F_1 hybrids by crossing a male-sterile line of onion with inbred lines in the S_2 were studied for 3 years (1990-92) at Vidra. Biometrical measurements of plant height, leaf breadth, number of leaves per plant and yield were made for both F_1 hybrids and their parents. Heterosis was clearly evident for plant height and yield, less for leaf breadth, and completely absent for number of leaves per plant. According to Janik et al. (1999), crop uniformity is considered a desirable character in modern agriculture because product uniformity is essential in marketing, uniformity in maturity permits crop scheduling, and uniformity in plant structure and maturation permits efficient mechanical harvest. Furthermore, crop uniformity is essential for maximizing yield, a little understood feature of uniformity that is expanded in this chapter. Production of F_1 hybrids of seed-propagated crops is a successful breeding technique because it exploits heterosis, promotes homogeneity in allogamous species and is a way for commercial breeders to control their product. The uniformity of hybrids has two dimensions: (1) genetic homogeneity–the presence of identical genotypes; and (2) genetic stability–phenotypic uniformity (homeostasis) in different environments. Methods to achieve genetic homogeneity are discussed, including vegetative or clonal propagation, selection, inbreeding and production of F_1 hybrids. The discussion on genetic stability includes ways in which to characterize stability, the genetics of stability, and the role of population buffering in competition and stability. Further information is presented on: maximizing genetic progress; strategies for achieving uniformity and stability in maize, tomato and onion; and uniformity vs. stability in subsistence agriculture. Netrapal et al. (1999) conducted a study during 1994-96 for the heterosis in line × tester cross of onion (*Allium cepa*) 3 lines (male-sterile parents) and 23 testers (fertile male parents) thus making a total of 26 inbred parents, and their 69 F_1's along with 6 controls quite a good number of F_1's showed desirable heterosis over the top parent for all characters except a few for maturity and neck thickness. All the characters revealed superiority of F_1's over the standard controls. Mostly the better performing F_1's also expressed higher heterosis over the better parent, top parent and standard control. The

F_1's (102-1 × 106) Smsms × 26-3-8-2op# × P 5, 75 Smsms × Early Grano, Pusa Red Smsms × Bahadurgarh Local × P 1, 75 Smsms × 31-7-60-0-p-3 # and (102-1 × 106) Smsms × 28-12-9 #-8 # emerged as better performing F_1's which gave 24.48-58.62% heterosis in yield over the top parent. For neck thickness (102-1 × 406) Smsms × 31-7-6-op-3-# × P 5 gave heterosis of − 1.91% over top parent. The hybrids 75 Smsms × Early Grano and (102 × 106) Smsms × 26-3-8-2 × P 5 gave more uniform bulbs than any other F_1 or the control. For total soluble solids the hybrids (102-1 × 106 Smsms) × SI 13 showed 6.0% heterosis over the top parents. For storage Pusa Red Smsms × SI 13 gave heterosis of 31.42% over the top parent.

Mani et al. (1999) crossed 5 red-skinned open-pollinated populations as males to 4 exotic yellow-skinned cytoplasmic male sterile inbred lines. The resulting 20 F_1 hybrids and their 9 parents were evaluated at 2 locations (Hawalbagh and Bhowali) in the Uttar Pradesh hills. The results indicated positive and significant heterosis over the better parent for bulb yield in the cross inbred 13 × L 43. In this study, the F1 between Inbred 13 × L 43 was used as a base material, advanced to the F_2 and subjected to 3 cycles of mass selection for bulb yield, skin colour, shape and size. An improved, high-yielding onion strain was thus developed and designated VL Piaz 3. It was released in 1993 for the hills of Uttar Pradesh. In trials at Hawalbagh, it produced a 20.6-47.4% higher yield over recommended controls. VL Piaz 3 combines a fairly good level of field tolerance of purple blotch and thrips, and a reduced tendency to bolting. It produces medium-sized, flat-globular red bulbs. EL-Sayed et al. (1999) evaluated the parents and F_1 hybrids from a half-diallel cross of onion for earliness (days from transplanting to harvest date), bulb ratio (neck diameter/bulb diameter), number of leaves per plant, bulb height and entire rings in trials conducted at the Onion Research Centre at Cairo. Based on the results obtained, information is derived on genetic variance, heterosis and heritability. Highly significant additive and non-additive gene effects were involved in the inheritance of all characters. The magnitude of mean square of the specific combining ability appeared small in comparison with that of the general combining ability indicating that additive type genetic variance is more important in control of these characters. A pronounced hybrid vigor was detected for most of the studied traits, indicating that non-additive gene effects also have a significant role in the inheritance of these characters.

POLLINATION CONTROL MECHANISMS

Pollination control: Pollination control in cross-pollinating species such as in onion is extremely difficult, considering each umbel has several hundred tiny individual flowers. Therefore, it is important to understand onion flowering habits and the inheritance of as many characters as possible to be efficient in breeding of the crop.

In case of normal open pollinated cultivar, selfing can be done only on a limited basis because inbreeding depression begins showing in the second generation. To make initial cross between two selections, the breeder has two choices. One is to hand emasculate stamens from one line, which is extremely difficult. The fact that the flowers open on an umbel for 2 or more weeks adds to the problem of making the cross. The second choice is to make a fertile × fertile cross, where two selections are caged together and then pollinated by hand or with insects such as common houseflies or bees. This method can efficiently be utilized if two selections are difficult enough so that the F_1 can be differentiated from two parents in the bulb stage. If they are different in that respect, seed should be saved separately from two plants and planted in separate progeny rows. The hybrid bulbs can then be identified and distinguished from bulbs resulting from selfs of the two original parents that were caged together. The F_1 bulbs are harvested and then caged together to produce F_2 progeny, or they can be used in back cross program.

Pollination requirements: Pollination in onion flowers occurs when pollen is transferred from the dehiscing anthers of one floret to a receptive stigma of another floret. Effective transfer of pollen between florets on an umbel or an individual plant can transpire through the action of an outside agent, but self-pollinating within the florets is impossible. Cross-fertilization between plants is common and even obligatory in the fertilization of male sterile onions used in hybrid seed production. Ven der Meer and Van Bennekom (1968) reported only 9% self-fertilization, and later, (1969), they concluded that seed set was less at lower temperature than higher ones.

The discovery of male sterility in onions (Jones and Emsweller 1936) made the production of hybrid onions possible under commercial conditions, and most of the onion seed produced now is hybrid seed. The procedure for utilization of male sterility in onions, which should be applicable to any crop plant in which male sterility is inherited in similar way, was shown in detail by Jones and Clarke (1943).

In the production of the hybrid seed, a male sterile row of a desired line to supply pollens to 3 to 10 rows of male sterile line (Franklin 1958), from which the hybrid seed should be obtained. Naturally the greatest volume of hybrid seed produced is desired, therefore, the male fertile or "bull" rows are kept at a minimum distance provided pollen is distributed sufficiently to set seed. Ericksen and Gabelman (1956) showed that pollen dispersal from appoint was logarithmic, with pollination at 7 feet from a source only one-half that occurring at 1 foot. To secure maximum seed set, the grower encourages pollen dispersal to maximum degree possible (Jones and Mann 1964).

MacGillivray (1948) showed that highest seed production occurred at Davis, California when plants received more than sufficient irrigation. Likewise, Howthorn (1951) obtained considerably higher seed yield with higher soil moisture. Nye (1970) reported that pollinator response to "wet" treatments was scarcely apparent, but use of nitrogen and phosphorus fertilizers caused decreased flower attractiveness.

Pollinators: Wind is not a factor in onion pollination (Erickson and Gebelman 1956). Insects are the primary vectors. When onion breeder wants to get sewed from a specific plant, they enclose the flower umbel within a bag or cage and introduce flies to transfer the pollen, or, if cross-pollination is desired, the umbels of the two lines are enclosed (Jones and Emsweller 1933). In large cage breeding work or pollination studies, honeybees are the primary agents used (Shrick et al. 1945, Moffett 1965, Walsh 1965, Bohart et al. 1970).

In commercial production of seed, the provision of and adequate number of flies is impractical so the industry depends upon honeybees as the primary agent. Bohart et al. (1970) reported 267 species of insect visitors on onion flowers, the most important of which are honeybees and drone flies. Of these, only the honeybees can be manipulated and used in large-scale onion seed production. Honeybees are effective pollinators of open pollinated onions because both pollen and nectar are available on all umbels. In hybrid seed production where male sterile plants are used, only the nectar collectors move freely pollen sterile to pollen fertile plants, making the necessary transfer of pollen from male parent to female parent. Honeybees then become less ideal pollinators of male sterile onion. Pollen collecting bees confine much of their activity to the pollen producing rows without adequately visiting and cross-pollinating the male sterile rows (Lederhouse et al. 1972). A strictly nectar collecting type of bee would be ideal because it would cross visit and effectively pollinate male sterile flowers. In the absence of this perfect type of bee, the grower can only try to compensate by having more honeybees present in the field. Then lack of intense attractiveness of onions to bees may cause the bees to neglect the crop, particularly if another highly attractive crop is in flower. The grower's only attractiveness is to make crop as attractive as possible with best cultural practices and to use a heavy population of bees. Even then, the seed yielding potential of crop may be never attained (Franklin 1970).

More research is needed on factors that affect attractiveness of onions to honeybees (Sanduleac 1969, Singh and Dharamwal 1970). Franklin (1970) noted that more placement of honeybee colonies in the onion field does not guarantee that the bees will work on onions. Although Nye et al. (1971) reported an average of 100 bees per 100 feet of male fertile rows and a maximum of 40 per 100 feet on male sterile rows, the number of honeybee visitors needed per onion plant, umbel, or linear feet of row has not been determined.

Stuart and Griffin (1946) used different rates and times of application of nitrogen on onions in greenhouse, and used honeybees to provide pollination. Their best production was 3.2 seed stalks per plant and 7.5 grams of seed per plant with a high nitrogen application from August 15 to January 1, low nitrogen during January-February (blooming), then high nitrogen until maturity.

Pollination recommendations and practices: As early as 1936, Shaw and Bourne indicated that onion seed growers might find it useful to provide themselves with a supply of bees. They did not go into details as to number of colonies, strength or location. Hamilton (1946) stated that a grower produced much more onion seed than he had in past after he rented 8 colonies of bees. Sanduleac (1961) stated that bees increase production of onion and leek seed in Romania 8 to 10 times, and he recommended about 2 colonies per acre. Le Baron (1962) stated that use of bees for pollination of onions in the Imperial Valley of California was a "must" and that two colonies per acre had given good results.

There have been no clear-cut guidelines on the use of bees for maximum onion seed production, and many beliefs based on limited observation have arisen. These include the size of colony cluster, its relative stage of development, and previous usage. The grower has learnt through experience that the use of honeybees is essential and is frequently frustrated by the erratic activity of the bees. They have generally adopted the practice of renting 5 to 10 colonies of bees per acre and having them placed in or adjacent to their seed fields at flowering time. One suggestion has been to have about 2 colonies per acre delivered when flowering is well started, then an additional two per acre at 3 to 4-day interval to take advantage of native bee behavior and maintain some level of nectar foraging activity throughout the blooming stage.

Much information is needed on the factors that influence the activity of bees on onion flowers because, as Franklin (1970) pointed out, the mere placement of colonies in the field does not guarantee that the bees will work on onions. Continuous nectar foraging activity is the essential factor in hybrid onion fields especially during the peak period of flowering.

HYBRID SEED PRODUCTION

The hybrid onion program originated in 1925 with the discovery by Dr. H. A. Jones of a bulb (15-53) in material of "Italian Red." Dr. Henery A. Jones, a former Professor of Vegetable Crops at University of California, Davis, and Research Horticulturist with US department of Agriculture and later on with the Desert Seed Company, El Centro, California is truly "Mr. Onion" having spent a lifetime of productive research with onion as quoted by Whitaker (1979). Bulb 13-53 was probably the single, most important onion bulb ever grown, because of its impact on onion breeding and breeding program in general. It was a great pleasure that 13-53 could be propagated vegetatively as it produced bulbils in the umbels placed in the seeds. Also male sterility was found to be stable under wide range of environmental conditions. The bulbils were saved and replanted to generate new plants which were crossed with other onion cultivars and the progenies were grown to understand the inheritance of the male sterile character.

The results proved that the male sterility was determined by an interaction between a nuclear gene and a cytoplasmic factor. The cytoplasmic factor was designated N for normal fertile cytoplasm and S for the sterile cytoplasm. The nuclear genetic condition was designated as Ms/- for the normal fertile cytoplasm and ms/ms for the sterile condition.

Male sterility can be maintained only when S msms is crossed as female parent with N msms as the pollen parent. The line N msms is also known as maintainer or B line. Havey (1999) reported that the primary source (S cytoplasm) of cytoplasmic-genic male sterility (CMS) used to produce hybrid onion (*Allium cepa* L.) seed traces back to a single plant identified in 1925 in Davis, California. Many open pollinated populations also possess this cytoplasm, creating an undesirable state of cytoplasmic uniformity. Transfer of cytoplasms from related species into cultivated populations may produce new sources of CMS. In an attempt to diversify the cytoplasm conditioning male sterility, the cytoplasm of *Allium galanthum* Kar. et Kir. was backcrossed for seven generations to bulb-onion populations. The flowers of galanthum-cytoplasmic populations possess upwardly curved perianth and filaments with no anthers, making identification of male-sterile plants easier than for either S- or T-cytoplasmic male-sterile onion plants. Mean seed yield per bulb of the galanthum-cytoplasmic populations was measured in cages using blue-bottle flies (*Calliphora eythrocephala* Meig.) as pollinators and was not significantly different from one of the two S-cytoplasmic male-sterile F_1 lines, a T-cytoplasmic male-sterile inbred line, or N-cytoplasmic male-fertile lines. Male-sterile lines possessing either the S galanthum cytoplasm were each crossed with populations known to be homozygous dominant and recessive at the nuclear locus conditioning male-fertility restoration of S cytoplasm and progenies were scored for male-fertility restoration. Nuclear restorers of male fertility for S cytoplasm did not condition male fertility for the galanthum-cytoplasmic populations. It is intended that these galanthum-cytoplasmic onion populations be used as an alternative male-sterile cytoplasm for the diversification of hybrid onion seed production.

Breeding of hybrids: The development of F_1 in onion requires the development of a male-sterile line (A line), an inbred maintainer for a line (B line) and a pollinator male line (C line). These lines have the following genetic constitution:

A line = S msms

B line = N msms

C line = N MsMs

The procedure of developing A and B lines by backcrossing has been out lined as follows:

Male-sterile line (A)		Male-fertile line No. 2 (B)
S msms	×	N msms
Fl S msms (50% No. 2)	×	N msms
BC$_1$, S msms (75% No. 2)	×	N msms
BC$_2$, S msms (87.5% No. 2)	×	N msms
BC$_3$, S msms (93.75% No. 2)	×	N msms
BC$_4$, S msms (96.87% No. 2)	×	N msms
BC$_5$, S msms (98.44% No. 2)	×	**N msms**

It may be possible to find male-sterile individuals within local adopted populations, and if so, appropriate pairs of male-sterile/maintainer lines can be developed from these. Alternatively, the breeder may start with known male-sterile lines, usually from unadapted cultivars made available to him by other onion breeders or genetic conservationists. These are converted to male sterile lines by crossing and back crossing with selected maintainer lines. Disregarding the original source of male sterile, the objective is same, i.e., the production of isogenic pairs of male-sterile/maintainer lines of agronomics value. The initial test crosses are grown to flowering and only those test crosses are used in further back crossing which are found to be male sterile. After 4-5 back crosses, the lines A (S msms) and B (N msms) become almost genetically identical except for sterility in line A and the other lines remain fertile. A line is maintained by crossing with B line and B line is maintained through selfing.

Development of pollinator line, i.e., C line (N MsMs) proceeds concurrently with that of male-sterile lines. Usually open pollinated selection/promising crosses are used as base population for this purpose. Generally 3 generations of selfing with field selection are carried out.

A schematic plan for breeding improved onion hybrids as outlined by Pike (1986) is given as:

Year	Procedure
1	Grow and select 100 bulbs, store, and plant, also grow supply of male-sterile bulbs for use in test crosses
2	Self selected bulbs and at the same time, test cross with known sterile
3	Grow out bulbs from self and F1 testcrosses, select, discard poor progeny rows and their F1 pair, store
4	Plant bulbs for seed production, observe sterility characters in F1 lines, if 100% sterile, self selections and make back cross to F1, discard pairs with fertile F1 lines
5	Grow out A and B lines as pairs, continue to select in B line side, save best bulbs from a line for next backcross
6	Self B line of selected progenies and make backcross to sterile side of pair
7	Grow bulbs and make final selection on the basis of B line side
8	Mass B line using 10-20 bulbs in cage while making the backcross to the sterile side of pair
9-12	At this point, begin seed to seed and continue through the fifth backcross using the same procedure as in the 8th year, several A and B lines should have been developed, begin making hybrid combinations for testing

Maintenance and multiplication of lines A, B, and C: The first increase of seeds of the lines A is done in insect proof cages followed by an increase in the open but not more than twice to obtain high degree of purity in stocks seeds of the parental lines. The seeds of the lines A and B can be increased in same cage while the line C must be grown separately in another cage. A nylon or wire net cage of 20 × 20 or 24 × 24 mesh measuring 6 m × 3 m × 2 m with a small door at one corner for entrance and exit is suitable for this purpose. This size of cage can accommodate 4 rows of bulbs spaced 60 cm apart to produce about 1-2.5 Kg of seeds. In the cage, bulbs of line A and B should be planted in alternate rows with equal number of bulbs of each. Before flowering the cage is put over the plants. The umbels should not touch the cage as it would allow the stigma to protrude and get contaminated with foreign pollen. About 3 days before the commencement of flowering a bee hive of medium size colony with a queen is placed inside the cage. Ten percent sugar is used as feeding medium. It is necessary to rogue out pollen bearing plants from A line and pollen sterile plants if any from B line every morning before the dehiscence of anthers. The C line can be multiplied in the same way in a separate cage.

For planting areas of commercial seed crop, it will be necessary to increase the seed of A, B, and C lines in open in isolation with an isolation distance of 3.2 km. The plot size is about 0.4 ha (Swarup 1991).

Production of hybrid seed: The hybrid seed is produced in the open isolated field. The bulbs of A and C lines are planted alternatively in a ratio of 4 rows of A line to 1 row of C line (Swarup 1991). The conventional ratio is 8:2. However, a higher planting ratio gives better seed yield. In UK 9:1 ratio has been found good. The flowering of A and C lines must synchronize. If needed synchrony can be achieved by:

1. Adjustment of storage temperature (warmer storage temperature within the range of 0-12°C causes earlier flowering), and
2. Adjustment of planting dates (early planting causes earlier flowering).

In large fields, 3-4 beehives/0.40 ha are placed to ensure large population of honeybees for cross-pollination. Daily rouging of pollen shedding plants and other off-types in A line in the morning before dehiscence of anthers is essential.

Frequently the seed from the pollinator row is discarded. The male plants are destroyed before the female is harvested to avoid contamination of the female seed with accidentally harvested seed from the male parent. If both the male and female parents are to be harvested the male should be harvested before the female and earlier than the optimal harvest time so that the more valuable hybrid seed can be harvested at its optimal time and free from contamination by accidentally harvested seed from the male.

Hybrid seed production often produces lower seed yields per unit of land occupied by the female line than the yields of the open pollinated cultivars. The process of stabilizing a male sterile line can reduce its vigor and fruitfulness due to inbreeding depression. Studies have attributed this due to:

1. Reduced number and size of umbels,
2. Reduced period of receptivity in individual flowers, and
3. Ovule abortion.

This tendency may be overcome by using single cross females without losing too much uniformity in the hybrid (Peters 1990).

Use of molecular tools in onion genetics/improvement: Use of molecular tools as an indirect assay of desired traits can substantially accelerate a breeding program. This would be a major benefit for a crop such as onion, which required 2 years per generation. The use of molecular techniques for onions is recent, but developing very rapidly. Onion is a diploid species with 14 chromosomes per cell, its nuclear DNA content per cell (referred to as genome size) is very high. DNA isolation is relatively easy, producing copious amounts of high quality DNA. The large size of onion genome suggested the need for use the PCR-based technique AFLP, rather than RFLP or the simpler PCR-based method of RADPs. AFLP works well in onion without any modification in protocol, yielding clean genetic fingerprints.

There are protocols within domesticated onions (*A. cepa*) for the production of haploid, then doubled haploids, though ovary culture. The acceleration in inbred development this approach allows is very attractive in a biennial crop, such as onion. We are interested in adapting it to allow production of doubled haploids with randomly introgressed portion of related wild genomes, starting *A. roylei*. The resulting lines would be extremely useful in genome studies, detection and transfer of QTL, or single major genes, affecting important traits such as the ability and timing of bulbing.

SEED PRODUCTION

Onion seed is usually produced in temperate and subtropical countries. In regions where high temperature prevails almost throughout the year, only the early bolting type of onion, requiring relatively low temperature exposure, can produce seed. Onion is a biennial crop for the purpose of seed production. In one season bulbs are produced from the seed and in second season bulbs are replanted to produce seed. Onion seed are poor in keeping quality and lose viability within a year. Therefore, it is essential to produce seed freshly and use the same for bulb production.

It is highly cross-pollinated crop, which is facilitated by protandrous nature of flowers. Its flowers are pollinated by honeybees. Two methods are employed in the production of the onion seed. The one used most commonly is the bud to seed method, which involves first producing bulbs as for the market and then planting them for seed production. While a smaller number of seed is produced by seed to seed method, generally medium sized bulbs are selected for planting seeds. The bigger the bulb size, the higher is the yield. However, very high sized bulbs if used will need a very high seed rate. A bulb size of 2.5 to 3.0 cm diameter may need 1500 kg to plant a hectare and may yield about 8850 kg of seed. Large sized bulbs of 3 to 4 cm diameter will need three times more seed bulb and may yield 1000 kg of seed per hectare. Bulbs are planted by first fortnight of October. The spacing depends on the size of the bulb. A closer spacing of 30 cm within row gives higher yields of seed than when bulbs are planted at 45 cm for variety "Pusa Red."

There are some varieties which do not store well sometimes someone wants to get seed within the same year. In such cases the seed to seed method is practiced. The seeds are sown in August and the seedlings are planted in September. Most of the bulbs give flowering stalks for seeds.

In India, seed to seed method is not popular. Sometimes bolters do appear in the bulb crop of onion, and the seeds produced from these direct bolters are considered unsuitable for raising the crop.

Advantages of this method are high seed yield, production of seed in poor keeping cultivars within storage, and less seed cost production by eliminating expenses on harvesting and replanting bulbs.

The isolation of onion seed fields should be planned well in advance. The greater the distance between the onion fields, less will be the extent of crossing. Other factors such as the direction of wind and weather conditions at pollination time influence the extent of crossing. Since it is a cross-pollinated crop, two cultivars may be kept at about 100 m apart to produce pure seed (Anon 1971).

The maturity of seed ready for harvest is indicated when fruits open and expose the black seed. Harvesting the seed at proper maturity is essential. Only fully ripe seed should be harvested. All the umbels do not mature at the same time. A field is considered ready for harvesting when about 10% of the heads have black seeds exposed. It is desirable to harvest seed at intervals. The seed heads are cut with 10-15 cm of stem attached.

An average seed yield of 8-10 quintals can be expected from a hectare. Higher yield up to 15 quintals are also obtained. For nucleus and foundation seed production the bulb to seed method should be followed because it provides opportunity for proper selection and rouging. However, seed to seed method will produce higher seed yield.

CONCLUSIONS

Onion (*Allium cepa*) is one of the most important crop among various *Alliums* grown in India and belong to family Amaryllidaceae. There has been a steady increase in area and production of onion in the last decade. Onion accounted for about 75% of total foreign exchange earning among fresh vegetables. Bulb and flowering of onion is initiated by environmental factors like temperature and photoperiod. The flower structure is called an umbel. The normal flower in onion is perfect, but genetic and cytoplasmic sterility variations were discovered. Onion is one of the pioneer crops in which heterosis was commercially exploited since about four decades. Although India is one of the leading onion producer, not much emphasis was given to the heterosis breeding in the past. One of the main component for exploiting heterosis in onion is the availability of male sterility. Pollination control in onion is extremely difficult, considering each umbel has several hundred tiny individual flowers. Therefore, it is important to understand onion flowering habits and the inheritance of characters to be efficient in breeding of the crop. Honeybees are the primary agent used for pollination. More research is needed on factors that affect attractiveness of onion to honeybees. The development of F_1 hybrids in onion requires the development of a male sterile line (A line) an inbred maintainer for a line (B line) and a pollinator line (C line). Hybrid seed is produced in the open isolated field. Although reasonable progress has been made so far in the onion and research, many important problems still have to be tackled. Rapid progress could be achieved through the use of simple *in vitro* and rapid multiplication methods.

REFERENCES

Anonymous. (1973). FAO Production Year Book.

Anonymous. (1971). Indian Minimum Seed Certification Standards, Central Seed Committee, Min. Food Agric. Comm. Dev. Co., New Delhi.

Aoba, T. (1967). Effect of different temperatures on seed. *J. Jap. Soc. Hort. Sci.* 36: 333-338.

Austin, R.B. (1972). Bulb formation in onions as affected by photoperiod and spectral quality of light. *J. Hort. Sci.* 47: 492-504.

Bohart, G.E., Nye, W.P. and Hawthorn, L.R. (1970). Onion production as affected by different levels of pollinator activity. *Agr. Expt. Sta. Bul.*, pp. 60, 482.

Currah, L. and Proctor, F.J. (1990). Onions in tropical regions. Natural Resource Institute Bulletin No. 35, pp. 35, NRI, UK.

El-Sayed, A.M., Atia, A.A.M., El-Hak, S.H.G., Azab, A.M. and Mohamed, H.Y. (1999). Studies on heterosis, gene action and combining ability of some traits in onion (*Allium cepa* L.). *Egyptian J. Hort.* 26: 85-95.

Ericson, H.T. and Gabelman, W.H. (1956). The effect of distance and direction on cross pollination in onions. *Amer. Soc. Hort. Proc.* 68: 351-357.

Franklin, D.F. (1958). Effect of hybrid seed production of using different ratios of male sterile and pollen rows. *Amer. Soc. Hort. Proc.* 71: 435- 439.

Franklin, D.F. (1970). Problems in the production of vegetable seeds. In the Indispensable Pollinators. Ark. Agr. Ext. Serv., Misc. Pub. 127, pp. 112-140.

Hamilton, R.P. (1946). Onions need bees. *Gleanings Bee Cult* .74: 23.

Havey, M.J. (1999). Seed yield, floral morphology and lack of male-fertility restoration of male-sterile onion (*Allium cepa*) populations possessing the cytoplasm of *Allium galanthum*. *J. Amer. Soc. Hort. Sci.* 124: 626-629.

Hawthorn, L.R. (1951). Studies on soil moisture and spacing for seed crops of carrots and onions. US Dept. Agr. Cir., pp. 26, 892.

Heath, O.V.S. and Hollies, M.A. (1965). Studies in the physiology of the onion plant. *J Expt. Bot.* 16: 128-144.

Janik, J., Coors, J.G. and Pandey, S. (1999). Exploitation of heterosis: Uniformity and stability. The genetics and exploitation of heterosis in crops. Proceedings of an international symposium, CIMMYT, Mexico, 17-22 August, pp. 319-333.

Jones, H.A. and Emsweller, S.L. (1936). A male sterile onion. *Amer. Soc. Hort. Sci. Proc.* 34: 582-585.

Jones, H.A. and Emsweller, S.L. (1933). Methods of breeding onions. *Hilgaria* 7: 625-642.

Jones, D.F. and Clarke, A.I. (1943). Inheritance of male sterility in onion and the production of hybrid seed. *Amer. Soc. Hort. Sci. Proc.* 43:189-193.

Jones, H.A. and Mann. L.K. (1964). Onions and Their Allies. Leonard Hill Book Co., London. p. 286.

Kato, T. (1964). Physiological studies on bulbing and dormancy of onion plant. III. Effects of external factors on the bulb formation and development. *J. Jap. Soc. Hort. Sci.* 33: 53-61.

Kononkov, P.F., Ustimenko, G.W. and Perez, A.P. (1969). Beitr Trop Subtrop Landw. *Trop. Vet. Med.* 183-192.

Le Baron, F.C. (1962). Onion seed, sample costs and production. Calif. Agr. Ext. Ser., Cost Data Sheet 22, Leaflet.

Lederhouse, R.C., Caron, D.M. and Morse, R.A. (1972). Onion pollination in New York. *New York's Food and Life Sci.* 1: 8-9.

Mac Gillivray, J.H. (1948). Effect of irrigation on the yield of onion seed. *Amer. Soc. Hort. Sci. Proc.* 51: 423-427.

Mani, V.P., Chauhan. V.S., Joshi, H.C. and Tandon, J.P. (1999). Exploiting gene effects for improving bulb yield in onion. *Indian J. Genet. and Pl. Breed.* 59: 511-514.

Moffett. J.O. (1965). Pollinating experimental onion varieties. *Amer. Bee J.* 105: 378.

Netrapal, and Singh, N. (1999). Heterosis for yield and storage parameters in onion (*Allium cepa*). *Indian J. Agric. Sci.* 69: 826-829.

Nye, W.P. (1970). Pollination of onion seed affected by environmental stresses. In the indispensable Pollinators, Report of the ninth pollination conference, Hot Springs. Ark. Oct. 12-15, 1970. Arkansas Agricultural Extension Service MP 127: 141-144.

Nye, W.P., Wailer, G.D. and Waters, N.D. (1971). Factors affecting pollination of onions in Idaho during 1969. *Amer. Soc. Hort. Sci. Proc.* 96: 330-332.

Pathak, C.S., Aghora. T.S., Singh, D.P. and Deshpande, A.A. (1987). Exploitation of heterosis in onion by using indigenous source of male sterility. National symposium on heterosis exploitation. Accomplishment and Prospects. Marathwada Agriculture University, Parbhani, India. Abstr. 53.

Pathak, C.S., Singh, D.P. and Deshpande, A.A. (1980). Annual Report of Indian Institute of Horticultural Research. 34-36.

Pathak, C.S., Singh, D.P. and Deshpande, A.A. and Sreedhar, T.S. (1986). Sources of resistance to purple blotch in onion. *Veg. Sci.* 31: 300-303.

Pathak, C.S., Singh, D.P. and Deshpande, A.A. (1987). A new type of cytoplasmic male sterility in onion. First All India Conference of Cytology and Genetics, Bangalore University, India, 1987. Abstr, 46.

Patil. J.A., Jadhav, A.S. and Rane, M.S. (1973). Male sterility in Maharashtra onion. *Res. J. Mahatma Phule Agric. Univ.* 4: 29-31.

Peters, R. (1990). Seed production in onions and some other *Allium* species. In H.D. Rabinowitch and J.L. Brewster (Eds.), Onions and Allied Crops, Vol. 1 Botany, Physiology and Genetics, CRC Press, Inc. Boca Raton, Florida. 161-176.

Pike, L.M. (1986). Onion Breeding. In M.J. Bassett (ed.), Breeding Vegetable Crops, AVI Pub. Co. Inc. Westport, Connecticut. 357-394.

Propandron, N. (1998). Revealing the heterosis phenomenon in some F1 onion hybrids obtained at the *Institutul de Cercetari pentru Legumicultura si Floriculture, Vidra. Anale-Institutul–de- Cercetari -pentru –Legumicultura- si- Floriculture- Vidra.* 15:45-50.

Sanduleac, E. (1961). The pollination of vegetable seed plants. *Apicultura.* 14: 25-26.

Sen, B. and Srivastava, S.N. (1957). Utilization of cytoplasmic male sterility in production of hybrid onion seeds. Proc 44th Indian Congr Part 2: 225.

Shaw, F.R. and Bourne, A.I. (1936). Insects pollinating onions. *Amer Bee J.* 76: 401-402.

Shrick, F.H., Douglass, J.R. and Shull, W.E. (1945). Experiments for control of the thrips initiated. Idaho Agr. Expt. Sta. Bul. 35, 264.

Singh, J.P. and Dharamwal, S.S. (1970). The role of honeybees in seed setting of onion at Pantnagar, District Nainital, U.P., India. *Indian Bee J.* 32: 23-26.

Stuart, N.W. and Griffin, D.M. (1946). The influence of nitrogen nutrition on onion seed production. *Amer. Soc. Hort. Sci. Proc.* 48: 398-402.

Swarup, V. (1991). Breeding Procedures for Cross Pollinated Vegetable Crops. ICAR, New Delhi, India: 118.

Vander Meer, Q.P. and Van Bennekom, J.L. (1968). Research in pollen distribution in onion seed fields. *Euphytica.* 17: 216-219.

Walsh, R.S. (1965). Pollination of onion plants by honeybees. *New Zeal. Beekeeper.* 27: 18-20.

Whitaker, T.W. (1979). The breeding of vegetable crops: Highlights of the past seventy five years. *Hort. Sci.* 14: 359-363.

Root Vegetable Crops

Pritam Kalia

SUMMARY. Significant heterosis for earliness and yield has been reported for root vegetable crops. Major root crops are carrot, radish, turnip and garden beet. These vegetables are used as cooked or as a salad. The hybrids in these crops are also gaining popularity all over the world. These are highly cross-pollinated crops. There are different systems like self-incompatibility and male sterility are available in these crops. Hybrid vigor is exploited commercially in these crops using these mechanisms. *[Article copies available for a fee from The Haworth Document Delivery Service: 1-800-HAWORTH. E-mail address: <docdelivery@haworthpress.com> Website: <http://www.HaworthPress.com> © 2004 by The Haworth Press, Inc. All rights reserved.]*

KEYWORDS. Root vegetables, carrot, radish, turnip, garden beet, crop biology, floral biology, heterosis, hybrid seed production

Root vegetables are characterized by prominent, fleshy underground structure, which, depending upon the crop may be a root, tuberous root, or a hypocotyl with taproot forming below it. Major root vegetables come from Umbelliferae (carrot), Brassicaceae (radish, turnip), and Chenopodiaceae (beet) families. These crops have variable food value (Table 1) and long storage life, which contributed to their nutritional importance. These develop suc-

Pritam Kalia is affiliated with the Division of Vegetable Crops, Indian Agriculture Research Institute, New Delhi 110012, India.

[Haworth co-indexing entry note]: "Root Vegetable Crops." Kalia, Pritam. Co-published simultaneously in *Journal of New Seeds* (Food Products Press, an imprint of The Haworth Press, Inc.) Vol. 6, No. 2/3, 2004, pp. 247-275; and: *Hybrid Vegetable Development* (ed: P. K. Singh, S. K. Dasgupta, and S. K. Tripathi) Food Products Press, an imprint of The Haworth Press, Inc., 2004, pp. 247-275. Single or multiple copies of this article are available for a fee from The Haworth Document Delivery Service [1-800-HAWORTH, 9:00 a.m. - 5:00 p.m. (EST). E-mail address: docdelivery@haworthpress.com].

TABLE 1. Nutritional constituents of the major root vegetable crops

Crop	Water (%)	Energy (cal.)	Protein (g)	Fat (g)	CHO (g)	A[b] (IU)	C[c] (mg)	Vitamins			Minerals					
								Thiamine (mg)	Riboflavin (mg)	Niacin (mg)	Ca (mg)	P (mg)	Fe (mg)	Na (mg)	K (mg)
Beet	87	43	1.6	0.1	9.9	20	10	0.03	0.05	0.4	16	33	0.7	60	335
Carrot	88	42	1.1	0.2	9.7	11,000	8	0.06	0.05	0.6	37	36	0.7	47	341
Radish	95	17	1.0	0.1	3.5	10	26	0.03	0.03	0.3	30	31	1.0	18	322
Turnip	92	30	1.0	0.2	6.6	Trace	36	0.04	0.07	0.6	39	30	0.5	49	268

Source: *National Food Review (1978), USDA.*
a = Data per 100 g sample.
b 1IU = 0.3 μ vitamin A alcohol.
c = Ascorbic acid.

cessfully in regions where cool night temperatures slow the respiration rate, enhancing retention of stored carbohydrates.

CARROT

Introduction

Carrot (*Daucus carota* L.; $2n = 2x = 18$), a member of Umbelliferae family, is a cool weather crop grown for its edible storage tap roots throughout the world and is most important of all the root crops. Carrot is widely grown both for fresh market and processing. It provides an excellent source of vitamin A and fiber in the diet. Black carrot is used for the preparation of *Kanji*, which is supposed to be good appetizer. Carrot has two distinct groups, i.e., tropical and temperate. The tropical types are annual, whereas, the temperate types are biennial. Carrot has wide range of variability for color from black, red to yellow and orange.

Besides carrot has medicinal properties as is apparent from antibacterial property of essential oils extracted from carrot roots. The essential oil from the carrot seed is useful in diseases of the kidney and in dropsy (Chopra, 1933; Kirtikar and Basu, 1935).

Origin and History

It is believed that carrot arose as a natural variation of the Eurasian wild flower Queen Anne's lace in Afghanistan in the region where the Himalayan and *Hindukush* mountains are confluent, and that it was domesticated also in Afghanistan and adjacent regions of Russia, Iran, India, Pakistan, and Anatolia (Vander Vossen and Sambas, 1993). Most wild forms of *Daucus* are found in the South-Western Asia and the Mediterranean, a few in Africa, Australia, and America (Banga, 1976). According to Matzhevitzh (1929), Afghanistan is the primary center of carrot because the largest diversity in morphological characters of this species has been found to occur in Afghanistan, the roots vary in the degree of ramification, fleshiness, and color ranges from white to anthocyanin red.

The cultivation of carrots may have begun 2000 to 3000 years ago as is evident from the indications emerged from seeds found by archaeologists in the Swiss Lake dwellings. Purple carrots from place of origin, together with a yellow variant, spread to the Mediterranean area and Western Europe in the 11th-14th centuries and to China, India, and Japan in the 14th-17th centuries. The orange types appeared as a chance mutation in the Netherlands during the 17th century and because of improved color and flavor became popular in Europe. Then Settlers brought it to North America, where Indians and colonists adopted it. It is said to have been introduced into India from Persia and were probably first grown in America in the Salem gardens about 1620 AD (Shoemaker, 1947).

Cytogenetics

The carrot is diploid with nine pairs of chromosomes. These are short but slightly variable in length having only 1 pg DNA per 1C nucleus. Approximately, 40 percent of this DNA is highly repeated. Of the nine pairs of chromosomes, four are metacentric, four are submetacentric, and one is satellited. According to Sharma and Bhattacharyya (1954), secondary constrictions on three chromosomes and a satellite on one have been noticed with a large gap separating the satellite and the remaining chromosomes in some stock. Generally, the rDNA of carrot has a high G+C content, but high A+T satellites have been noticed in some stock.

The DNA polymorphism of carrot observed in a limited population so far is about 25 percent for cDNA and genomic clone and 70 percent of the clones are polymorphic across the entire *Daucus* genus (Rubatzky et al., 1999). There are 22 species of *Daucus* most of which have 11 pairs of chromosomes except for two with nine pairs. A modified squash technique for karyotypic analysis developed in carrot (Xing Jim et al., 1994) was also found to be effective in five other species of Umbelliferae.

Genetic Resources

The modern orange carrots have a narrow genetic base considering that they all have been derived from a few 18th century Dutch cultivars. The exploitation of the genetic variation existing in wild *Daucus* germplasm in the Mediterranean and South-Western Asian region has begun only recently. The germplasms collections of carrot and other species are relatively small. About 5600 accessions are held worldwide with over 1000 of these at the Vavilov Institute in Russia (VIR) (Frison and Serwinski, 1995). The genetic resources unit of Horticultural Research International, Wellesbourne, and Warwick (UK) is also maintaining small working collection. The most accessions of which are *Daucus carota*, but samples of *D. broteri*, *D. glochidiatus*, *D. gnacilis*, *D. hispidifolius*, *D. involcratus*, *D. littoralis*, *D. montividensis*, and *D. muricatus* are also held. Approximately, 800 accessions are maintained in USDA collection at the North Central Regional Plant Introduction Station in Ames, Iowa. In these collections, 95 percent are *D. carota* and the other species are *D. aureus*, *D. broterii*, *D. capillifolius*, *D. crinitus*, *D. durieua*, *D. glochidiatus*, *D. guttatus*, *D. littoralis*, *D. muricatus*, and *D. pusillus*. Besides, France, The Netherlands and Japan also hold small working collections.

Breeding Objectives

1. Early and high root yield.
2. Combining desirable characters of both European and Asiatic groups, especially high carotene content and ability to set seeds is plains.
3. Developing F_1 hybrids using cytoplasmic male sterility.
4. Desirable uniform root size, shape (cylindrical), dark orange external and internal color, i.e., uniform in xylem and phloem, top-root ratio, i.e., small tops and smooth heavy tender roots.
5. Improvement in quality, especially carotene content, flavor, texture, high sugar and dry matter in roots.
6. Resistance to cracking and breaking of the root during harvesting and post harvest handling.
7. Broad shouldered, uniform tapering or stump rooted thin and self-coloured slow bolting carrots.
8. Tolerance to environmental stresses and wider adaptability.
9. Resistance/tolerance to diseases, insect pests and nematodes, especially leaf blight (*Alternaria* and *Cercospora*), black rot, powdery mildew, bacterial soft rot, carrot yellows, caterpillars, carrot fly.
10. Resistance/tolerance to defects such as excessive secondary root development, splitting, secondary growth and cavity spot.

Crop Biology

Plant Characteristics

Carrot is an annual or biennial erect herb. At final vegetative state, it is 20-50 cm tall and 120-150 cm at flowering.

Root

The edible portion of carrot root is actually an enlarged fleshy taproot. This is straight, conical to cylindrical, 5-50 cm long and 2-5 cm in diameter at top, orange, reddish violet, yellow or white. The core of mature roots is usually somewhat higher in color than the phloem, and top of the root is often green. In case of good quality colorless cultivars, the core is minimum and cortex is maximum. Besides the core is deeply pigmented so that cortex and core are evenly colored.

Leaves

Leaves are pinnately compound with long petioles often sheathed at its base numbering 8-12. These grow in a rosette, glabrous, green with 2-3 pinnate leaf blades and the segments are divided into often-linear ultimate lobes.

Floral Characteristics/Biology

Flowering stalks in carrot are few to many branched, each ending in an umbel, which is a compound inflorescence. The development of the umbel begins with a broadening of the floral axis and internode elongation (Borthwick et al., 1931). Each umbel comprises of 50 or more umbellate, each of which has approximately 50 flowers. The primary umbel bears more flowers at maturity usually over 1000, whereas secondary, tertiary, and quaternary umbels bear successively fewer flowers. The first, second, and other order umbels usually flower at an interval of 8-12 days from each other and anthesis in a single umbel is completed in 7-9 days. The floral development is centripetal and arrangement is spiral. Thus, the first mature flowers are on the outer edges of the outer umbellates.

Braak and Kho (1958) revealed that though the primary umbel consists mainly of bisexual flowers, but male flowers can occur frequently (between the edge and centre of umbellet) in subsequent umbels. The pollen from flowers at the centre of an umbellet is larger and more frequently fertile than that from peripheral flowers (Nair and Kapoor, 1973). Carrot has protandry and the petals separate and the filaments begin unrolling to release the anthers at anthesis. After straightening of filament, the pollen is shed and the stamen is quickly abscised. Thereafter, the petals open fully and the style elongates. The carrot has a split style, which separates when the flower is receptive to pollination. The petals of male fertile plants fall soon after the split stigma is receptive. However, the petals of petaloid, but not of brown anther, male sterile plants are persistent until the seed ripens. The carrot flowers are epigynous with five small sepals, five petals, five stamens and two carpels. The mature flower and developing fruit from it are 2 mm long. Each carpel has two ovule primordia during early stage of development, but only the lower one continues to grow. Borthwick (1931) found that carrot embryo sac is monosporic (developing from the chalazal macrospore) and 8-nucleate.

For hybridization, as soon as the first bud in an umbel opens, the whole umbel of the female parent is enclosed in a muslin bag. The anthers are removed from the early opening outer flowers in the outer whorl of the umbellets until enough have been emasculated. This process should be completed before any stigma becomes receptive. The unopened central florets in the emasculated umbellets and all late flowering umbellets are removed, leaving the female parent inflorescence with only emasculated flowers. This umbel is then enclosed in a small cloth cage with a pollen bearing umbel from the selected male parent. To ensure artificial cross pollination, the male umbel is gently rubbed over the emasculated umbel for few days daily in the morning or live house flies and pupae are introduced to ensure a continuing supply of active pollinators during the full period of stigma receptivity. In case the pollen parent possesses some dominant marker genes, which can distinguish hybrids in the seedling stage or later at the root stage, then it is not necessary to emasculate the flowers. The roots of male and female parents can be planted very close to each other, about 20 cm apart, and cover a few umbels each of the male and female parents in the some cloth bag before anthesis. The umbels of male and female parent should be gently rubbed against each other or houseflies be introduced into the cloth bag to effect cross-pollination during peak flowering.

Heterosis

It is a well-proven fact that F_1 hybrids show a great promise in getting uniform and high yields in crops. The first report of heterosis in carrot upon crossing of two inbred lines was made by Poole (1937). Inbreds producing roots averaging 12.8 and 24.9, respectively, produced hybrids with roots weighing 80.5 g upon hybridization. The degree of sterility in carrot depends more on cultivar than meteorological conditions and male sterility was mostly confined to primary umbels. The hybrids obtained from male sterile forms yielded 34.4-44.9 percent more and contained higher amount of dry matter and sugars compared to intervarietal pollination. Katsumata and Yasui (1965) reported that F_1 hybrids among the cultivars Yokono, male sterile Sansun and Kuroda possessed good root form, high carotene content and were high in quality. Most of the F_1 hybrids exhibited heterosis for carotene content. The male sterile forms of the cultivars Nantes 14 and Chantenay 2461 were successful in the production of heterotic hybrids (Litvinova, 1979) and heterosis was 20-22 percent higher than in hybrids produced with out using CMS (cytoplasmic male sterility). These F_1 hybrids showed better uniformity of morphological characters. The heterosis breeding in carrot has been facilitated by the cytoplasmic male sterility (CMS) which is of two types, viz., (i) brown anther type, in which the anthers degenerate and shrivels before anthesis, based on S-cytoplasm and at least two recessive genes with complementary action and (ii) petaloid, in which five additional petals replace anthers, based on S-cytoplasm and at least two dominant genes with complementary action. In hybrid development petaloid steriles are employed more widely and then the brown anther type. If genetic and environmentally stable brown anther steriles were available, then they would be preferred over petaloids because of their higher seed yielding potential. Basically, there are three lines in heterosis breeding, namely the male sterile, male fertile sister line and the pollinator line which is male fertile and has a good combining ability with the male sterile line. The male sterile and the pollen parent lines are inbred for several generations for attaining uniformity. The loss in vigor in these can be restored by hybridization.

The hybrids in carrot are normally three way crosses, (A×B) ×C, because the hybrid vigor in a single cross F_1 female seed parent normally results in much greater seed production than that of inbred male sterile parent. Single cross hybrids, A×B, are on an average more uniform than three way crosses. Moreover, they do not require an extra year to produce F_1 seed parent stock. So the single crosses can be used, if their productivity is adequate. Reduced uniformity of three way crosses can be overcome if backcrosses are utilized as seed parents in hybrids as a result the final product attains the form [(A×B) ×B] ×C. Although compared to three way crosses, it consumes an additional year of seed parent production, but it permits utilization of less similar seed parent inbreds, A×B, than what is required.

Inheritance of Cytoplasmic Male Sterility

Cytoplasmic inheritance of petaloidy, first described by Thompson, in carrot lines initiated from a male sterile wild carrot plant found near Orleans, Massachusetts (USA) in 1953. It was named the Cornell cytoplasm and used to produce the majority of hybrid carrots in the United States (Goldman, 1996). Thompson (1962) developed a complex model for inheritance of pollen sterility from the study of wild carrot (petaloid type) along with the brown anther material from Welch and three additional brown anther sources from Gabelman and concluded that there are at least two and probably three duplicate dominant maintainer genes and an epistatic restorer operating in cytoplasm of these material. The useful maintainer line would therefore have to be free from the restorer and homozygous dominant at one of the Ms loci. The studies of Morelock (1974) supported the hypothesis that petaloid sterility in wild carrot cytoplasm is controlled by two dominant nuclear genes (15 fertile: 1 sterile in F_2). The brown anther type is controlled by two recessive as evident from 15 sterile:1 fertile in domestic cytoplasm.

Pollination Control Mechanisms

Carrot is an out breeding species due to being protandrous. It has no self-incompatibility system, but inbreeding is severe. The distinctive umbels and floral nectaries attract insects, which are most responsible for performing pollination. The occurrence of two distinct genetic cytoplasmic types of male sterility (CMS), i.e., brown anther and petaloid type has provided a system for control over pollination for commercial production of hybrid carrot cultivars. The brown anther type was first discovered by Welch and Grimball (1947) and this report was followed in 1953 by the discovery of a sterile wild carrot (petaloid type) by Munger of Cornell University in USA.

Hybrid Seed Production

Hybrid cultivars have the advantage of relatively uniform roots and have been produced by the use of two systems as described by Riggs (1987). The first system is with cytoplasmic male sterility (CMS) in which the pollen does not develop beyond the microspore stage, sometimes referred to as 'brown another form.' The other type is petaloid form in which the five anthers are transformed into petaloid structures during their early development and do not produce early pollen. The final morphology of the petaloid anthers varies from petal-like to filamentous (Eisa and Wallace, 1969). According to Riggs, most commercial F_1 carrot hybrids are produced from the petaloid CMS. Production of parental lines and the hybrid seed by either system should be done in accordance with the instructions of the maintenance breeder. Less insect activity because of smaller petals on the male sterile flowers of the seed producing lines results in to low seed yield. Seed producers frequently use colonies of bees to supplement the natural level of pollinating insect activity when producing hybrid carrot seed.

Seed production of F_1 hybrid cultivars is based on cytoplasmic male sterility (CMS) of one of the parent-inbred lines. The number and arrangement of pollen parent plant rows relative to seed parent rows in fields varies depending mainly on inbred characteristics and grower practices. Generally the ratio of female to pollinator rows as reported by Takahashi, 1987, is from 2:1 to 4:1. But a common male:female ratio is 4:1 which is often grown is an 8:2 arrangement with four two-row beds of female alternating with a single two-row bed of the pollen parent in an isolated field for production of F_1 hybrid seeds. The seeds are harvested from male sterile line and the pollinator plants are removed before collecting the seeds from the male sterile female parent.

Maintenance of Inbred/Pure Lines

For effecting individual plant pollinations/self-pollinations for maintenance of inbred/pure lines, the roots should be planted at 60 × 50 cm spacing and cages of 50-80 cm long wire cylinders, 40-60 cm diameter, covered with muslin cloth and open at each end are used. These are placed over umbels to be pollinated, and tied tightly with a wire closure around seed stalks at the bottom, just below the umbels. Similarly, top is also tied tightly and flies are added through a tube inserted in the top or side. Self-pollination and sib mating for maintenance of inbred/pure lines and varieties, respectively, can also be made in an insect free area by rubbing umbels of parental stocks together or moving pollen with brushes or by hand. These can be maintained by controlled pollination under mesh screen cages of sizes 1 m × 2 m to 8 m × 30 m depending on plant numbers with the help of honeybees or flies added to distribute pollen. Besides, inbred/pure lines and varieties can also be maintained by planting in isolation.

Seed Production

Carrot is an insect pollinated out breeder. Its flowers are typically bisexual, but pollen shed and stigma receptivity are not synchronized and thus do not facilitate self-pollination. It has an extended period of flowering, where umbels produced over a long period. Flowering within each umbel also takes a prolonged time as a result seed development and maturation also show a variation. This variability will obviously pose problem for seed producers as well as for growers using the seed in order to achieve uniform production of storage roots.

For large-scale seed production of carrot, seed to seed method is followed where the crop is sown and remains *in situ* to flower the following spring and seed is harvested in the late summer of the year following sowing. Generally, spacing for seed-to-seed plantings ranges from 50 to 90 cm between rows and about 3-5 cm with in rows. In root-to-seed method, the stecklings are lifted and replanted before the onset of winter after selection of desirable types incase of tropical types, whereas incase of temperate types, in snow bound areas, the roots/stecklings are stored in trenches and replanted in the following spring after snow melts. The stecklings are spaced at 75-90 cm distance in rows and 20-30 cm in plants. Isolation distance for commercial carrot seed crop should be 1000 m and for basic seed it should be at least 1600 m.

Rouging in case of seed-to-seed production method is carried out for early bolters and those with atypical foliage characters. In case of root-to-seed method, remove the plants with atypical foliage and those bolting the first year. Secondly, after the roots have been lifted ascertain their trueness to type, according to root shape, color and size. All those, which do not conform to quality, should be discarded.

RADISH

Introduction

The radish (*Raphanus sativus* L., 2n = 2x = 18), a member of the family Brassicaceae, is a cool season, fast maturing, easy to grow annual or biennial herbaceous plant that are grown for their roots. Radish is mainly grown for its thickened fleshy root used mainly as *salad*. Its young leaves are also consumed after cooking. Although radish is grown and consumed worldwide, it contributes little to the nutrition.

Origin and History

The area of maximum diversity for radish runs from the eastern Mediterranean to the Caspian Sea to China and still more to Japan. Radish is a crop of ancient cultivation in the Mediterranean (before 2000 BC), from where it spread to China in about 500 BC and to Japan in about 700 AD (Kasem-Piluek and Beltran, 1993). Radishes were a common food in Egypt before the building of the pyramids, and they were highly regarded by ancient Greeks. Roman writings described several forms and colonies of radish at the beginning of the Christian era. It was introduced to the new world by Columbus (Peirce, 1987).

Radish is believed to have evolved from *Raphanus raphanistrum*, a widely distributed weed in Europe. More than one source is thought to have contributed to it, which is evident from the ecological and morphological differences existing in different regions of the world. The important wild species of radish which might have contributed to the present day cultivated radish are: *Raphanus raphanistrum* Linn., *Raphanus maritimus* Smith, *Raphanus landra* Morett., and *Raphanus rostratus* DC. These are found in Mediterranean region. The ecological classification of radish cultivars comprises five main varieties, viz., *Raphanus sativus* var. *niger* (Mill) Pers.; var. *radicula* DC.; var. *raphanistroides* Makino; var. *candatus* (L.), and var. *oleifer* Netz. All these intercross freely among each other as well as related wild species.

Cytogenetics

In *R. sativus* and *R. raphanistrum*, Karpechenko (1924) observed 2n = 18 chromosomes. Radish showed 2n = 18 at mitosis and n = 9 at meiosis except in case of artificial tetraploids (2n = 36). Karyotypic analysis revealed three tetrasomic and disomic chromosomes. Richaria (1937) reported that since radish chromosomes are extremely small, i.e., about 1.5-3.5 μm in length at mitosis so detailed karyotypic analysis is in progress. Wang and Luo (1987) and Wang et al. (1989) indicated valuable information for karyotypic analysis of *Brassica* species from Giemsa banding techniques, which may be applicable to chromosome studies in radish. Kaneko (1980) found a spontaneous haploid plant of radish in the progenies of nucleus restored radish (2n = 18) bred by successive backcrossing to *Raphanobrassica* (2n = 36, RRCC), and exhibited the chromosome configuration of (3-0) II + (3-9)I at MI in PMCs. One hexavalent formation by the secondary association of bivalent chromosomes was reported by Maeda and Sasaki (1934) from Shogoin-Daikon cultivar of radish. The basic genome (x = 9) of radish chromosome complements therefore, appears to comprise some homologous and non-homologous chromosomes. Kato and Fukuyama (1982) proved from mitotic study of interspecific hybrids in a phylogenetic relationship of radish that R genome of Japanese radish (*R.*

sativus L., 2n = 18) identified with that of Seiyo-no-Daikon (*R. raphanistrum* L. subsp. *landra*, 2n = 18) because their F_1 hybrids normally formed nine bivalents at MI in PMCs.

Crop Biology

Plant Characteristics

Leaves

Leaves are alternate, glabrous to sparingly hispid, lower leaves in a radical rosette, petioles 3-5.5 mm long, leaf blades oblong, oblong-ovate to lyrate pinnatifid. These are 3-5 jugate with a round or ovate terminal lobe, 5-30 cm long. Higher leaves are much smaller, short petioled, lanceolate-spathulate, and subdentate.

Root

The primary root and the hypocotyls develop in to edible portion of radish root. Radish roots vary greatly in size, shape and other external characters as well as the time period they remain edible. Depending upon cultivar, the length may vary from 2.5-90 cm, shape from oblate to long tapering and the external color may be white or different shades of scarlet with some red cultivars having white tip. Various flesh colored genotypes have also been developed (Kalia, 2003).

Inflorescence

This is typical terminal erect, long, many-flowered raceme. The flowers 1.5 cm in diameter, fragrant, small, white, rose or lilac in color with purple veins in bractless racemes, pedicel up to 2.5 cm long; sepals 4 erect; petals 4, clawed, spathulate, 1-2 cm long, stamens 6, tetradynamous and style 3-4 mm long.

Siliqua

It is cylindrical 3-7 cm long, up to 1.5 cm in diameter, consisting of 2-several superposed joints, lower joints very short and seedless, whereas upper one(s) much la rger terete, spongy and divided into 2-12 one seeded compartments, long conical seedless beak and indehiscent. In case of *Raphanus caudatus*, the pod length is usually 20-30 cm.

Seed

These are ovoid-globose and about 3 mm in diameter. Upon maturity, seed are yellowish at first, turning brown with age.

Floral Biology

The buds borne in racemes on the main stem and its branches open under the pressure of the rapidly growing petals. This process starts in the afternoon, and usually the flowers become fully expanded during the following morning. The anthers open a few hours later being slightly protogynous. Under normal environmental conditions, the anthesis starts at 8 a.m. and 12 noon. The pollen fertility is considerable high at the time of anthesis and one day before anthesis. Fertility is reduced, a day after anthesis. The stigma becomes receptive two days before and remains up to four days after anthesis. However, pollination on the day of flower opening gives better fruit set (Siddique, 1983).

Breeding Objective

1. Early maturity and high root yield.
2. Non-pithy blunt ended cylindrical roots.
3. High pungency and nutrients.
4. Slow and late bolting with less foliage.
5. Tolerance to heat, drought and wet situations, and cold hardiness, etc.
6. Resistant/tolerant to alternaria blight, white rust, downy mildew, grey leaf spot, soft rot, yellows, radish mosaic virus, and aphids.

Heterosis

Pal and Sikka (1956) obtained high yielding hybrids, which gave 30-60 percent higher yield than the better parent. Significantly higher yield than better parent, earliness for 50 percent root formation and long sized roots were observed in radish hybrids (Singh et al., 1970; Singh 2003).

Self-incompatibility and male sterility phenomenon in radish easen exploitation of hybrid vigor. Self-incompatibility system is of sporophytic type in which the papillae on the stigma surface disrupt the pollen tube penetration. Male sterility has been reported by Ogura (1968) in Japanese radish, which is governed by the interaction of a recessive gene 'ms' and S-cytoplasm (Bonnet, 1970). The hybrid breeding in radish, thus, became feasible using either incompatibility or male sterility. The first hybrid radish was developed by Frost (1923) who observed that the crosses between selfed lines were very vigorous and usually exceeded the better parent in root size and all plant characters. Radish superiority of F_1 hybrids were superior over parental lines for earliness and productivity. Heterosis for root diameter and root length has also been observed.

Pollination Control Mechanisms

Radish is a cross pollinated due to presence of self-incompatibility. It is entomophilous root vegetable where pollination is greatly affected by insect activity, especially bees. Pollination is normally limited by environmental conditions, especially temperature, during the flowering season. Wild honeybees mainly pollinate radish and wild flower flies. During rainy winter and low temperature insect pollination is rare. With the discovery of male sterility in radish (Ogura, 1968), it became possible to have a control over pollination. Male sterility in radish was controlled by the genetic interaction between cytoplasmic and nuclear genes. There are two pairs of ms genes in the nucleus, the genotype of the male sterile plants being Sms_1ms_1 ms_2ms_2 and the maintenance line Nms_1ms_1 ms_2ms_2. Selfing can be accomplished by bud pollination.

Hybrid Seed Production

Both the self incompatible line and male sterile lines are being used for producing hybrid seed, commercially (Singh et al., 2001). The scheme to produce economically large amounts of F_1 hybrid seeds by Ito and associates at the Takii Nursery Co. (Haruta, 1962; and Ito, 1954). Single crossing, three way crossing and double crossing are three methods of hybrid seed production in radish (Haruta, 1962). However, double-crossing method is, probably, the most effective means of hybrid seed production in radish and this method consists of different breeding steps varying from I to VIII.

Breeding materials are prepared in step I by observing groups of available plants for morphological and ecological traits under usual cultivation conditions. Selection for the following two or three generation is carried out. In step II, pedigree selection is carried out by diallel cross testing among available groups for one generation.

Parental combination is determined in step III by diallel cross testing among superior strains that are evaluated by pedigree selection in step II for one generation. In step IV, fixation of horticultural traits is enhanced in each strain for few generations. The evaluation of incompatible genes and individual plant selection by self pollination are carried out in parallel with comparisons of combining ability by diallel crossing among strains and fixation of parental strains are increased. Promising cross combinations between strains not only carrying different self-incompatible genes but also possessing higher combining ability and more intense economic traits should be selected.

Progeny testing and evaluation of seed production is performed for one generation in step V, whereas in step VI foundation seeds are mainly produced by bud pollination and selected parental strains are tested for self-incompatibility and cross compatibility. In step VII, open pollination is performed with plantation of parental strains is 1:1 ratio for adequate production of original seeds. The production of marketing seeds by open pollination among seed parents in step VIII.

The field for hybrid seed production should be 500-1000 m apart or even more to avoid out crossing. The proportion of the two parental lines in mixed planting is influenced by the level of self-incompatibility, vigour in growth of seed parents, plant posture and quality and quantity of pollen. The seed parents are planted at a ratio of 1:1 where F_1 hybridization is between two strains with the same level of traits. Maximum rate in mixed planting is 1:3. The ratio of female to male parent rows is usually 1:1 (Takahashi, 1987).

Maintenance of Inbred/Pure Lines

Due to existence of self-incompatibility in radish, selfing can be accomplished by bud pollination. The flower buds should be pollinated 2 days prior to their opening by their own pollen from previously bagged flowers of the same plant when the plant has about 30 percent flowering than just at the commencement of flowering. At this stage self-incompatibility is not active. Usually about 8 to 10 flower buds were pollinated in each inflorescence branch to ensure better seed set and the unopened young flower buds at the terminal end are removed. An alternative to conventional bud pollination method for overcoming self-incompatibility in radish by the application of carbon dioxide at relatively high concentration (3-5%) was given by Ito (1981). Normally CO_2 at this concentration is applied for 2 hours in the evening of every other day and this treatment is continued for 3-4 weeks. By this method, self-fertilized seeds could be obtained in pollination at flowering and sufficient original seeds of inbreds/pure lines as parents of F_1 hybrids maintained (Kimura and Fujita, 1988).

Seed Production

There are two seed production systems followed in radish open pollinated cultivars. These are seed to seed and root to seed systems. The seed to seed system is used for commercial seed production, if the stock seed is of higher quality, whereas root to seed system is preferred for raising nucleus seeds. In this system, the roots are harvested when fully mature. True to type root stecklings are given up to half cut and two-third shoot cut before replanting. This method results in a higher seed yield (Jandial et al., 1997). The selection and roguing are done on the basis of foliage characters, root characters (color, shape, size, flesh color, core size, pungency, etc.) and bolting time. Diseased, misshapen, undersized, and other undesirable roots are discarded. Hairy, forked, early as well as late bolters are also removed. Then the selected roots after giving suitable root and shoot cut, are planted at 50-90 cm distance in rows and 20 cm apart with in rows. An isolation distance of about 1000 m is essential from other cultivars.

In case of seed-to-seed system, there are three stages at which roguing should be carried out, viz., (a) at marketable maturity of root for their relative size, shape, color, proportions of each color on bicolored cultivars, solidity, (b) at stem elongation: remove early bolters, stem color off types, wild radish types, and (c) at flower bud and very early at start of anthesis for flower color.

For root-to-seed system, used for basic seed production, the root selected on the basis of external morphology should also be examined for internal solidity. For this Watts (1960) described an immersion technique in which the radish roots put in a bucket of water. The solid roots sink and selected, whereas pithy float and are discarded. The solid selected roots are then planted and grown for seed production.

TURNIP

Introduction

Turnip (*Brassica rapa* L.; $2n = 2x = 20$), a member of family Brassicaceae, is a cool season, frost hardy crop mainly grown for its fleshy roots. Turnip is used in roasts, stews, soups, casseroles, and as a boiled or cooked vegetable or sliced in salads.

Origin and History

Its cultivation was probably the first attempt by Celts and Germans (Sturtevant, 1919). It was described in ancient Greek times of Alexander the Great, where empire included the Middle East and Persia, from where it must have found its way to East Asia. In Europe and Japan, a well-defined, polymorphic group of vegetable turnips were created, independently, by the 18th century. Turnip has been grown for nearly 4000 years and has spread all over the world from its original home in temperate Europe. It reached Mexico in 1586, Virginia in 1610, and New England in 1628. It was introduced in Canada in 1540 by Cartier and in Virginia in 1609 (Shoemaker, 1949). Turnip is thought to have two main centers of origin, viz., Mediterranean area primary centre for European types and Eastern Afghanistan with adjoining area of Pakistan is considered to be another primary center with Asia Minor,

Transcaucasus and Iran as secondary centers. The parents of cultivated turnip are found wild in Russia, Siberia, and Scandinavia.

Genetic Resources and Cytology

Major germplasm collections are present in the gene banks in Japan, the UK and the USA. Breeding work in turnip is almost at halt excepting for the maintenance of existing cultivars marketed by the major International Japanese and European seed companies.

Frandsen (1945) reported that *Brassica rapa* hybridizes readily with *B. napus* to produce triploid hybrids (3n = 29) which are almost sterile. Spontaneous and artificial amphidiploids (2n = 58) (Frandsen and Winge, 1932; McNaughton, 1973; and Olsson, 1963) might be used to introgress useful characteristics from one species to another.

Inheritance of Color and Phenotypic Variants

The genetics of flesh and skin color in both turnip and Swede was studied by Davey (1931) who reported that white flesh is dominant to yellow in both species and is determined by a single locus in turnip and by two homologous loci in Swede. Skin color in turnip is determined by two independent loci conditioning the presence or absence of green or red pigmentation, respectively. In both the cases, the allele for colored roots is dominant and if both dominant alleles are present the root has a purple phenotype (Brar et al., 1969). In Swede also the skin color inheritance is similar to turnip except for that environment has a greater effect upon intensity of pigmentation. Cours and Williams (1977) reported that characters, viz., cream corolla (cr), light yellow corolla (ly), dark yellow corolla (dy), cupped petal (cup), apetalous (pl), puckered leaf (pkl), anthocyaninless hydathode (ahd), anthocyaninless bud tip (ab), anthocyaninless anther tip (aa), and anthocyaninless style tip (as) are governed by a single recessive gene, whereas rolled petal margin (Ropm) was controlled by a single dominant gene. Similarly, polypetalous (Pp) was also dominant but strongly influenced by environment. However, striped petal (sp) was exclusively maternally inherited and red petal margin (rpm) preferentially maternally inherited. Gene interaction and linkage study revealed that genes for flower color were non allelic and independently inherited. However, cream was epistatic to light yellow, dark yellow and orange; light yellow epistatic to dark yellow and dark yellow epistatic to orange. Anthocyaninless anther tip was found closely linked to anthocyaninless style tip, but there was no linkage between the gene for yellow green plant (yg2) and any of the three-flower color genes cr, ly, and dy.

Crop Biology

Plant Characteristics

Root: Turnip has stout taproot often fusiform to tuberous. The fleshy thickened underground portion of turnip is actually the hypocotyls, the color and shape of which vary depending upon cultivar. A distinct taproot and secondary roots arise from the lower part of the swollen hypocotyls. Thickening begins in the central part of the hypocotyl, followed by the upper and the lower parts. Turnips are variable in shape, from flat through globose to ellipsoid and cylindrical, blunt or sharply pointed, flesh white, pink or yellow, apex white, green, red, pink or bronze.

Leaf: Leaves are very variable, depending upon cultivar, growing in a rosette during the vegetative stage. Basal leaves are more or less petioled, bright green, lyrate-pinnati-partite, dentate, crenate or sinuate with large terminal lobe and up to 5 pairs of rather small lateral lobes. Lower cauline leaves are sessile, clasping and pinnatified. Upper cauline leaves are sessile, clasping, undivided, glaucous, entire to dentate.

Inflorescence: It is a loosely corymbiform raceme with open flowers over topping the buds. Pedicel in 1-3 cm long and sepals yellow green petals yellow, clawed, 6-11 mm long; stamens 6, tetradynamous; carpels 2, superior ovary.

Fruit: It is siliquae, linear, 4-10 cm × 0.2-0.4 cm, beak 0.5-3 cm long. Seeds 20-30, globose, 1-1.5 mm in diameter, dark brown with a fine distinct reticulum.

Floral Biology

Turnip is insect pollinated, which visit flowers in search of nectar. This is secreted by two active nectaries situated between the two short stamens and the ovary. The process of flower bud opening starts in the afternoon, and usually the flowers open fully the next morning. The peak period of anthesis is between 8 to 9 a.m. Dehiscence of anthers begin with flower opening with maximum between 10 a.m. and 12 noon. Pollen fertility is maximum at and a day before anthesis. The stigma becomes receptive two days before and remains receptive for two days after anthesis.

Breeding Objective

1. Earliness in attaining marketable size.
2. High yield.
3. Stump rooted vari.eties with thin taproot and non-branching habit.
4. Slow bolting and no pithiness.

5. Resistance/tolerance to white rust, phyllody, club root, powdery mildew, turnip mosaic virus, cabbage root fly, and turnip root fly.

Heterosis

Improvement for uniformity is on important breeding objective in turnip. Wit (1966) demonstrated heterosis for number of characters including dry matter yield of root and leaf in this crop. Several F_1 hybrid cultivars in turnip have been bred by Japanese breeders (Anon, 1980). However, their acceptance had been slow in USA (Rubatsky, 1981). Heterosis for root yield, root length, root diameter and number of leaves per plant was observed by Pathania et al. (1987).

Pollination Control Mechanisms

There are two pollination control mechanisms in turnip, viz., sporophytic self-incompatibility system and cytoplasmic male sterility (CMS). Effective self-incompatibility avoid the need for emasculation and crosses can be made by enclosing flower heads from two compatible plants in a cellophane or muslin bag with blowflies as pollinators. Ohkawa (1985) demonstrated occurrence of cytoplasmic male sterility.

Hybrid Seed Production

A distance of 1000 m to avoid out crossing must separate the fields for F_1 hybrid seed production. Takahashi (1987) suggested that the ratio of female to male parent is usually 1:1 while utilizing cytoplasmic male sterility as well as seed parents planting in case of self-incompatibility.

Maintenance of Inbred/Pure Lines

Self-incompatibility does pose a problem in producing selfed seed of inbreds/pure lines, but this can be overcome if style is pollinated after removing the stigma. Besides, Monteiro et al. (1988) suggested that salt (NaCl) solution and carbon dioxide could also be used to over come self-incompatibility that will facilitate self-seed production, thus maintenance of inbred/pure liens.

Seed Production

Turnip cultivars are grouped into two categories from seed production point of view, viz., (a) Annual/Asiatic/tropical type, (b) Biennial/European/temperate type. The latter produces seeds only in temperate climate area available in hills whereas the former types can be grown for seeds in plains as well as hills. Seed-to-seed or *in situ* and root to seed or transplant method are used for seed production in turnip. Generally, *in situ* method gives higher seed yield than the transplant method, but this method can be practiced only if quality nucleus seed is and to raise the commercial seed crop. Root to seed method is followed for basic seed production, where only desirable selected roots are used in order to maintain high quality standards of the seed.

Roguing for seed-to-seed method is carried out firstly at early vegetative stage, i.e., before the swelling of the roots to check for leaf type, color and relative height and secondly at the start of anthesis to check for flower color and size. In case of root-to-seed method, roguing is carried out besides at early vegetative stage when roots are lifted for storing/re-planting, depending upon local winter climate and custom to check for root shape, relative size, color of root and shoulder. When the replanted crop reaches anthesis roguing is done based on flower color and size. Selected roots are then planted at 50-90 cm distance amongst rows and 30-40 cm within rows after giving two-third cut to the top protecting growing point and half cut to root. Isolation distance of about 1000 m is maintained from other cultivars and related crops.

GARDEN BEET

Introduction

The garden beet (*Beta vulgaris* L. 2n = 2 x = 18), a member of the family Chenopodiaceae, is grown for a fleshy root, a marketable product composed of hypocotyl and crown. The roots of garden beet are eaten boiled either as a cooked vegetable or cold as a *salad* after adding oil and vinegar. A large chunk of the commercial production is processed into boiled and sterilized beets or to pickles. Young leaves are eaten as potherbs in Indonesia and Japan. Root and leaves also have a medicinal value against infectious and tumours, and garden beet juice is a popular health food. Betanins, obtained from the roots, are used industrially as red food colorants, e.g., to improve the color of tomato paste strains with strikingly colored, large leaves are from as ornamentals.

Origin and History

It was first described in 1558 in Germany and referred to as Roman beet and in England in 1576. In 1800, it was introduced in USA and become known as garden beet. According to Burkill (1935), beetroot was introduced in India in remote time and then via sea route it was taken by the Arabs to China in 1850 AD.

Beet thought to have originated in Europe and North America and spread eastward from the Mediterranean area. Its wild forms occur around the Mediterranean, Asia minor, and the Near East. The present day cultivated garden beet has probably originated from *Beta vulgaris* L. ssp. *maritima*, a variable species of the Mediterranean, possibly by means of hybridization with *Beta patula*, a closely related species of Portugal and the Canary Island. *Beta vulgaris* L. ssp. *maritima* has been found to grow wild on seashores in Britain and through Europe and Asia to the East Indies.

Genetic Resources

Genetic material collected from breeding work in sugar beet and fodder beet can be used in breeding vegetable beets. Collections of wild and cultivated beta material from eastern Mediterranean and secondary centers of diversity are kept by sugar beet breeders in western Europe and North America, Since diversity is eroding rapidly, therefore *Beta* genus has a high priority for IBPGR.

Inheritance of Color

Color is an important characteristic feature of garden beetroot as it plays an impressive role in adorning the salad plate. Watson and Gabelman (1984) studied the genetics of pigments betacyanine, betaxanthine, and sucrose concentrations in roots of table beet and found their quantitative inheritance, whereas qualitative pattern was observed by other researchers (Holland and Dowker, 1969; Keller, 1936; and Pederson, 1944). Two loci determined qualitative color differences in which R locus has five alleles, viz., R, R^t, r, R^p, and R^h, whereas Y locus had three alleles, viz., Y, Y^r, and y.

The R-Y-genotype determines a phenotype with red roots, hypocotyls and petioles while in the presence of the recessive rr genotype Y-gives yellow roots, hypocotyls and petioles. The R-yy genotype roots are white with red hypocotyls. The double homozygous recessive rryy had yellow hypocotyls and white root. In the presence of Y allele, the R^H allele produces a red hypocotyl. R^t governs striped petiole independent of the genotype at the Y locus and R^p provides pink color to roots, hypocotyls and petioles. The Y^r allele determines pigment production in the roots only. Thus, R-Y^r-plants have red roots and rrY^r-plants have yellow roots but both have green tops, while R^t-Y^r-plants have red roots and striped petioles. The dominance relationship in alleles, viz., R^t>R>r and Y>Y^r>y was reported by Keller (1936). The ratio of violet betacyanine to betaxanthine determines the quantitative differences in beetroot colour. Roots with high, medium, and low betacyanine to betaxanthine ratio have violet, red, and orange color, respectively. Watson and Gabelman (1984) reported highly significant GCA and SCA for both pigment concentrations and pigment ratio. However, triallelic system at the R locus with in complete dominance was found controlling the ratio of betacyanine to betaxanthine (Wolyn and Gabelman, 1989).

Crop Biology

Plant Characteristics

Garden beet is a highly variable, robust, erect, usually biennial herb.

Root

The main root is long, stout, tapered, side-roots forming a dense, extensive root system in the top 25 cm of the soil. The hypocotyls and the upper part of the main root are conspicuously swollen, being globular, flattened, cylindrical or tapering, adventitious roots occur in two opposite rows on the lower part. The swollen root consists of alternating layers of strongly colored conductive tissue and light coloured spongy tissue.

Leaf

These grow in a basal rosette and have long petioles. Leaves are alternate, often ovate and cordate, 20-40 cm long, margins wavy. Leaf tissues puckered between nerves, subglabrous, green, and dark green or red, often shiny.

Inflorescence

It has a long, paniculate, more are less open spike, 50-150 cm long. Flowers are greenish, sessile, bisexual, usually 2-3 (-5) together, subtended by minute bracts. Perianth 5 partite, turning thicker at base as fruits ripen. There are 5 stamens, 1 celled ovary, superior, surrounded by a disk and pistil short with 2-3 stigmas. Fruit is single seeded, enclosed within the swollen corky perianth bases, 3-7 mm in diameter, 1-6 fruits adhering in groups called glomerules or seed balls. These also referred to as multiple beet seed or multigerm. If the fruit is formed from a single ovary, it is termed as single germ beet seed or monogerm. The seeds are kidney shaped brown, 1.5-3 mm in diameter and 1-5 mm thick.

Floral Biology

It is allogamous and anemophilous. The anthesis starts at 7 a.m. and continues up to 5 p.m., but the peak hours are from 11 a.m. to 1:00 p.m. Anther dehiscence begins at 8:00 a.m. and continues up to 6:30 p.m, however the peak period is from 12:30 p.m. to 2:30 p.m. depending upon temperature and humidity. High temperature and low humidity favors anthesis as well as anther dehiscence. As the stigma receptivity starts 8 hrs before anthesis and reaches peak just after anthesis and remains for 8 hrs after that.

Breeding Objective

1. Breeding high yielding varieties having dark red, uniformly colored roots with absence of any internal white rings.
2. Uniform shape and size of roots, and slow bolting habit.
3. Developing varieties of spherical, flattened spherical, cylindrical or conical root shapes depending upon regional preferences.

Heterosis

Heterosis breeding is at infancy in beetroot. Although with the introduction of CMS, the production of hybrids has been facilitated. Male sterility was governed by the segregation of a gene X in S-cytoplasm, with fertility dominant to sterility (Bliss, 1965). Another gene Z with partial male fertility and complete dominance over male sterility was also reported but it was independent of and hypostatic to X. Gabelman (1974) and Gabelman (1974) reported that monogerm character and cytoplasmic male sterility (CMS) were transferred from sugar beet.

For hybridization, if dominant marker is available it can be used to avoid emasculation of female parent and the selfed plants can be rogued either at seedling stage or at root stage. If the pollen parent does not possess any marker genes it would be necessary to emasculate the flowers before crossing since these are hermaphrodite. During pollination care should be taken to avoid contamination from foreign pollen by wind. The bags covering the plants of the male and female parents should not be opened on a windy day and pollination must be done when the air is still, preferably in a glasshouse or plastic cage (Swarup, 1991).

Pollination Control Mechanisms

Beetroot is predominantly wind pollinated. Its pollens are dust like and produced abundantly. For isolation of plant for breeding work pollen proof conditions are required. Since for hybridization, emasculation process in cumbersome, therefore mass pollination of selected root was used for development of improved cultivars of beetroot. The transfer of CMS system from sugar beet has now facilitated this process.

Hybrid Seed Production

Sowing seed parent and pollen parent in 1:1 ratio can produce the commercial F_1 hybrid seed in fields isolated from other compatible crops or varieties by at least 1000 m, since beetroot is an anemophilous. Zoning scheme can also be employed to confine seed production of different hybrids in separate geographical areas. Hybrid seed production in beetroot is feasible only if monogerm character and cytoplasmic male sterility (CMS) system is available. Dominant marker, if available, can also be made use of for the purpose, but then selfs need to be thinned.

Maintenance of Inbred/Pure Lines

Selfing is necessary for maintenance of inbreeds/pure lines and are for this purpose, the entire plant or a few flowering branches are enclosed in a thick muslin cloth bag or preferably a Kraft-paper bag so that pollen does not get blown away by wind. The bags must be shaken once or twice doily to ensure better seed-set and must not be opened on windy days.

Seed Production

The seed to seed and root to seed are the basic methods of seed production for beetroot. The seed to seed method is normally only used for commercial seed multiplication while root to seed is used for nucleus or basic seed production. This method allows for inspection and roguing of roots.

In seed-to-seed method sowing is normally done late compared to root to seed method as the plants remain is the field throughout the winter; therefore this system is not for areas where there is a problem of server winter. Generally, August to September is the main sowing time in the northern hemisphere. The seed should be sown at 60 cm distance between the rows and the 30 cm distance between plants should be maintained by thinning. Alternatively, sowing can also be carried out in a four-row bed system with 25-30 cm between the rows in a 110 cm wide bed to give an optimum plant density of 200 plants per square meter. The stecklings can then be transplanted in the early spring, but where there is a problem of soil being wet as in some Mediterranean areas and also dry temperate areas in India, late autumn transplanting is adopted. The main roguing in this method is done at lifting and replanting, although plants that bolt prematurely can be removed before lifting. At lifting roguing is done to discard plants showing incorrect leaf shape and/or color, premature bolting, incorrect root shape, seed borne pathogens, etc., and same process in repeated at replanting so as to avoid escapes.

The root-to-seed method is in two stages. In the first stage roots are produced like that of production of beet for the market. Seed is sown during July-August in such areas where growing conditions are satisfactory into the autumn. The seedlings are singled as soon as possible following emergence so that root develops its characteristic shape. If monogerm or rubbed seed has been sown at proper spacing then thinning is not necessary.

During autumn lifting of roots is carried out and desirable selected roots are stored. Lifting is completed before the on set of damaging frosts, and thus the timing depends upon local climates. Mechanical damage should be avoided. The lifting and root handling becomes easier if the plants are topped before lifting and care is taken that crown is not damaged.

The root is then stored either in suitable buildings or storage in the field in pits. Although, most beetroot seed producers have developed suitable shed systems, but field storage is still used in some areas:

1. Storage in protected structures: The advantage of this storage is that the air temperature can usually be controlled when necessary, thus avoiding over heating and frost damage. The optimum temperature for storage is 4-5°C and optimum relative humidity is 80-90 percent. A stacked tray or crate system is very suitable and can be coordinated with field operations at lifting and planting times.

2. Field storage: There are different versions of field storage like clamps and pits, etc., on well drained sites, the selected roots are arranged in pyramids or ridges. The roots are stacked 60-200 cm above ground level. The piles of roots are covered with straw and then a wooden plank is placed on the pit plugging the sides with soil. The wooden plank must have a hole on one corner where in plastic pipe or straw funnel or chimneys can be erected from aeration point of view so as to reduce the risk of condensation.

In the spring as early as the local conditions become favorable, the roots are transplanted in rows 100 cm apart and 30 cm with in rows. The roots must be set upright with their crowns at finished soil level. The soil around root must be made firm as this process reduces drying out and assist in early establishment of new fibrous roots. In specialized seed producing areas, the seed companies supply the roots to the growers who produce the seed on contract basis.

In this method, roguing is carried out before cutting tops for lifting and plants showing incorrect leaf color and/or morphology, early bolters and plants showing symptoms of seed borne pathogens are removed. In case of lifted roots, discard roots that are not true to type. Shape, size, crown, base and surface corkiness should be taken into consideration. At replanting, the roots showing storage diseases should be discarded. In case of bolting plants, plants showing incorrect leaf shape, color, vigor, and seed borne pathogens.

Since beetroot is a highly cross-pollinated crop and its pollen is wind borne over relatively long distances and sufficient isolation should, therefore, be ensured. Between same types of cultivars, e.g., red globe isolation of at least

500 m should be maintained and between different types of cultivars (e.g., between red globe and cylindrical types), an isolation of 1000 m is necessary. Beetroot also has cross compatibility with other subspecies of *Beta vulgaris*, viz., spinach beet, Swiss chard, sugar beet, mangold, etc., therefore, adequate isolation of these different seed crops has to be ensured (George, 1999). Zoning scheme is generally used for this purpose in order to confine seed production of each of the different species in separate geographical areas. There is a variation with respect to minimum isolation requirements between different types of *Beta vulgaris* vis a vis authority or scheme. Inspection for interior root characteristics is essential by cutting out a thin wedge of root flesh before replanting. Dark (1971) described and recommended the discard strip technique to be used when high genetic quality is required is seed and pollination contamination is suspected.

REFERENCES

Anon. (1980). Swedes and turnips. Grower guide No. 14. Grower Books, London, 92.

Banga, O. (1976). Evolution of crop plants (Ed. N.W. Summands). Longman, New York, USA.

Bliss, F.A. (1965). Inheritance of male sterility in table beets (*Beta vulgaris* L.). *Diss. Abstr.*, 25: 6901.

Bonnet. A. (1970). Eucarpia CRNA, P. Versailles, pp. 83-88.

Borthwick. H.A. (1931). Development of macro gametophyte and embryo of *Daucus carota*. *Bot. Gaz.*, (Chicago), 92: 23-44.

Braak. J.P. and Kho, Y.O. (1958). Some observations on the floral biology of the carrot (*Daucus carota* L.). *Euphytica*, 7: 131-139.

Brar, J.S., Gill, H.S. and Nandpuri, K.S. (1969). Inheritance of qualitative characters in turnips (*Brassica rapa*). *J. Res. Panjab Agric. Univ.*, Ludhiana, India. 6: 907.

Burkill, I.H. (1935). A dictionary of the economic products of the Malay Peninsula, Crown Agents, London.

Chopra, R.N. (1933). Indigenous drugs of India, The Art Press, Calcutta.

Cours, B.J. and Williams, P.H. (1977). Genetic studies in *Brassica campestris* L. *Cruciferae Newsletter*, 2.

Dark, S.O.S. (1971). Experiments on the cross-pollination of sugarbeet in the field. *Journal of the National Institute of Agricultural Botany*, 12: 242-246.

Davey, V. McM. (1931). Colour inheritance in swedes and turnips and its bearing on the identification of commercial stocks. *Scott. J. Agric.*, 14: 303.

Eisa, H.M. and Wallace, D.H. (1969). Morphological and anatomical aspects of petaploidy in carrot. *Journal of the American Society for Horticultural Science*, 94: 545-548.

Frandsen, K.J. (1941). Contribution to the cytogenetics of *Brassica napus* L., *Brassica campestris* L., and their hybrid the amphidiploid *Brassica napocampestris* Arsskr. K. Vet.-Landbohojsk, 59.

Frandsen, K.J. and Winge, O. (1932). *Brassica napocampestris*, a new constant amphidiploid species hybrid. *Hereditas*, 16: 212.

Frison, E.A. and Serwinski, J. (1995). Directory of European Institutions holding Crop Genetic Resources Collections, IBPGRI, Rome.

Frost, H.B. (1923). Heterosis and dominance of size factors in *Raphanus*. *Genetics*, 8: 116-153.

Gabelman, W.H. (1974). Breeding F_1 hybrid table beets, In: XIX Proc. XIX International Horticultural Congress, Warsaw, 11-18 Sept. 1974 Vol. IB, Antoszewski, R., Harrison, L. and Nowosielski, J., eds., 669. ISHS, Wageningen, Netherlands.

Gabelman, W.H. (1974). F_1 hybrids in vegetable production, In: Proc. XIX International Horticultural Congress, Warsaw, 11-18 Sept. 1974 Vol. III, Antoszewski, R., Harrison, L. and Nowosielski, J., eds., 419. ISHA, Wageningen, Netherlands.

George, Raymond A. T. (1999). Vegetable seed production (2nd edn.). CABI Publishing, Wallford, Oxon OX10 8DE, UK.

Goldman, H. (1996). A list of germplasm release from univeristy of Wisconsin carrot breeding program, 1964-94. *Hort. Sci.*, 31: 882-883.

Haruta, T. (1962). Studies on the genetics of self and cross incompatibility in cruciferous vegetables. *Rep. Agric. Nagaoka farm Takii Co.*, 2: 1.

Holland, H. and Dowker, B.D. (1969). The breeding of Avonearly, a red beet variety resistant to bolting. *J. Hortic. Sci.*, 44: 257.

Ito, S. (1954). On the breeding system in cruciferous vegetables. *Rep. Agr. Nagaoka Farm Takii Seed Co.*, 1: 1.

Ito, T. (1981). Feasible self-seed production methods in self-incompatible crucifer lines. Proc. 1st Int. Symp. on Chinese Cabbage, AVRDC, Taiwan, pp. 345-355.

Jandial, K.C., Samnotra, R.K., Sudan, S.K. and Gupta, A.K. (1997). Effect of steckling size and spacing on seed yield in radish (*Raphanus sativus* L.). *Environ. Ecol.*, 15(1): 46-48.

Kaneko, Y. (1980). Haploid plant of *Raphanus sativus*. *CIS No.* 28: 21.

Karpechenko, G.D. (1924). Hybrids of *Raphanus sativus* L. × *Brassica oleracea* L. *J. Genet.*, 14: 375.

Kato, M. and Fukuyama, T. (1982). Production of *Raphanus sativus* 'Aokubi-miyashige' with *R. raphanistrum* 'Seiyo-nodaikon' cytoplasm. I. Process nucleus substitution and selection in the 4th-6th generations. *Bull. Exp. Farm Coll. Agri. Ehime Univ.*, 4: 29.

Katsumata, H. and Yasui, H. (1965). *Engei Shikonjo hokoku/Bull. Hort. Res. Sta. Ser. D,Kurume*, 3: 79-112.

Keller, W. (1936). Inheritance of some major colour types in beets. *J. Agric. Sci. (Camb.)*, 52: 27.

Kimura, A. and Fujita, Y. (1988). The original seed production by CO_2 application, in seed production of vegetables means of high techniques. *Res. Veg. Seed Prod. Seibundo-Shinkosha*, Tokyo, pp. 246.

Kirtikar, K.R. and Basu, B.D. (1935). Indian medicinal plant (Lolit Mohan Basu) Allahabad.

Litvinova, M.K. (1979). Tsitoplazmatich, muzhsk. Steril'nost, i selektsiya rast kiev, Ukranian SSR, pp. 200-203.

Matzkevitzh, V.I. (1929). The carrot of Afghanistan. *Bull. Appl. Bot. Genet. Pl. Breed.*, 20: 517-557.

Maeda, T. and Sasaki, T. (1934). Chromosome behaviour in the pollen mother cells of 'Shogoin daikon' and 'Nerima daikon', the horticultural varieties of *Raphanus sativus* L. *Jpn. J. Genet.*, 10: 78.

McNaughton, I.H. (1973). *Brassica napocampestris* (2n = 58), I. Synthesis, cytology, fertility and general considerations. *Euphytica*, 22: 301.

Monteiro, A.A., Gabelmen, W.H. and Williams, P.H. (1960). The use of sodium chloride solution to overcome self-incompatibility in *Brassica campestris. Cruciferae Newsetter*, 13: 122.

Morelock, T.E. (1974). Influence of cytoplasmic source on expression of male sterility in carrot, *Daucus carota* L. Ph.D. Thesis, University of Wisconsin, Madison, USA.

Nair, P.K.K. and Kapoor, S.K. (1973). Pollen morphology production of *Daucus carota* L. *J. Palynol.*, 9: 152-159.

Ogura, H. (1968). Studies on the new male sterility in Japanese radish with special reference to the utilization of this sterility towards the practical raising of hybrid seeds. *Mem. Fac. Agr. Kagoshima Univ.*, 6: 39.

Ohkawa, Y. (1985). Occurrence of cytoplasmic male sterility in *Brassica compestris* and comparison with that of *B. napus. Bull. Nat. Inst. Ag. Sci.* Japan D. 36, 1.

Olssson, G. (1963). Induced polyploids in *Brassica* in recent plant breeding research, Svalof, 1946-1961, Akerberg and Hagbag, Eds.

Oyen, L.P.A. (1993). *Beta Vulgaris* L. In: Plant resources of South-East Asia (Eds. J.S. Siemonsma and Kasem Piluek), No. 8., Vegetables, Pudoc Scientific Publishers, Wageningen.

Pal, B.P. and Sikka, S.M. (1956). Exploitation of hybroid vigour in the improvement of crop plants, fruits and vegetables. *Indian J. Genet.*, 16: 98-104.

Pathania, N.K., Rattan, R.S. and Thakur, M.C. (1987). Heterosis and combining abilities in turnip (*Brassica rapa* L.). *Veg. Sci.*, 14: 161-168.

Pederson, A. (1944). The colours of beets (*Beta vulgaris* L.). *K. Vet-of-Landbokojsk Arsskr.*, p. 60.

Peirce, L.C. (1987). Vegetables: Characteristics, production, and marketing. John Wiley & Sons, New York, USA.

Piluek, K. and Beltran, M.M. (1993). *Raphanus sativus* L. In: Plant Resources of South-East Asia (Eds. J.S. Siemonsma and Kasem Piluek), No. 8, Vegetables, Pudoc Scientific Publishers, Wageningen.

Poole, C.F. (1937). Year book USDA, 379.

Richaria, R.H. (1937). Cytological investigation of *Raphanus sativus, Brassica oleracea* and their F_1 and F_2 hyrbids. *J. Genet.*, 34: 19.

Riggs, T.J. (1987). Breeding F_1 hybrid varieties of Vegetables. In: Hybrid production of selected cereal oil and vegetable crops (Eds. Feistritzer, W.P. and Kelly, A.F.) FAO, Rome, pp. 149-173.

Rubatsky, V.E. (1981). An overview of California cruciferae production, present and future concerns. Proc. Eucarpia Cruciferae Conference, 16-18 September, A°s, Norway, 68.

Rubatzky, V.E., Quires, C.F. and Simon, P.W. (1999). Carrots and related vegetable umbelliferae. CAB International, Wallingford, UK.

Sharma, A.K. and Bhattacharyya, N.K. (1954). Further investigations on several genera of umbelliferae and their relationship. *Genetica* 30: 1-68.

Shoemaker, J.S. (1947). Vegetable growing. John Willey and Sons, Inc. New York, USA.

Shoemaker, J.S. (1949). Vegetable growing (2nd edn.). John Willey and Sons, Inc. New York, USA.

Siddique, B.A. (1983). Studies on floral biology and morphology of *Raphamus sativus* L. *Acta Botanica Indica*, 11: 150-154.

Singh P.K., Tripathi S.K. and Somani K.V. (2001). Hybrid seed production of radish (*Raphanus sativus* L.). *J. New Seeds* 3(4): 51-57.

Singh P.K. (2003). Personal Commn.

Sturtevant, E.L. (1919). Notes on edible plants, Ed. U.P. Hendrick, New York. Agric. Exp. Sta.

Swarup, V. (1991). Breeding procedures for cross-pollinated vegetable crops (2nd Edn). ICAR, New Delhi, India.

Takahashi, O. (1987). Utilization and seed production of hybrid vegetable varieties in Japan. In: Hybrid production of selected cereal oil and vegetable crops (Eds. W.P. Feistritzer and A.F. Kelly). FAO, Rome, pp. 313-328.

Thompson, D.J. (1962).Studies on inheritance of male sterility in the carrot, *Daucus carota* L. *Proc. Am. Soc. Hortic. Sci.*, 78: 332-338.

Toxopeus, H. (1993). *Brassica rapa* L. In: Plant resources of south-east Asia (Eds. J.S. Siemonsma and Kasem Piluek), No. 8, Vegetables, Pudoc Scientific Publishers, Wageningen.

Van der Vossen, H.A.M. and Sambas, E.N. (1993). *Daucus carota* L. In: Plant resources of south-east Asia (Eds. J.S. Siemonsma and Kasem Piluek), No. 8, Vegetables, Pudoc Scientific Publishers, Wageningen.

Wang, X.-H. and Luo, P. (1987). Studies on the karyotypes and C-banding patterns of Chinese Kale and cabbage. *Acta Bot. Sinica*, 29: 49.

Wang, X.-H., Luo, P. and Shu, J.J. (1989). Giemsa N-banding pattern in cabbage and Chinese Kale. *Euphytica*, 41: 17.

Watson, J.F. and Gabelman, W.H. (1984). Genetic analysis of betacyanine, beta-xanthine and sucrose concentrations in roots of table beet. *J. Am. Soc. Hort. Sci.*, 109: 386.

Watts, L.E. (1960). The use of a new technique in breeding for solidity in radish. *J. Hortic. Sci.*, 35: 221-226.

Welch, J.E. and Grimball, E.L. (1947). Male sterility in carrot. *Science*, 106: 594.

Wit, F. (1966). Quality plant material. *Veg.* 13: 315-310.

Wolyn, D.J. and Gabelman, W.H. (1989). Inheritance of root and petiole pigmentation in red table beet. *J. Heredity*, 80: 33.

Xing Jin, H., Tang-Lesheng, He, X.J. and Tang, Z.S. (1994). Cytological study of six species of carrot. *J.S.W. Agricultural University*, 16(5): 488-491.

Garden Pea

Kevin McPhee

SUMMARY. Pea has been grown for food and feed since its domestication the Mediterranean region in approx. 7000 B.C. It serves as an excellent source of protein, fiber, minerals and vitamins especially in developing countries. Nitrogen fixation by *Rhizobium leguminosarum* in symbiotic relationship with pea is a primary benefit to inclusion in crop rotations. Several biotic and abiotic stresses constrain production worldwide. Genetic resistance is available for many of the diseases and when coupled with cultural control methods can be effectively controlled. Heterosis and male sterility are present in pea, however, to date no hybrid cultivars have been developed or commercialized due to difficulty associated with cross pollination. Seed production and maintenance of inbred pure lines is through bulk production or single plant selection in disease free fields isolated by at least 30 m from neighboring pea fields. Clean equipment, routine field inspection and use of electric eye technology further increase seed purity. Wide adaptation and use of modern technology in cultivar improvement ensure that pea will continue to be an important crop worldwide. *[Article copies available for a fee from The Haworth Document Delivery Service: 1-800-HAWORTH. E-mail address: <docdelivery@ haworthpress.com> Website: <http://www.HaworthPress.com>]*

KEYWORDS. Pea, genetic diversity, crop biology, heterosis, pollination control mechanisms, seed production, inbred lines, hybrid seed production

Kevin McPhee is affiliated with the USDA/ARS and Crop and Soil Sciences Department, Washington State University, Pullman, WA 99164-6434.

[Haworth co-indexing entry note]: "Garden Pea." McPhee. Kevin. Co-published simultaneously in *Journal of New Seeds* (Food Products Press, an imprint of The Haworth Press, Inc.) Vol. 6, No. 2/3, 2004, pp. 277-288; and: *Hybrid Vegetable Development* (ed: P. K. Singh, S. K. Dasgupta, and S. K. Tripathi) Food Products Press, an imprint of The Haworth Press, Inc., 2004, pp. 277-288. Single or multiple copies of this article are available for a fee from The Haworth Document Delivery Service [1-800-HAWORTH. 9:00 a.m. - 5:00 p.m. (EST). E-mail address: docdelivery@haworthpress.com].

INTRODUCTION

Pea (*Pisum sativum* L.) originated in the Near East and Mediterranean regions and has been grown for food and feed since early Neolithic times (Zohary and Hopf, 1973). Evidence from carbonized remains indicate that pea has been cultivated with cereal crops such as wheat and barley since its domestication in 6-7000 B.C. Limited historical evidence makes it difficult to determine the exact location of domestication; however, soon after domestication pea and other legumes spread to other parts of the world with the movement and activities of man (Marx, 1977). The nutritional value and ability to store it for extended periods has contributed to widespread production of the pea.

The genus *Pisum* is a member of the family Papilionaceae tribe Viciae and is composed of two species, *P. sativum* L. and *P. fulvum* Sibth. and Sm. After a significant amount of study and discussion *Pisum sativum* has been further divided to include several subspecies, *P.s.* ssp. *sativum*, *P.s.* ssp. *elatius*, *P.s.* ssp. *humile*, *P.s.* ssp. *arvense*, and *P.s.* ssp. *hortense*. The literature indicates that ssp. *elatius* and ssp. *humile* are the progenitors of the garden pea *P.s.* ssp. *sativum* (Zohary and Hopf 1973). *Pisum s.* ssp. *arvense*, includes the field pea and the 'Austrian' winter pea, both possessing colored flowers and variously pigmented seeds. Subspecies *hortense*, the garden pea, is harvested at the succulent immature stage and consumed fresh or preserved through canning or freezing.

Pea continues to serve an important role in modern agriculture as a nitrogen fixing rotational crop with cereals. The high protein concentration contributes favorably to diets of people in many developing countries. World production of dry pea is estimated at 10.6 million MT from 5.9 million hectares (FAO 2001). In addition, an estimated 7.2 million MT of fresh peas for canning and freezing are produced on approximately 876,000 ha worldwide (FAO 2001). The United States accounts for less than 2% of the world's dry pea crop and only 17% of the fresh pea crop. Canada is the largest producer of dry peas at 25% of world production (FAO 2001). Dry pea production in the US is centered in the Pacific Northwest (PNW) states of Washington and Idaho, however, production in the midwest states of North Dakota and South Dakota, Nebraska and eastern Montana has increased steadily in recent years and promises to exceed production in the PNW. Fresh peas for canning are produced primarily in Washington, Oregon, Idaho, Wisconsin, Minnesota, and New York and several other states also have significant production.

Genetic diversity within the genus *Pisum* has resulted in a myriad of uses for the crop. Dry peas are used for animal feed and for human consumption in soups and processed snacks. Some minor uses of the dry pea are as a cover crop, green manure (plowed prior to seed development) and bird feed. Minor uses of the dry pea include extraction of starch and protein for food processing and industrial applications. The fresh pea is primarily consumed fresh from the garden, canned or as a frozen product.

The most widely produced pea in the world is the dry edible yellow pea although many other types are produced for a variety of uses. Primary uses are in animal rations while a small fraction is used in human diets. It is difficult to estimate the quantity used for human food due to limited reporting from many developing countries, but estimates approach 40-50%. The dry green pea is produced primarily for human consumption in soups and snacks. The fresh green pea can either be frozen, canned or consumed fresh. The Austrian Winter pea is another type of dry pea which when harvested at maturity can be sold for seed or more commonly used as an ingredient in bird feed. Due to its excessive biomass production, the Austrian Winter pea is commonly used as a green manure crop to improve soil tilth and fertility. A rather minor pea type produced for the snack industry in the Orient is the marrowfat pea. It is also harvested dry and is characterized by extremely large seed size and an irregularly shaped, dimpled seed surface.

CROP BIOLOGY

Pea is a cool-season crop best suited to production in the temperate regions. In the warmer areas of the tropics pea is restricted to production in the cooler highlands. Temperatures between 7 and 24°C are suitable for plant growth, but optimum yields are achieved between 13 and 21°C (Duke, 1981). Temperatures above 32°C during flowering often result in abortion and reduced seed set. Genetic diversity within *Pisum* has afforded development of genotypes adapted to a broad production area worldwide and with a wide range of uses. The pea crop is produced primarily under rainfed conditions, but can also be irrigated. Irrigation is most common for the fresh market crop.

Spring sowing of pea in the US begins in late March and is completed by mid-May. England, France, and other European countries begin sowing the pea crop in February. Fields for fresh pea production are sown on a delayed schedule to extend the harvest season and optimize the capacity of each processing plant. Optimum sowing density for the dry pea crop is 90 plants m^{-2} and slightly higher for the fresh pea. Cultivars with sufficient winterhardiness to survive harsh winter conditions have also been developed. Fall-sown peas are sown at densities near 120 plants m^{-2} from mid-September to early October depending on moisture availability.

Pea plant growth begins with imbibition and germination of the seed. Emergence is hypogeal and characterized by the radicle protruding through the testa. Soon after the radicle begins growth the plumule, sometimes called the "plumule hook," begins to elongate upward. The plumule remains bent or hooked during emergence to protect the apical meristem. As the plumule approaches the soil surface and detects light it straightens and begins to turn green from the synthesis of chlorophyll. Once straightened, the first leaf becomes visible.

Numerous soil-borne pathogens attack the seed during germination causing seed rot and damping off resulting in reduced plant populations. The testa surrounding the cotyledons is relatively "leaky" and easily damaged due to its fragility. This results in loss of solutes from the seed attracting soil-borne microorganisms particularly *Pythium* spp. and *Rhizoctonia solani* which then colonize the seed piece and reduce viability. Effective seed treatments are available and widely used to protect the seed from attack by these pathogens. Certain soil limiting microelements are often included with the fungicidal treatments, for example, molybdenum in the PNW.

Growth habit in pea has been studied extensively and the predominant type for modern cultivars is quite different from that of the early cultivars. Early cultivars had long vines and a normal leaf type, i.e., two to three pairs of leaflets at each node (Figure 1). Due to the tendency of pea plants to lodge and make harvest difficult alternate forms were sought. Vine length was reduced by the incorporation of *le* which reduces internode length and produces a dwarf plant type. When the afila (*af*) mutation which converted leaflets to tendrils was discovered in the mid-1960s (Goldenberg, 1965) breeders began combining the tendrilled phenotype with the shortened vine and found that the new plant type remained upright through harvest. The mechanism of lodging resistance involves greater stem strength of the shortened internodes and improved mutual support from intertwining of the tendrils of neighboring plants. The upright canopy provides additional benefits through greater air movement

FIGURE 1. Standard and modern plant types showing reduced vine length and conversion of leaflets to tendrils.

through the canopy which is believed to reduce the incidence of foliar disease, particularly sclerotinia white mold.

The reproductive stage of growth begins with bud formation and the onset of flowering. Flowering begins as early as 20 days after sowing (DAS) and can be as late as 60 DAS. Pod development is divided into three stages, flat pod, pod fill and maturity. Reproductive growth of the dry pea crop is often terminated by the onset of terminal drought approximately 90 DAS. This is a characteristic of production in the rainfed areas. Harvest maturity occurs at 105 to 110 DAS. The fresh pea crop on the other hand is harvested from 50 to 70 DAS and is determined by the physiological stage of the developing pea seed. A tenderometer which measures the firmness of immature pea seeds is used to determine the maturity of the pea and harvest timing.

A primary benefit to inclusion of pea in crop rotations is the ability of the pea plant to host the di-nitrogen fixing bacteria, *Rhizobium leguminosarum*. The symbiotic relationship between the pea plant and *R. leguminosarum* provides the pea plant with most of the nitrogen required for growth. Residual nitrogen from the pea residue and decomposing nodules benefits subsequent cereal crops and reduces the amount of nitrogen fertilizer which must be applied. The nitrogen benefit is one of the main reasons for using of pea as a green manure plow down crop.

There are several insect pests of pea including the pea aphid (*Acyrthosiphon pisum* (Harris), Homoptera: Aphididae), pea leaf weevil (*Sitona lineatus* L., Coleoptera: Curculionidae), pea seed weevil (*Bruchus pisorum* L., Coleoptera: Bruchidae), seedcorn maggot (*Delia platura* (Meigen), Diptera: Anthomyiidae), wireworm, armyworm species (Lepidoptera: Noctuidae), grasshopper, several looper species (Lepidoptera: Noctuidae, subfamily Plusiinae), flower thrips (*Frankliniella* ssp.), pea leaf miner (*Liriomyza huidobrensis* (Blanchard), Diptera: Agromyzidae), and pea moth (*Cydia nigricana* (Fabricius), Lepidotera: Tortricidae) (Table 1). Control of these insects is primarily through insecticidal sprays at appropriate times during the growing season. Many of the insects do not cause economic losses in most years and are not monitored, however, the pea leaf weevil, pea seed weevil and aphids frequently cause economic losses and are monitored and sprayed routinely. The pea aphid can cause serious yield loss if allowed to increase to large enough numbers and they vector many viruses which cause disease and depending on severity can devastate the crop.

The primary virus diseases of pea include pea seed borne mosaic virus, pea enation mosaic virus, bean (pea) leaf roll virus, pea streak virus, red clover vein mosaic virus, and several others. Virus diseases of lesser importance include bean yellow mosaic virus, faba bean necrotic yellows virus and pea early browning virus (Kraft and Pfleger, 2001). The virus diseases commonly cause chlorosis and distortion of foliage, stunted growth, and plant death. Plants

TABLE 1. Insect pests and disease pests of pea.

Insects	Virus	Soil-Borne Fungi	Foliar Fungi	Bacteria
Pea aphid	Pea enation mosaic virus	Fusarium wilt	Downy mildew	Bacterial blight
Pea leaf weevil	Pea seedborne mosaic virus	Fusarium root rot	Powdery mildew	Brown spot
Pea seed weevil	Bean leaf roll virus	Aphanomyces root rot	Ascochyta blight complex	
Pea leaf miner	Pea streak virus	Pythium damping off	Botrytis gray mold	
Thrips	Red clover vein mosaic virus	Rhizoctonia root rot		
Seedcorn maggot	Bean yellow mosaic virus	Thielaviopsis root rot		
Wireworm	Faba bean necrotic yellows virus			
Armyworm	Pea early browning virus			
Grasshopper				
Loopers				
Pea moth				

which do survive to reproduce often have reduced yield and the seed are often misshapen, small and have poor vigor. Genetic resistance to pea seed borne mosaic virus and pea enation mosaic virus has been identified and bred into many cultivars currently grown. Resistance to bean leaf roll virus and red clover vein mosaic virus is available in the germplasm, but these diseases are not considered a high priority in many production regions.

Numerous fungal pathogens, both soil borne and foliar, attack peas (Table 1). The soil-borne pathogens of economic importance in the U.S. include Fusarium wilt caused by *Fusarium oxysporum* Schlechtend.:Fr. f. sp. *pisi* (J.C. Hall) W.C. Snyder and H.N. Hansen races 1, 2, 5, and 6; Aphanomyces root rot caused by *Aphanomyces euteiches* Drechs., Fusarium root rot caused by *Fusarium solani* (Mart.) Sacc. f.sp. *pisi* (F.R. Jones) W. C. Snyder and H.N. Hansen, *Pythium* spp., Rhizoctonia and Thielaviopsis root rot caused by *Thielaviopsis basicola* (Berk. and Broome) Ferraris. Genetic resistance is available in pea for Fusarium wilt. A single gene for each has been identified and when possible incorporated into new cultivars. Most cultivars are resistant to race 1 and some are resistant to race 2. Resistance to races 5 and 6 is only needed in western Washington and British Columbia, Canada. Race 2 is becoming more problematic in all pea production regions of the U.S. PNW and elsewhere.

Tolerance to Aphanomyces and Fusarium root rot have been reported to be quantitatively inherited and consequently difficult to select for. Molecular markers for the resistance genes have been identified for each of these diseases but have not been adopted for screening segregating populations in breeding programs (C.J. Coyne, personal communication; Cargnoni, Weeden and Gritton, 1994).

Seedling diseases of pea have been successfully controlled through seed treatment with captan and metalaxyl. *Pythium* ssp. primarily attack the seed prior to emergence resulting in the seed becoming mushy and discolored. Attack after germination results in the plumule and radicle appearing water soaked and translucent. Rhizoctonia infects the hypocotyl and epicotyl of young seedlings. Water-soaked lesions first appear and coalesce eventually killing the plant. Seed rot and seedling damping off reduce plant stand and may reduce yield. Cultural control practices include reduced soil compaction, long crop rotation, avoiding poorly drained fields and clean tillage.

Fungal pathogens affecting the foliar portion of the pea plant include downy mildew (caused by *Peronospora viciae* (Berk.) Casp.), *Ascochyta* spp., and botrytis gray mold (caused by *Botrytis cinerea* DC.) in the warm and humid production areas and powdery mildew (caused by *Erysiphe pisi*) which is adapted to the warmer and drier areas. Genetic resistance is available for downy mildew and powdery mildew and many resistant cultivars are available. Powdery mildew can severely reduce crop yield and quality by coating the foliage reducing photosynthetic potential and preventing the plant from drying down at maturity. Seeds within the pod can become blemished if pod infection is severe and mycelium penetrate the pod wall. Severe disease development results in excessive spore production which can be a health hazard to workers and can cause combine fires during harvest. Genetic resistance to the *Ascochyta* spp. is complicated and is likely controlled by quantitative trait loci.

HETEROSIS

Heterosis has been recognized in pea since the early studies of Mendel (Mendel, 1866). More recent reports on heterosis in pea have been given by Krarup and Davis (1970), Gritton (1975), and Sarawat et al. (1994). Heterosis in predominantly self pollinating species is of interest, but difficult to apply in practice without sufficient means to achieve cross pollination. Studies of F_1 hybrids in pea have been conducted with the intent of capitalizing on heterosis in commercial seed production and for the potential of using data from F_1 hybrids to identify superior populations for subsequent selection and cultivar development. To date no known hybrid cultivars have been developed or commercialized.

The primary traits which have been evaluated for heterotic response are yield and its components, days to flower, duration of flowering period and plant height. Very similar relative heterotic values for pods per plant (32, 31 and 38%), seeds per pod (10, 8 and −1.4%), and seed weight (7, 1 and −2.1%) were obtained from each of the three studies, Krarup and Davis (1970); Gritton (1975); and Sarawat et al. (1994), respectively. Heterosis for seed yield in pea compared to the mid-parent value ranged from 30 to 56% depending on the growing environment and was greater in poor environments compared to the better environments (Sarawat et al., 1994). This effect is primarily due to increases in the number of pods per plant (Sarawat et al., 1994; Gritton, 1975).

In order for hybrid peas to be useful they must out perform the best pure line and not simply exceed the mid-parent value or even the highest parent. All the F_1s studied by Gritton (1975) were superior to the best parent from the cross which, although his results are only applicable to the specific lines studied, may indicate that hybrids superior to the best inbred lines could be possible if elite material were used to generate the hybrid.

General and specific combining ability are both important in the expression of heterosis (Lejeunehenaut et al., 1992; Sarawat et al., 1994). Gritton (1975) concluded that although high yielding inbred lines could be produced based on the significance of additive effects, it would be difficult to develop pure lines equal to F_1s due to the importance of non-additive effects in the F_1. Inbreeding depression beginning in the F_2 was reported in all studies. It can be concluded that heterosis is present in pea and, with an economical method of generating F_1 hybrids, the beneficial effects could be exploited in field production systems.

POLLINATION CONTROL MECHANISMS/ HYBRID SEED PRODUCTION

The pea flower is cleistogamous resulting in pollen shed occurring twenty-four hours prior to the flower opening and nearly complete self-pollination. Therefore, hybrid genetic lines must be generated via pollen sterility mechanisms. Pollen (male) sterility in pea had not been reported to occur naturally until Singh and Singh (1995) characterized a spontaneous mutant with a single recessive gene conferring male sterility. Prior to this discovery, several male sterile mutants were generated using a variety of mutagenesis treatments (Klein, 1969; Nirmala and Kaul, 1991; Myers and Gritton, 1988). All male sterile genes which have been reported act as recessive traits and nearly all the mutants have full female fertility (Kaul, 1988).

Myers and Gritton (1988) studied many of the available male sterile mutants and determined through allelism tests that nine distinct mutants existed (Table 2). Others may have existed at one time, but have been lost (Myers and Gritton, 1988).

Pollen grain formation is complicated and requires the precise functioning of numerous genes. Nirmala and Kaul (1991) divided microsporogenesis into three stages: pre-meiosis, meiosis, and post-meiosis. Pre-meiosis involves cell separation and DNA synthesis. Post-meiosis begins once the cells separate from the pollen mother cell, the callose wall is formed and the cells prepare to undergo mitosis. Many of the male sterile (ms) genes which have been reported act during the post-meiotic stage. Sterility of the male gametophyte occurs through a variety of mechanisms, beginning during pre-meiosis through post-meiotic events, however, most cases of sterility involve post-meiotic events (Kaul, 1988; Kaul and Nirmala, 1989) (Table 3).

MAINTENANCE OF INBRED/PURE LINES

Maintenance of inbred and pure lines is accomplished using bulk increases or regenerating breeder seed through single plant selection. Bulk increases should be grown in fields free from disease, with a known crop history and iso-

TABLE 2. Origin, mutagen treatment, and source of nine non-allelic male sterility genes in *Pisum*. Adapted from Myers and Gritton (1988).

Gene Symbol	Originating Line	Mutagen Treatment	Source
ms-2	'Dippes gelbe Viktoria'	X-ray	W.Gottschalk
ms-3	'Dippes gelbe Viktoria'	X-ray	W.Gottschalk
ms-5	CSC8617 × 'New Season'*af*	EMS	E.T. Gritton
ms-6	CSC8617 × 'New Season'*af* and New Line Early Perfection' × 'New Season'*af*	EMS	E.T. Gritton
ms-7	CSSC8221 × 'New Season'*af*	EMS	E.T. Gritton
ms-8	'Juneau'	Sodium azide	F.J. Muehlbauer
ms-9	M392 × 'Fenn'	EMS	D. Auld
ms-10	'Juneau'	EMS	F.J. Muehlbauer
ms-11	'Juneau'	EMS	F.J. Muehlbauer

TABLE 3. Gene action of several male sterile mutants in pea. (Modified from Kaul, 1988.)

Mutant Identifier	Gene Action
ms	Archesporial tissue formation inhibited
msk_1	During sporogenous tissue formation
msk_2	During leptotene
38B, 69	During pachytene
msk_3, msk_4, 195B	During late prophase-I
71A	During All
98A	After All
67, 78, 395	During tetrad formation
33C, msk_5, msk_6, ms_5, ms_7, ms_8, ms_9, ms_{11}	After microspore liberation
503	After exine formation

lated by modest distances from other pea crops. Single plant selection requires that 100 to 200 individual plants be selected from field plots or grown in the greenhouse. In the case of greenhouse grown plants each plant is inspected and off-types discarded. Progeny rows from each single plant are grown under field conditions, inspected and off-types discarded. At this point all uniform plots can be bulked for increase and distribution or grown an additional generation in larger plots to confirm uniformity.

Production in arid regions is common due to reduced incidence of disease. Routine inspection of the fields must be done to identify disease and to rogue all off-type plants. Sowing and harvest equipment should be thoroughly cleaned to remove all residual seed. Greater purity of the seed can be accomplished using an electric eye to remove off-type seed. A high level of seed purity can be maintained if diligence and care are practiced at each step during the increase.

SEED PRODUCTION

Seed production of pea is most successful in more arid regions due to reduced incidence of disease. The predominantly self-pollinating habit of pea allows relatively short isolation distances of 30 m or less to minimize out crossing. However, where possible greater distances should be used. Great care must be taken to ensure high quality of the harvested crop. Moisture content less that 13% is optimum, but may be up to 20% if drying facilities are available. In extremely dry conditions where seed moisture content may reach 9-10% or when harvest is delayed beyond the optimum, extreme care must be taken to adjust the combine to minimize mechanical damage to the seed.

Special care is taken to verify the absence of pea seed borne mosaic virus as well as other foliar fungi which may be transmitted via the seed. Of particular interest are the causal organisms of the Ascochyta blight complex, *Ascochyta pisi*, *Mycosphaerella pinoides*, and *Phoma medicaginis*. Southern Idaho and the Palouse region of the US PNW are the primary areas of seed production for the fresh pea (wrinkled) industry while seed for smooth dry edible peas is produced in the PNW as well as the Midwest states.

Pea seed can be stored for one to two years at ambient temperature without serious loss of viability. Long-term storage, > 2 years is best at low humidity and temperatures near 5-7°C. Storage conditions with high humidity and high temperatures reduce seed viability.

CONCLUSION

The garden pea is an important agricultural crop worldwide and serves as a protein source in many developing countries especially those where meat products are not available or are not consumed due to social or religious issues. A wide range of genetic diversity in the genus *Pisum* has allowed the crop to be adapted to a wide production area spanning from 60 N to 60 S latitude and representing elevations from sea level to 2000 m above sea level and average temperatures from 16 to 35°C. The genetic diversity of pea is also represented in the types of pea produced and their varied uses.

Benefits to agricultural production are numerous but center around the synergistic effects of rotation with cereal crops. Examples include improved soil fertility primarily through nitrogen fixation in association with *Rhizobium leguminosarum*, allowing control of grassy weeds and disrupting cereal disease cycles. An additional benefit is that the pea crop appears to be well adapted to increasingly popular direct seeding systems. Stubble from the previous crop provides support to the pea crop allowing greater ease of harvest. In addition, the small amount of residue produced from the pea crop helps offset the large amount of residue produced by the cereals.

Production of pea is through the use of pure lines due to the predominant self-pollinating habit. Male sterile mutants have been created through mutagenesis and found spontaneously in nature, however their application in creating hybrids for field production has not been adopted despite the presence of significant hybrid vigor. This is primarily due to difficulty in achieving cross pollinations. Until significant genetic improvement using existing breeding procedures and production systems is no longer possible hybrid seed production is not likely to be considered.

Seed production is accomplished through contracts with selected growers at sufficient, but modest isolation distances. Disease-free areas in relatively dry and arid environments are most common for increasing seed stocks.

REFERENCES

Cargnoni, T.L., N.F. Weeden and E.T. Gritton. 1994. A DNA marker correlated with tolerance to Aphanomyces root rot is tightly to *Er-1*. Pisum Genetics, 26:11-12.

Duke, J.A. 1981. Handbook of legumes of world economic importance. Plenum Press, New York, New York. pp. 200-205.

FAO, 2001. http://apps.fao.org.

Goldenberg, F.B. 1965. Afila, a new mutation in pea (*Pisum sativum* L.). Boletin Genetica, Instituto de Fitotecnia Castelar, Argentina. pp. 27-28.

Gritton, E.T. 1975. Heterosis and combining ability in a diallel cross of peas. Crop Science, 15:453-457.

Kaul, M.L.H. 1988. Male sterility in higher plants. Springer Verlag, Germany, p. 1005.

Kaul, M.L.H. and C. Nirmala. Cytogenetical basis of male sterility. Plant science research in India. pp. 251-264.

Klein, H.D. 1969. Male sterility in *Pisum*. The Nucleus, XII (2):167-172.

Kraft, J.M. and F.L. Pfleger. 2001. Compendium of pea diseases and pests. 2nd Edition. APS Press, St. Paul, Minnesota, US. pp. 67.

Krarup, A. and D.W. Davis. 1970. Inheritance of seed yield and its components in a six-parent diallel cross in peas. Journal of the American Society of Horticultural Science, 95:795-797.

Lejeunehenaut, I., G. Fouilloux, M.J. Ambrose, V. Dumoulin and G. Eteve. 1992. Analysis of a 5-parent half diallel in dried pea (*Pisum sativum* L.) I. Seed yield heterosis. Agronomie, 12(7): 545-550.

Marx, G.A. 1977. Classification, genetics and breeding. *In:* J.F. Sutcliffe and J.S. Pate (Eds.) Physiology of the garden pea. Academic Press, New York, NewYork. pp. 21-44.

Mendel, G. 1966. Experiments in hybridization. [Translation by Royal Horticultural Society of London, Harvard University Press, Cambridge, 1963] Verh. Naturforsch Ver. Brunn 4.

Myers, J.R. and E.T. Gritton. 1988. Genetic male sterility in the pea (*Pisum sativum* L.): I. Inheritance, allelism and linkage. Euphytica, 38:165-174.

Nirmala, C. and M.L.H. Kaul. 1991. Male sterility in pea I. Genes disrupting pre- and post-meiosis. Cytologia, 56:587-595.

Sarawat, P., F.L. Stoddard, D.R. Marshall and S.M. Ali. 1994. Heterosis for yield and related characters in pea. Euphytica, 80:39-48.

Singh, B.B. and D.P. Singh. 1995. Inheritance of spontaneous male sterility in peas. Theoretical and Applied Genetics, 90:63-64.

Zohary, D. and M. Hopf. 1973. Domestication of pulses in the old world. Science, 182:887-894.

Advances in Watermelon Breeding

Tarek Kapiel
Bill Rhodes
Fenny Dane
Xingping Zhang

SUMMARY. The continuous discovery of best possible combiners and their outcome has replaced the superior varieties by hybrids in watermelon. Nowadays several types of watermelons are being marketed–some are having red flesh, some with yellow, some with white flesh. Even with the differences in its shape and size, some are round, some are oval, some are square. Some are with seeds some are seedless. This all happened because of the efforts of crazy mind and fast changing market needs. The introduction of new watermelon genes and marker genes have equipped us to deal with pest and pathogens. Beyond that, using the genes of a rootstock of a related cucurbit, we can combat soil pests as well as adverse environmental conditions. Use of male sterility can reduce the cost of hybrid seed. Seed production requires good cultural practices and timely harvest. Triploid hybrid seed production varies considerably. *[Article copies available for a fee from The Haworth Document Delivery Service: 1-800-HAWORTH. E-mail address: <docdelivery@haworthpress. com> Website: <http://www.HaworthPress.com> © 2004 by The Haworth Press, Inc. All rights reserved.]*

Tarek Kapiel and Bill Rhodes are affiliated with the Clemson University, Department of Horticulture, Clemson, SC 29634 USA.

Fenny Dane is affiliated with the Department of Horticulture, Auburn University, Auburn, AL 36849 USA.

Xingping Zhang is affiliated with Syngenta Seeds, Inc., Woodland, CA 95695 USA.

[Haworth co-indexing entry note]: "Advances in Watermelon Breeding." Kapiel, Tarek et al. Co-published simultaneously in *Journal of New Seeds* (Food Products Press, an imprint of The Haworth Press, Inc.) Vol. 6, No. 4, 2004, pp. 289-322; and: *Hybrid Vegetable Development* (ed: P. K. Singh, S. K. Dasgupta, and S. K. Tripathi) Food Products Press, an imprint of The Haworth Press, Inc., 2004, pp. 289-322. Single or multiple copies of this article are available for a fee from The Haworth Document Delivery Service [1-800-HAWORTH, 9:00 a.m. - 5:00 p.m. (EST). E-mail address: docdelivery@haworthpress.com].

http://www.haworthpress.com/web/JNS
© 2004 by The Haworth Press, Inc. All rights reserved.
Digital Object Identifier: 10.1300/J153v06n04_01

KEYWORDS. Watermelon breeding, grafting, diploid hybrids, triploid hybrids, controlled pollination, male sterility, hybrid seed production

INTRODUCTION

The watermelon (*Citrullus lanatus*) has undergone several facelifts as well as some authentic changes during the past 20 years. The first change came when the open pollinated (OP) varieties gave way to hybrid varieties. Even though OP varieties were inbred varieties maintained by isolation, some variability in fruit size and shape did appear from time to time. As cultural practices such as plastic mulch, closer spacing, drip irrigation and more precise fertilization maximized the environment for watermelon production, it became clear that the hybrids were yielding more than the inbreds. The grower could afford to pay more for seed, and seed companies could afford to increase research efforts. However, many of the first modifications to the crop were largely cosmetic.

More emphasis was placed on marketing and postharvest handling. The popularity of watermelon in the U.S. waned sharply in the 80s, and producers, shippers and supermarkets realized the necessity of carefully harvesting mature fruit and taking care to handle, ship, and display the watermelon properly. A smaller fruit became popular for smaller families, and supermarkets began to sell slices of watermelon. Then the so-called triploid "seedless" watermelon (Figure 1), long dormant in the western market, emerged as a real competitor with seeded watermelon varieties.

The outbreaks of the bacterial disease cryptically known as "blotch," although very expensive to growers and the industry, resulted in the improvement of sanitation in seed production, and bacterial detection on contaminated seed and plants. We also discovered that triploids were more resistant to *Acidovorax avenae* than diploid varieties. Unfortunately, liability costs associated with watermelon seed drove up their cost.

Triploid seed production for the seedless watermelon has improved. Good triploid seed lots give more than 90% germination under carefully controlled environments. Seed companies start to prime triploid watermelon seed to further improve germination and uniformity. The cost of triploid seed production is more and the retail cost of triploid seed remains high. The introduction of marked male sterility into seed production systems will reduce the cost of triploid seed production (Zhang and Rhodes, 2000).

The continued discovery and introduction of new watermelon genes and marker genes have increased our ability to deal with pests and pathogens. Genome information for the watermelon has increased exponentially in the last decade. Beyond that, we can use the genes of a rootstock of a related cucurbit

FIGURE 1. Triploid or seedless watermelon.

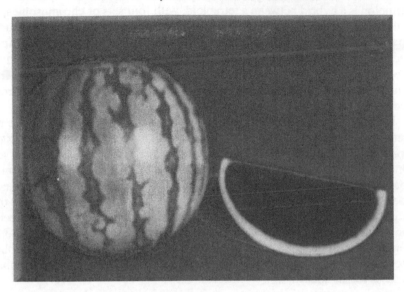

to combat soil pests and pathogens as well as adverse environmental conditions. We can choose a superior genotype and clone it for large-scale production without the problems of seed production.

Pollination is an important problem to address, especially for triploid production. Because cucurbits require so much pollination, this family has already served as a warning that pollination may soon become a limiting factor in food production.

CROP BIOLOGY

Germination

Germination of the traditional OP seed was of little concern. Several inexpensive seed were planted per hill to insure a good stand. If soil temperatures were too cold for good germination, hills or entire fields were replanted. However, the adoption of hybrid seed, several times more expensive than OP seed, and especially triploid seed, encouraged the use of transplants and the development of transplant houses. Transplant survival at lower temperatures than seed could germinate insured that the crop was in place as early as possible, and earlier harvest dates were realized.

The triploid seed posed yet another problem. The triploid ("seedless") watermelon is a hybrid between a tetraploid, with four sets of chromosomes, and a diploid, with only two sets. The seed has a small triploid embryo, and air spaces inside a thick, hard tetraploid seed coat. Only a few years ago, triploid seed were associated with low seed yield, poor germination, poor seedling establishment and high seed cost. Nonetheless, high yields of pest resistant fruits with tough rinds, high sugar, and crisp, brightly colored flesh virtually free of seed were reasons enough to work toward improvement of the seed.

First and foremost, triploid seed production in a low humidity climate with long hot days and cool nights insures seed with more robust embryos. Extraction and cleaning of fresh triploid seed is more critical than with diploid seed. Unlike diploid seed, the more perishable triploid seed, with the extra air spaces, are usually not fermented to remove bits of flesh. Seed are washed as soon as they are extracted from flesh. It is critical that the seed are dried thoroughly and quickly. Dryers with heater and force air flow are highly recommended for drying triploid watermelon seed. Under less than optimal storage conditions, triploid seed do not have the longevity of diploid seed. Curiously, the germination of older triploid seed can be enhanced by a simple tumbling process for an hour or so.

Moving the germination process inside germinating chambers or greenhouses helped a great deal but increased the cost of plant establishment in the field. Higher soil temperature and more stable moisture level than usually encountered in the field were necessary to achieve optimal germination. Germination of triploid seed and survival of transplants was increased from 70 to 80% to more than 90% under optimal conditions.

Sachs (1977) reported that diploid watermelon seed do not germinate below 15°C. He reported that diploid watermelon seeds could be primed for low-temperature germination by pretreatment at 30°C or 35°C by imbibing the seeds in salt solutions prior to sowing. The best treatment was 2-3% KNO_3 for 6 days, and primed seed could be dry stored for 20 weeks. In a related species, *Cucumis melo*, Edelstein et al. (1995) noted that varieties differ in their ability to germinate at low temperature, and this difference in germination is due to seed coat structure and oxygen availability. Thorton's (1968) work with watermelon supported this view. Rhodes et al. (1992) noted improved germination in certain watermelon types after 8 years storage, but the hypothesis that aging increased oxygen availability through a modified seed coat, enhancing germination, was not tested. Thorton discovered cutin deposits in the nucellar layers (inner membrane) of diploid watermelon seed and related that to dormancy.

The germination of triploid seed requires precisely control of the temperature and moisture. Sowing triploid seed in moist (but not wet) soilless soil mixture and placing the seedling flats in a humid germination room with temperature maintained at 30-32°C for 48 to 72 hrs allow triploid seed to germinate quickly and evenly. Seedling flats are then moved to greenhouse with temperature maintained at 25°C till soil emergence. Only light water is applied before seedlings emergence. The seed coat often remains attached to the cotyledons after soil emergence. The removal of the tetraploid seed coat that often sticks to the triploid cotyledons is necessary to insure normal growth of the young seedling in need of ample high quality light. Apparently, horizontal seed orientation at planting (Maynard, 1989) and vigorous washing immediately after germination helped to remove the seed coats. Planting depth at about 2 cm is also critical to reduce the seed coats problem and maintain stable moisture level around the seed during germination. The cotyledons are often bizarre, and the seedling develops normally only after true leaves emerge.

Scarification of the seed coat near the cotyledon end of the embryo to allow the entry of oxygen significantly improves the rate and percentage germination. Interestingly, triploid seed germination can be improved by tumbling, which does not actually scarify the seed but may disrupt the membrane that surrounds the embryo. Because the triploid seed germinates well lying flat in the germination medium, a mechanical protocol which placed seed flat in the same orientation in a germination tray, followed by simultaneous penetration of all the seed coats at the cotyledon end would serve to prepare all the seed in the tray for germination in the soil or soilless medium.

Although scarification of triploid seed is a routine practice in Japan and China, no reports on predisposing triploid watermelon seed to germination have been found. Because the threat of watermelon fruit blotch has discouraged the greenhouse industry from growing triploid transplants and the value of each triploid seed is so great. We studied pregermination enhancement of triploid watermelon seed. We compared germination (radicle emergence) within the same seed lots that were tumbled with various scarifying agents with germination of seed that were not treated at all. We concluded that tumbling treatments did increase the rate of germination. We also noted that 4-year-old seed that did not germinate well responded much more to tumbling treatments than did 2-year-old seed. If the tips of the 4-year-old seed were manually removed with needle nose pliers, these seed germinated as well as 2-year-old seed. By orienting each seed where a reciprocating knife can penetrate the cotyledon end of the triploid seed, it should be possible to provide oxygen and a systemic inducer to the embryo just before seeding in seedling flats. Seed companies that are already using seed scarification treatments on triploid seed may be able to further enhance these treatments. The grower may be able to direct seed in the field after treatment with a savings of $100 or more per acre due to increased germination. If older seed can now be planted with some assurance of germination, seed companies should be able to recover more of their production costs.

We also enhanced the germination of triploid watermelon seed (Sterling F1 triploid hybrid obtained from Hollar Seeds Co.) by clipping the seed. We used common fingernail clippers to remove a portion of the seed coat above and opposite the radical end, resulting in an opening ~ 7 × 5 mm prior to germination. Seeds were arranged on the upper part of the germinating papers, rolled and placed in a vertical position in water at 30 + 0.1°C in a controlled incubator (model 815 Precision Scientific low temperature incubator, Precision Scientific, Inc.). Our results indicated that the percentage of seed germination using the previously described method was significantly higher (97%) than the control (71%) (Figure 2).

It seems likely that germination enhancement treatments could also be applied simultaneously with treatments to eliminate seed-borne diseases. For example, Hopkins (1994) found that 1% HCl added to freshly extracted watermelon seed almost eliminated watermelon fruit blotch. However, HCl treatment has adverse effects on the germination of triploid seeds. A commercial food disinfectant Tsunami 100 can significantly reduce the watermelon seed contamination risk by treating the freshly harvested watermelon seed before drying. This disinfectant has no effects on germination of both triploid and diploid watermelon seeds at proper concentration and treating periods (Lovic, 2001).

FIGURE 2

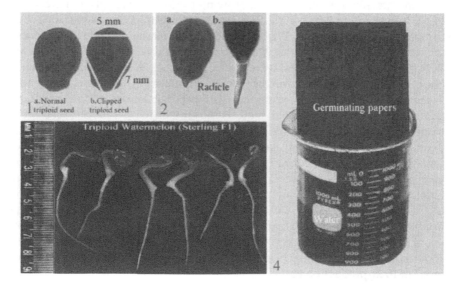

GRAFTING

Introduction. Growing grafted vegetables was first launched in Japan and Korea in the late 1920s by grafting watermelons to gourd rootstocks (Lee, 1994).

Currently, grafting is routine in Korea, Japan, and some Asian and European countries where land use is very intensive and the farming area is small. In Japan and Korea, where land use is intensive and the availability of new farmland is scarce, almost 95 percent of the watermelon crop is grafted before being transplanted to the field or greenhouse. At present, most of the watermelons [*Citrullus lanatus* (Thunb.) Matsum. & Nakai] in Korea and Japan are grafted before being transplanted to the field or greenhouse. In 1992, Japan cultivated almost 24,000 hectares of grafted watermelon seedlings in the field, and over 3,000 hectares in the greenhouse (Lee, 1994).

Although grafted plants are widely used in the United States for a variety of orchard and vineyard crops (e.g., apples, grapes), growing grafted vegetables is seldom practiced in the United States or in other western countries where land use is not intensive.

Watermelons are grafted with either gourd stocks (*Lagernaria siceraria* or *Cucurbita ficifolia*) or the interspecific hybrid *C. maxima* × *C. moschata*. In humid southeast Asia, citron is found to be a better rootstock for watermelon. Watermelon grafts can produce more, larger fruit for a longer season. The fruit size of watermelons grafted to rootstocks having vigorous root systems is often significantly increased compared to fruit from intact plants, and many growers are practicing grafting mainly for this reason. The benefits also include resistance to soil-borne diseases, increased growth at low temperatures, increased tolerance to salt or wet soil, increase in nutrient uptake, and extended duration of economical harvest time (Lee, 1994).

Scion and Rootstocks Being Grafted. Scions derived from many watermelon varieties as well as from well-established micropropagation systems were grafted on different rootstocks. In Japan, *in vitro*-produced shoot tips were used as scion material for grafted seedlings, but domestic demand of triploid seedling was not great enough over a short market period to justify *in vitro* propagation.

If tissue culture scions are uniform enough for the type of grafting robots described by Kurata (1994), grafted seedlings may be competitive in larger markets, such as the United States. Grafted tetraploid (parent) plants can produce manyfold the number of triploid seed produced by nongrafted plants, thus meeting the demands of the growers for more seed at lower prices. (Triploid seed currently cost several hundred dollars a pound.)

Watermelons are commonly grafted to various rootstocks, especially for cultivation in greenhouses or plastic houses; numerous rootstocks also have been developed. Watermelons are commonly grafted to gourd [*Lagemaria siceraria* (Mol.) Stand1.] or to interspecific hybrids [*C. maxima* Duch. × *C. moschata* (Duch.) Duch. ex Poir.]. Wild citron watermelons are found to be a better rootstock for commercial watermelons in hot and humid southeastern Asia. However, many rootstocks having distinctive characteristics are available today, and growers select the rootstocks they think are the most suitable for their growing season, cultivation methods (field or greenhouses), soil environments, and the type of crops and cultivars.

Several new rootstocks are being developed. For example, bur-cucumber (*Sicyos angulatus* L.) collected near Andong, Korea, showed good compatibility with cucumbers and watermelons for early summer growth and good resistance to nematodes (Lee, 1994) (Table 1).

Grafting Methods. Various grafting methods have been developed and growers must choose their favorite methods based on experiences and preferences. Grafting cucurbitaceous crops is commonly done when scion and rootstock seedlings are young, i.e., before the outgrowth of the first true leaf between the cotyledons. Grafting methods vary considerably with the type of scion and rootstock being grafted, and the sowing time for scion and stock seeds vary with grafting method. The "hole insertion grafting" would be convenient for watermelons because of their small seedling size compared to the size of stock seedlings, such as gourd and squash (Lee, 1994). However, cut grafting is popular for watermelon in Korea (Oda et al., 1997) (Figure 3).

In the past, the growers themselves routinely carried out grafting, but it is now rapidly shifting to cooperative operations owing to the efficiency, ease of postgraft care, and recent expansion of the commercial seedling industry (such as sales of plug-grown seedlings). Even though the single-edged razor blade is still the most widely used grafting knife among farmers today, many other de-

TABLE 1. Rootstocks, major grafting methods, and purpose of grafting watermelon plants.

Popular rootstock species[z]	Grafting method[y]	Purpose[x]
Gourd (*Lagemaria siceraria* var. *hispida*)	1	1, 2
Interspecific hybrids[w]	1, 2	1, 2, 3
Wax gourd (*Benincasa hispida* Cogn.)	1, 3	1, 2
Pumpkin (*Cucurbita pepo* L.)	2, 3	1, 2, 3
Squash (*Cucurbita moschata* L.)	1, 2	1, 2, 3
Sicyos angulatus	2	4

[z]The names of the numerous varieties within the same species were not listed.
[y]Grafting methods: 1 = hole insertion; 2 = tongue approach; 3 = cleft grafting.
[x]Purpose of grafting: 1 = fusarium wilt control; 2 = growth promotion; 3 = low-temperature tolerance; 4 = nematode resistance.
[w]Many interspecific hybrids are commonly obtained by fertilized ovule culture *in vitro*.
Adapted from Lee, 1994.

FIGURE 3. Cut grafting method; (a) Watermelon scion is tapered from the hypocotyl; (b) Rootstock is prepared to make a hole by removing leaf and buds; (c) Making a hole in rootstock's hypocotyl; (d) Scion is inserted into the hole; (e) Complete graft with adventitious roots one week after transplanting.

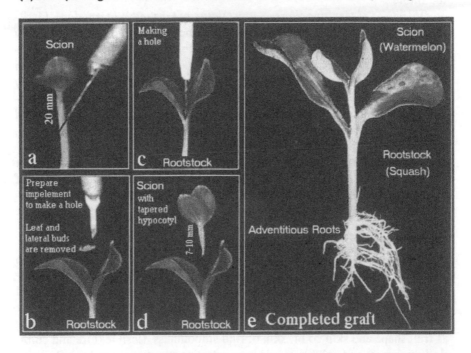

vices, such as specifically designed knives, clips, tubes, or glue, have now been developed for easier grafting and postgraft care (Lee, 1994).

In Japan, Korea, and Taiwan, watermelon buds are grafted from young watermelon plants onto young rootstocks of cucurbit relatives (e.g., *Lagenaria siceraria*) by sophisticated grafting technology. Hundreds of millions of plants are produced for the industry in these countries. Semi-automatic grafting machines and/or fully automatic grafting machines using robots were developed (Itagi, 1992; Ito, 1992; Kobayashi, 1991). Kurata (1994) described four methods of robot grafting in detail. By using the most sophisticated machine, grafting efficiency could be significantly increased from the present 150 seedlings/hour per expert. The labor required for intensive postgraft care, mostly 7 to 10 days of careful management, could be markedly reduced by using specifically designed conditioning chambers (Lee, 1994).

Recently, the tube grafting method has been developed for plugs (Oda et al., 1997). This is popular for the manual grafting of watermelon plants. Grafting robots and healing chambers have been developed, and are used in nurseries producing grafted plugs. Because grafting gives increased disease tolerance and vigor to crops, it will be useful in the low-input sustainable horticulture of the future.

Objectives of Grafting Watermelon (Value of Grafted Watermelon). Grafting's early purpose was to avoid or reduce the soilborne disease caused by *Fusarium oxysporum*, but the reasons for grafting, as well as the kinds of vegetables grafted, have increased dramatically. The purpose of grafting has been greatly expanded from reducing infection by soilborne diseases caused by pathogens such as *Fusarium oxysporum* Schlect. to increasing low-temperature and salt and wet-soil tolerance, enhancing water and nutrient uptake, and increasing plant vigor and extending the duration of economical harvest time, among other purposes (Lee, 1994).

With regard to reducing the need for soil fumigation, the primary use of grafting will be to increase disease and nematode resistance through the use of selected rootstock with known resistance to soilborne pests. Root diseases particularly wilt; devastate intensive watermelon cropping regions worldwide. Watermelon grafted to rootstocks of cucurbit relatives are resistant to root diseases.

Watermelon grafts can produce more, larger fruit for a longer season. In Egypt we have observed that the fruit size of watermelons grafted to rootstocks having vigorous root systems is often significantly increased compared to fruit from intact plants, and many growers are practicing grafting mainly for this reason. However, in grafted cucumbers, other quality characteristics, such as fruit shape and skin color, skin or rind smoothness, flesh texture and color, rind thickness, and soluble solids concentration reputedly can be influenced by rootstock (Choi et al., 1992; Kang et al., 1992). Even though these are usually regarded as cultivar-specific hereditary characteristics, they can apparently be greatly influenced by the rootstock (Kang et al., 1992). The effects of rootstocks on fruit quality in cucumber are reputedly often detrimental (Choi et al., 1992).

Triploid watermelon seed does not germinate in cold, wet soils as well as diploids; germination is slower due to reduced embryo size and thicker seed coat; fissures on the seed coat provide safe harbor for fungal spores; and triploid fruit set is later than most diploid cultivars. Because of these problems producers often transplant rather than direct-seed seedless watermelons, and grafting may provide an alternative solution to overcome these problems to obtain a successful protocol for mass propagation of seedless watermelons. Also, a successful protocol for clonal propagation of triploid watermelon combined both tissue culture and grafting technique (Wang et al., 1980; Gao et al., 1983).

In Japan the machine-grafted watermelons, if intensively managed, produce from last frost in the spring to first frost in the fall. Contrast this to plants in the southeastern U.S., which are dead in July. In Egypt, hand grafts are widely used (personal observation).

Conclusions. Grafting currently is used in commercial agricultural production to achieve higher yielding field and greenhouse crops, repair damaged sections of a plant, increase temperature or salinity tolerance, produce higher yielding varieties, and extend the duration of economical harvest time. Research is being conducted to identify disease resistant germplasm to be used as rootstocks in grafting. Breeding multi-purpose rootstocks and developing efficient grafting machines and techniques will undoubtedly encourage increased use of grafted seedlings. Grafted tetraploid (parent) plants can produce manifold the number of triploid seed from nongrafted plants, thus meeting the demands of the growers for more seed at lower prices. Large-scale commercial production of vegetable seedlings is expanding rapidly in many developed countries, and this will lead to an increased commercial supply and use of grafted vegetable seedlings throughout the world (Lee, 1994). Grafting is extremely laborious and time-consuming, and growers are trying to reduce the labor input required. Attempts have been made to mechanize grafting operations. As the grafting operations and the healing of grafted plants become easier, grafting vegetable crops may become popular all over the world. Since plants gain disease tolerance and vigor by grafting, grafting of vegetables may be useful in the low-input, sustainable horticulture of the future (Oda, 1999).

INDUCED SYSTEMIC RESISTANCE

In some plants, researchers have demonstrated that it is possible to induce a valuable level of resistance to a broad range of pests and pathogens. The "immunization" need not be made with a pest or pathogen but may be a simple compound such as salicylic acid or 2,6-dichloroisonicotinic acid. However, inoculations of large populations of plants become expensive and time-consuming, and the cost/benefit ratio may quickly fade.

Micropropagated triploid watermelon could become competitive with triploid seedlings. The triploid seedling is derived from a very expensive hybrid seed. This expensive seed must be germinated under rigorously controlled conditions, and even then, may not germinate well. The time necessary to grow a seedling to the transplant stage in the greenhouse is shorter than the time necessary to develop a transplant from tissue culture, but the gap is being closed. Micropropagated triploids may be made more competitive with transplants from seed in a simple way: impart protection to the micropropagule *in vitro* when it is very fast, very precise, very effective, and very cheap to treat hundreds of thousands of micropropagules with a few dollars worth of inoculum.

Watermelon transplants have a number of challenges during their first few weeks in the field when induced resistance would be most effective. Cucumber beetles, foliar pathogens, soil pathogens, late frosts, and desiccation are some of the challenges that young watermelon transplants face. It is conceivable that transplant shock, as well as many of the other dangers that exist for the plant just out of tissue culture, can be minimized by inducing the plant's own resistance mechanisms.

Zehnder et al. (1998) have noted that plants have evolved both constitutive and inducible mechanisms for protection against insect and disease pests. Most genes exploited for resistance have been constitutive (always on) rather than inducible. Inducers include agents that stimulate production of phenolics (Carrasco et al., 1978), compounds as simple as salicylic acid (Malamy et al., 1990; Metraux et al., 1990), saccharin (Siegrist et al., 1998) and phosphate (Orober et al., 1999), more complex chemicals such as 2,6-dichloroisonicotinic acid (Kessman et al., 1994; Colson and Deverall, 1996) and even bacteria (Alstrom et al., 1991; Liu et al., 1995; van Peer et al., 1991; Wei et al., 1991; Zehnder et al., 1998). Salicylic acid is a natural inducer (Hammerschmidt and Smith-Becker, 1999). Kuc (1999) has reviewed what is known about structure and function and notes that, to date, a chemical structural feature has not been reported as being essential to induce resistance.

Van Loon et al. (1998) have reported that in their Pseudomonas-Arabidopsis system, the plant growth-promoting bacteria (PGPR) have an O-antigenic side chain of the outer membrane lipopolysaccharide that is one of the inducing determinants. This system requires jasmonic acid and ethylene-dependent signaling at the site of resistance (Knoester et al., 1999; Loon et al., 1999). They suggest that salicylic acid induction of resistance via the systemic acquired resistance (SAR) pathway in radish involves another mechanism. Others (Wees, S.C.M. et al., 1999) have corroborated these additional pathways to induced resistance. Myer et al. (1999) showed that nanogram amounts of SA produced by *Pseudomonas* activated the SAR pathway in bean.

Cucumber has been a model for studies of inducible resistance (Liu et al., 1995; Orober et al., 1999; Wei et al., 1991; Widders et al., 1992). Their relatedness to watermelon is significant because the cucurbits exploit cucurbitacin metabolism in their constitutive and inducible resistance to insects (Chambliss et al., 1966). One might expect that related inducible systems exist in watermelon, and that these systems could be most ideally activated during the preparation of grafted seedlings for transplants.

Systemic acquired resistance can be induced *in vitro* (Siegrist et al., 1998) expanding its value for micropropagated and grafted triploid watermelon.

Bioassays such as that described by Han et al. (2000) for screening rhizosphere microorganisms for SAR induction should prove very helpful.

GENOMICS

Citrullus lanatus var. *lanatus* (Thunb.) Matsum & Nakai is the progenitor of the cultivated watermelon. *C. lanatus* is one of four known diploid (n = 11) species that belong to the xerophytic genus *Citrullus* Schrad. ex Eckl. & Zeyh., found in the temperate regions of Africa, Central Asia, and the Mediterranean (Whitaker and Davis, 1962; Jeffrey, 1975; Whitaker and Bemis, 1976), and has a genome size of 4.2×10^8 bp (Arumuganathan and Earle, 1991).

The 1999 gene list from the Cucurbit Genetics Cooperative Report (Rhodes and Dane, 1999) provides an update of the known genes in watermelon. Partial sequencing and database comparisons of randomly selected vegetative leaf cDNA clones of commercial F1 hybrid watermelon 'Lucky' have been reported by Ok et al. (2000). Database comparisons indicated that 51.9% of the watermelon leaf cDNA clones revealed a high degree of similarity to other plant genes, while 43.3% did not have a database match and require further analysis. The expressed sequence tag clones (704 ESTs) were divided into the following ten categories depending on gene function: primary metabolism (66 clones), amino acid synthesis and processing (44), secondary metabolism (7), membrane and transport (26), signal transduction (17), DNA, RNA related and gene expression (18), cell wall and metabolism (7), cell division (2), defense (19) and others (Table 2). Further cDNA screening should facilitate the isolation of many horticulturally important genes in watermelon.

The construction of a detailed genetic map of watermelon (*C. lanatus* var. *lanatus*) has been hindered by its narrow genetic base with respect to isozymes and DNA polymorphism (Navot and Zamir, 1986; Hashizume et al., 1996; Lee et al., 1996; Zhang et al., 1994; Hawkins et al., 2001; Levi et al., 2000). In a survey of 26 allozyme loci in 550 cultivated watermelon accessions, very little variation was found (Navot and Zamir, 1987), but significant divergence was detected in wild *Citrullus* forms using simple sequence repeat length polymorphic markers (Jarret et al., 1997). Similar results were indicated using polymorphic RAPD markers. Genetic similarity values among *C. lanatus* var. *citroides* accessions varied from 71-90.5%, among *C. lanatus* var. *lanatus* PI's from 75-96% and among watermelon cultivars from 92.8-98.3% (Levi et al., 2000, 2001).

The first watermelon map, which included 19 protein coding loci, was derived from a cross between *C. colocynthis* and the watermelon cultivar 'Mallali' (Navot and Zamir, 1986). The map was extended to 22 isozyme loci and the locus for bitterness and flesh color, and comprised 7 linkage groups

TABLE 2. Summaries of watermelon ESTs with significant similarity to known genes. Reprinted with the permission of Springer from Ok, S., Chung, Y.S., Um, B.Y., Park, M.S., Bae, J.M., Lee, S.J., and Shim, S.J. (2000). Identification of expressed sequence tags of watermelon (*Citrulus lanatus*) leaf at the vegetative stage, *Plant Cell Reports*, 19: 932-937. .

Putative identity	Accession No.[a]
Primary metabolism	
3-Ketoacyl-coA thiolase	AA660039 (2)
3-Methylcrotonyl-CoA carboxylase	AA660057 (2)
ACC oxidase	AI563157 (1)
ACC synthase	AI563272 (1)
Acetyl-CoA acyltransferase	AI563047 (1)
Acid phosphatase precursor	AA660161 (2)
Aconitate hydratase	AA660145 (1)
Acyl-CoA dehydrogenase and epoxide hydrolase	AA660045 (3)
Adenosylmethionine-8-amino-7-oxonon-anoate aminotransferase	AA660037 (3)
Amidophosphoribosyltransferase	AA660094 (1)
AMP-binding protein	AA660131 (2)
Carbonic anhydrase	AA660102 (2)
Chlorophyll A/B binding protein	
CAB-21	AA660064 (1)
CP29	AI563142 (1)
CAB-36	AA660163 (1)
CAB-37	AA660116 (1)
CAB-151	AI563254 (2)
Cinnamoyl-CoA reductase	AI563265 (2)
Cytochrome b5	AA660129 (1)
Cytochrome c oxidase subunit 1	AI563078 (2)
Cytochrome cl, heme protein	AI563065 (2)
Cytochrome P450-like TBP	AI563252 (1)
Deoxycytidylate deaminase	AI563080 (2)
Electron transfer flavoprotein beta unit	AI563129 (2)
Ethylene-forming enzyme-like dioxygenase	AI563096 (2)
Ferredoxin	AI563241 (2)
Formate dehydrogenase precursor	AA660126 (1)
Fructose-bisphosphatase	AI563089 (2)
Fructose-bisphosphate aldolase	AI563093(1)
Fumarate hydratase	AI563262 (1)
Glucosyltransferase	AI563099 (2)
Glutamate dehydrogenase 2	AI563225 (1)
Glyceraldehyde-3-phosphate dehydrogenase	AI563173 (1)
Ketol-acid reductoisomerase	AI563215 (2)
Lipase	AA660109 (2)

Putative identity	Accession No.[a]
Lipase (lysophospholipase)	AA660096 (3)
Malate dehydrogenase	AA660028 (2)
Malate oxidoreductase	AI563049 (1)
Monodehydroascorbate reductase	AA660034 (2)
NADH dehydrogenase	AA660051 (2)
N-Carbamyl-L-amino acid amidohydrolase	AA660138 (2)
Oxalyl-CoA decarboxylase	AI563232 (3)
Oxygen-evolving enhancer protein 2	AA660030 (1)
Oxygen-evolving enhancer protein 3	AA660156 (2)
Phenylalanine ammonia lyase	AI563248 (1)
Phospho-2-dehydro-3-deoxyheptonate aldolase 1	AA66OtO8 (2)
Phosphoglycerate kinase	AI563095 (2)
Photosystem I accessory protein E	AI563120 (1)
Photosystem I P700 apoprotein AI	AI563072 (1)
Photosystem I protein psa L	AI563046 (2)
Photosystem I psaH protein	AI563128 (2)
Photosystem I reaction center subunit III	
Precursor	AA660032 (1)
Photosystem II 10 kDa polypeptide	AA660025 (1)
Photosystem II oxygen-evolving complex protein 1	AA660158 (1)
Photosystem II oxygen-evolving complex protein 2	AI563177 (1)
Photosystem II oxygen-evolving complex protein 3	AI563234 (3)
Photosystem II protein psbK	AI563267 (1)
Signal transduction	
ADP-ribosylation factor	AA660148 (1)
Calreticulin	AA660017 (1)
Ethylene receptor	AI563052 (1)
G-protein beta subunit	AA660061 (2)
Phosphoenolpyruvate carboxykinase	AI563063 (1)
Protein kinase (Arabidopsislputative receptor)	AI563246 (2)
Protein kinase (novel serine/threonine)	AA660087 (2)
Protein kinase (receptor-like)	AA660112 (1)
Protein kinase (shaggy-like)	AA660105 (1)
Protein kinase AFC 2	AA660015 (2)
Protein kinase Xa2I	AI563098 (3)
Protein kinase YAK1	AA660097 (3)
Protein phosphatase 2C ppH1	AA660004 (2)
Protein phosphatase PP2A catalytic subunit (serine/threonine)	AA660155 (1)
Protein phosphatase pp-x isozyme (serine/threonine)	AA660123 (2)
Ran binding protein I	AA660005 (2)
Steroid binding protein	AI563210 (2)

TABLE 2 (continued)

Putative identity	Accession No.[a]
DNA, RNA related and gene expression	
Alfin-I	AI563088 (2)
Ankyrin-like protein	AI563040 (2)
Cys3His zinc-finger protein	AI563208 (1)
DNA-binding protein	AA660085 (2)
DNA-binding protein G2p	AI563197 (1)
DNA-damage-repair/toleration protein	AA660003 (3)
Endonuclease PI	AA660099 (3)
EREBP-4	AA660094 (2)
HD-ZIP protein	AI563062 (2)
Histone HI	AI563143 (2)
Histone H3.2	AA660088 (1)
Homeobox-leucine zipper protein hat22	AA660125 (1)
Polypyrimidine tract-binding protein I	AI563235 (3)
Ribonucleoprotein	AA660019 (1)
RNA helicase	AI563193 (2)
RNA-binding protein RZ-1	AI563146 (1)
SCARECROW homolog	AA660090 (2)
Transcription factor	AI563116 (3)
Cell wall and metabolism	
Beta-galactosidase precursor	AI563107 (2)
EDPG precursor	AA660101 (2)
Extensin-like protein	AI563152 (4)
(hydroxyproline-rich)	AI563104 (2)
Glycoprotein	AI563103 (2)
Pectin methylesterase	AI563150 (3)
Pectinesterase	AI563153 (2)
Ribophorin I homologue	AI563050 (1)
Rubisco activase	AA660135 (2)
Rubisco small subunit	AA660114 (1)
Sedoheptulose-1,7-bisphosphatase chloroplast precursor	AA660026 (2)
Thioredoxin-f	AI563275 (1)
Transketolase	AI563214 (1)
Uricase 11	AI563243 (2)
Amino acid synthesis and processing	
40S Ribosomal protein S2	AI563127 (1)
40S Ribosomal protein S5	AA660095 (1)
50S Ribosomal protein L3	AA660031 (2)
50S Ribosomal protein L35	AI563242 (3)

Putative identity	Accession No.[a]
60S Ribosomal protein LIB	AA660110 (2)
60S Ribosomal protein L23-1	AA660016 (1)
60S Ribosomal protein L24	AI563148 (1)
60S Ribosomal protein L34	AA660127 (1)
60S Ribosomal protein L37A	AI563135 (1)
Alanine aminotransferase	AI563086 (2)
Alanyl-tRNA synthetase	AI563076 (3)
Aminoacyl-tRNA synthetase	AI563108 (1)
Asparagine synthetase	AI563114 (2)
Chaperonin 10	AA660089 (2)
Chaperonin 60 beta subunit	AI563217 (2)
ClpC protease	AI563160 (1)
Cyclophilin	AA660049 (1)
Cysteine synthase	AI563124 (1)
Elongation factor (EF-TuB)	AA660106 (1)
Glutamine synthetase root isozyme 4	AI563060 (1)
Glutamyl-tRNA synthetase	AI563051 (3)
Glutathionine reductase cytosolic	AA660006 (1)
Heat shock protein	AA660075 (1)
Nucellin	AI563122 (3)
Peptide chain release factor 2	AI563239 (2)
Peptidyl-tRNA hydrolase	AI563056 (3)
P-Protein-like protein	AI563121 (1)
Protease (trypsin-like)	AI563071 (1)
Protease inhibitor (cysteine)	AI563159 (3)
Protease inhibitor (Rcmti-V)	AI563140 (3)
Protease inhibitor (serine)	AI563053 (3)
Protease inhibitor Trichosanthes trypsin	AI563213 (2)
Protease inhibitor 11	AI563224 (3)
S-adenosyl methionine synthetase	AI563202 (1)
S-adenosylmethionine decarboxylase	AI563200 (1)
Spermidine synthase	AI563105 (2)
Threonine synthase	AI563068 (2)
Translation initiation factor (eIF-4A.6)	AA660065 (1)
Translation initiation factor (eIF-4E)	AA660115 (2)
Translation initiation factor SUI 1	AA660143 (1)
Tyrosyl-tRNA synthase	AA660100 (2)
Ubiquitin	AI563151 (1)
Ubiquitin protein ligase E3	AI563218 (2)
Ubiquitin-conjugating enzyme E2	AI563249 (1)

TABLE 2 (continued)

Putative identity	Accession No.[a]
Secondary metabolism	
Acetoacetyl-coenzyme A thiolase	AI563074 (3)
Chromoplast-specific carotenoid-associated protein	AI563106 (2)
Cycloartenol synthase	AI563211 (1)
Geranylgeranyl hydrogenase	AI563245 (2)
Glutathionine s-transferase	AA660029 (3)
Lipoxygenase	AA660152 (3)
Loxc homologue	AI563203 (2)
Membrane and transport	
ABC transporter	AI563287 (3)
Acyl-binding/lipid-transfer protein isoforml	AI563066 (3)
Adenine nucleotide translocator	AI563236 (2)
ATPase B subunit V-type	AA660024 (1)
ATPase beta-subunit	AI563144 (3)
ATPase C subunit	AA660033 (3)
ATPase C subunit V-type 16 kDa proteolipid chain	AA660091 (1)
ATPase gamma-subunit	AI563256 (1)
ATPase metal-transporting P-type	AI563041 (3)
ATPase mitochondrial FI subunit	AI563274 (1)
ATPase P-type 4	AI563169 (1)
Coatomer, beta-prime subunit	AI563087 (3)
Cyclic nucleotide-regulated ion channel	AI563184 (2)
Lipid-transfer protein	AA660082 (2)
Mitochondrial phosphate transporter	AI563176 (1)
Monosaccharide transport protein	AA660083 (1)
Outer plastidial membrane protein	AA660042 (2)
Oxoglutarate malate translocator	AI563181 (1)
Probable membrane protein	AA660162 (3)
Protein translocase	AI563115 (2)
Pyrophosphate-energized vacuolar membrane proton pump	AA660118(1)
Scarlet protein	AA660060 (2)
Tetracycline transporter-like protein	AA660062 (2)
Umecyanin	AA660011 (2)
Water-stress induced protein	AA660147 (3)
Xylose permease	AA660154 (1)
Cell division	
Chloroplast FtsH protease	AI563064 (1)
UDP-glucose glucosyltransferase	AI563082 (2)

Putative identity	Accession No.[a]
Defense	
Aquaporin	AI563055 (3)
Catalase	AI563227 (1)
DnaJ protein	AI563180 (2)
DnaJ-1 protein	AA660002 (1)
ERD15 protein	AI563233 (2)
Ferritin	AI563100 (1)
Fis I	AI563255 (2)
HinI	AI563209 (3)
Imidazole glycerol-phosphate dehydratase	AA660081 (2)
In2.1 protein	AI563162 (2)
Jasmonate induced protein	AI563130 (3)
Manganese superoxide dismutase	AI563126 (2)
Metallothionein-like protein	AI563198 (1)
Mlo protein	AI563090 (2)
Oxidoreductase P2	AI563205 (1)
Polyphenol oxidase	AI563244 (3)
Selenium-binding protein	AA660150 (1)
Sti	AI563174 (1)
Trehalase	AI563125 (3)
Others	
21 kDa protein precursor	AI563069 (2)
AP2 domain containing protein	AI563257 (1)
Apoptosis protein MA-3	AI563110 (1)
AT103 protein	AI563042 (2)
CoL 2	AA660070 (1)
CONSTANS	AI563221 (3)
Cuc 2	AA660001 (2)
Diminuto	AI563189 (2)
Dormancy-associated protein	AI563216 (1)
Drosophila couch potato protein	AI563113 (1)
21 kDa protein precursor	AI563069 (2)
AP2 domain containing protein	AI563257 (1)
Apoptosis protein MA-3	AI563110 (1)
AT103 protein	AI563042 (2)
CoL 2	AA660070 (1)
CONSTANS	AI563221 (3)
Cuc 2	AA660001 (2)
Diminuto	AI563189 (2)
Dormancy-associated protein	AI563216 (1)
Drosophila couch potato protein	AI563113 (1)

TABLE 2 (continued)

Putative identity	Accession No.[a]
EST gbIN65759	AI563133 (3)
Glycosylatable polypeptide	AI563079 (1)
GRPF1	AI563268 (2)
Haemophilus influenzae permease	AA660093 (3)
HvBI2D homolog	AI563179 (2)
Hypothetical protein (different 26)	AI563261 (2)
IAA induced protein ARG2	AA660063 (1)
KIAA0005	AI563237 (3)
Lac Z-pho C	AA660071 (2)
NAM	AI563260 (2)
Peroxidase	AI563266 (2)
Putative protein (different 4)	AI563228 (2)
Putative small subunit	AA660111 (2)
Senescence-associated protein senl	AI563212 (2)
Sipl protein	AA660084 (1)
Small glutamine-rich tetratricopeptide	AI563094 (3)
SRGI protein	AI563058 (3)
TI F9.10	AI563247 (3)
T'7N9.3	AA660079 (1)
TCTP protein	AI563270 (1)
Unknown function protein	AI563043 (2)
Unknown protein (different 8)	AI563253 (2)
Unnamed protein (different 2)	AI563172 (2)
Yeast cat8 regulatory protein	AI563201 (3)
Haemophilus influenzae permease	AA660093 (3)

[a] The numbers in parentheses indicate range of similarity with high scoring segment pair (HSP): (1), 91-100%; (2), 71-90%; (3), 51-70%; and (4), less than 50%

spanning 354 cM. Hashizume et al. (1996) constructed a 524 cM linkage map spanning 11 linkage groups in a backcross population of a cultivated Japanese *C. lanatus* line and a wild African form. This map contained 58 RAPD markers, one isozyme, one RFLP and 2 morphological markers segregating into 11 linkage groups. Half of the RAPD markers on one linkage group displayed high levels of segregation distortion. Recently linkage maps were constructed in populations segregating for resistance to race 1 and 2 of *Fusarium oxysporum* f. sp *niveum* (Hawkins et al., 2001; Levi et al., 2001). Hawkins et al. (2001) used F_2 and F_3 lines to examine the linkage relationships between different molecular markers (isozymes, RAPDs, SSRs), several morphological traits and race 1 and race 2 Fusarium wilt resistance. Since a high number of RAPD markers showed segregation distortion in both generations or were unlinked, only 2 linkage groups could be detected in the F_2 covering 113 cM, and 5 linkage groups in the F_3 covering 139 cM. Distorted markers were subsequently mapped one marker at a time and clustered on one large linkage group. Even though 320 RAPD primers were tested for linkage to Fusarium wilt race 1 or 2 resistance, only loose linkages could exist between three RAPD markers and race 1 resistance and four RAPD markers and resistance to race 2 (Hawkins et al., 2001). Xu et al. (1999) reported close linkages between OP P01$_{700}$ and resistance to Fusarium wilt race 1 in PI 296341FR, which was subsequently converted into a SCAR marker, and should be useful for marker-assisted selection for breeding Fusarium race 1 resistant cultivars (Xu et al., 2000). Levi et al. (2001) used a BC$_1$ population [(PI-296341-FR × NHM) × NHM] to construct a linkage map, which contains 155 RAPD markers segregating into 17 linkage groups. Since many of the linkage groups only contain a few RAPD markers, which are not transferable across populations, and large chromosomal regions of the watermelon genome are not yet covered, the construction of a detailed linkage map will require several different strategies such as the use of the published watermelon expressed sequence tags (Ok et al., 2000).

HETEROSIS

Diploids

In the beginning of this chapter, the watermelon was introduced as an acceptably inbred crop. Unlike corn, watermelon can inbred indefinitely without inbreeding depression (Porter, 1933). Crimson Sweet, an OP variety developed by Charles V. Hall in Kansas about 1960 is still popular in many parts of the world. It has been a parent of many varieties released since that time. However, Singletary and Moore (1985) reported that hybrids outyielded OP cultivars 40-60% and were more uniform. Kalloo (1993) and Zhang and Wang (1987) reported that watermelon hybrids demonstrated heterosis with respect to yield, number of fruit, size and weight of fruit, fruit uniformity, number of female flowers, earliness and total soluble solids. More obvious heterosis was achieved with hybrids under better cultural conditions-drip irrigation, black plastic, and closer spacing. The fact that hybrid diploids exhibit more heterosis than inbred diploids is suggested by the eventual adoption of more expensive hybrids despite the availability of OP seed. In the largest watermelon producing country China, almost 100% of its 1.2 million hectares of commercial watermelon production uses hybrid varieties. Commercial watermelon production in Japan and Korea has long been using hybrid varieties. Almost all of the new variety releases are hybrids in the United States, although OP varieties are being used in some commercial fields.

Triploids

Tri-X-313 was the first commercially introduced triploid hybrid in the United States in late 80s. It is still one of the most popular triploids in North America, Europe, and Israel. Most of the late triploid hybrid releases from different seed companies are in the class of Tri-X-313 with some minor modification. This type of triploid hybrids are from a gray skin and globe shaped tetraploid female parent and a striped and elongated diploid male parent. The presence of the third genome in the triploid hybrids resulted in a tougher rind, a smaller fruit, and strong plants. The triploid is generally more resistant to many diseases, has a fruit with a longer shelf life and soluble solids higher than the midparent. Most of the current triploid hybrids over-yield diploid hybrids due to more fruit per plant and longer harvest periods. Big commercial growers in west coast of the U.S. primarily grow seedless watermelon. The diploid watermelons are only grown as the pollenizer of seedless watermelon. The big seedless watermelon growers are even looking for diploid watermelons that do not produce harvestable fruit and produce pollen in a long period of time to use as pollenizer for the triploid watermelon.

The potential of triploid hybrids is under explored and used. The diversity of triploid varieties is much less than the diploids. Most of the commercial triploid hybrids are derived from very few essentially derived tetraploids. The choice of the tetraploid maternal parent can clearly enhance the superiority of the triploid (Henderson, 1977). It takes more time and efforts to create a usable tetraploid because of the low fertility, difficult fruit set, and hollow heart defects associated with many tetraploids. Low fertility is now only a problem for propagating the tetraploid parent but also a severe problem for producing triploid hybrid seed. Hollow heart and hard seed coats are two more major challenges for developing superior triploids. Some tetraploids produce more hollow heart and hard seed coats triploids. However, the combination of tetraploid and diploid parents is also important. Due to the special defects that triploid hybrids might have, extensive variety evaluation is highly recommended before commercialization.

POLLINATION CONTROL MECHANISMS

Open Pollination

Open pollination and open pollinated (OP) varieties have been mentioned before. Isolation of a planting of an inbred line one mile or more will insure that insect pollination will occur only among the individuals in the isolation block. Because watermelon is monoecious, self- and cross-pollinations occur in such a field. Outcrosses occur at the rate of 37% (Singletary and Moore, 1965). With a ratio of 7-8:1 male/female flowers opening on a single morning, fertilization of a single female flower with pollen from a number of males in the same vicinity is insured, given a minimum number of bees. Fruit set may not occur unless the female flower receives several visits from the bumblebee and many more visits from the honeybee (Stanghellini et al., 1997). Some varieties of watermelon, e.g., Giza in Egypt, and some mutants, e.g., *tendrilless* (Rhodes et al., 1999) exhibit an abundance of andromonoecious flowers that produce some viable pollen. This morphology facilitates self-pollination (Rhodes, personal observation).

Hand Pollination

Pollen from a single plant can be collected by hand and distributed to a single female flower on the same plant or a different plant with the exclusion of foreign pollen (Rhodes and Zhang, 1999). This process requires careful covering the day before the flowers are biologically scheduled to open and careful hand pollination on the morning that the flowers open, followed by replacement of the cover.

Female flower buds that will open the next morning are identified the afternoon before they open by their coloration and position on the vine. The most mature female flower buds are located toward the end of a branch of the vine. The bud to be pollinated the next day has begun to yellow but remains firm to the touch. If the bud is not firm and the petals are yellow, the bud has opened, has been pollinated, and then closed back up. This flower will likely become a fruit and should be removed from the vine. To insure fruit set, females on the same branch, all female flowers that are open pollinated should be removed from the vine during the selection of unopened female flowers. The female flower bud which will open next morning is protected from unwanted pollen by a cover to keep out all insect pollinators. A glassine bag, with a wire twist tie, or even a paper cover created by rolling a inch-wide strip of paper around the end of a pencil, could be used. Plastic tape can secure the bloom but is difficult to use and induces more floral abortion than the other methods. It is essential to ensure that a given cover will be sufficient to prevent insects from getting into the female flower. The cover must be easily removed for pollination the next morning and be easily reinserted over the flower after pollination. Male flowers also are located the afternoon before pollination the next morning. Male flowers, like female flowers, begin to yellow before opening the next day and bloom in sequence along the vine. Male flowers that will open in sequence with the female flower are often on the second node below a prospective female flower. For a controlled hand pollination, these flowers will be covered the evening before they open or can be collected in late afternoon and kept on moist sand until pollination the next morning. The male flowers collected shed pollen earlier than the flowers in the field if they are kept in a warm place in the light. Alternatively, male flower buds may be collected in the early morning before they open. These procedures save the labor involved in covering individual male flowers. When selecting males for a self-pollination, it is necessary to insure that the male used is from the same plant. Therefore, it is best to select a male that is on the same vine as the prospective female. The male flower is more easily closed than the female flower, and care must be taken not to damage the female.

In a large field, finding the flowers that have been covered or closed for the next morning's pollination is facilitated by some kind of marking system. Wire flags, small pieces of bamboo or stiff fence wire segments cut long enough to see above the foliage can be used.

A hand pollination, cross or self, is made in this way. On the morning of the pollination, male flowers are taken to the female flower. The cover is carefully removed from the female flower. In early morning, the petals of the female may not open readily, and may need to be gently pried open. If the bud is still hard, it may need to be marked and covered for pollination the following day. With the female flower open, the petals of the male flower are pulled back or removed to expose the anthers. If the anthers do not appear to have shed any pollen, the male flower should be discarded, and a more suitable male should be used. The surface of the pollen-covered anthers are rolled over the surface of the stigma, to insure that the stigma is completely covered with pollen. The female flower is carefully recovered and tagged.

Tagging can identify the parents, the pollination date, and the pollinator. The following information, in order, could appear on the tag. The first item on the tag is the identification of the female flower. The identification of the female is followed by a crossing symbol, either an X for a cross or an X within a circle for a self-pollination. If the pollination is a cross, then the X is followed by the identification of the male. If the pollination is a self, then the identification follows the circled X. The next item on the tag is the date of the pollination. The date allows an evaluation of earliness of fruit production. The last bit of information on the tag is the initials of the person who did the pollination. Knowing the pollinator is quality control on workmanship. The tag should be placed around the stem just above the female flower. Thus, the tag will remain next to the developing fruit. Sometimes the stem tag is defaced or lost, and it is hard to check until harvest. An added precaution is to place an identical tag on the wire flag used to mark the site of the female flower.

Pollination should begin between 6:30 and 7 a.m. and is often futile after 11 a.m. Most successful pollinations are made by 9 a.m. Covering of flowers should begin in the late afternoon.

When the branches of two plants overlap, tagging mistakes can be made, and the plants are probably too close together to obtain maximal fruit set. Therefore, it is necessary to thin plants to arrive at a uniform stand of one plant per hill, with hills given a more generous spacing than a production field. However, commercial hybrid seed production fields use higher density to increase seed yield. Each hill may have two or more plants per hill at the beginning of the season but should only have one at pollination. Thinning should take place in two phases. First, the hill should be thinned to the two largest plants when the plants have reached the four to five leaf stage. The next thinning should take place when the vines begin to run. At this time, the hills should be thinned to one plant. In observational and comparison blocks, hills should be thinned to the three healthiest plants while in the expanded cotyledon stage and two plants at the four or five true-leaf stage.

Male Sterility

No cytoplasmic male sterility has been reported in watermelon (Robinson, 1999). Watts (1962) reported the first genic male-sterile mutation in watermelon in an X_2 population of Sugar Baby irradiated with gamma rays. This trait was tightly linked with glabrousness. It is now designated *gms* (Rhodes and Dane, 1999). Fertile male flowers do occur very late in the development of the mutant and partially male fertile glabrous lines can be developed (Rhodes, 1991). Love et al. (1986) introduced the gene into tetraploids, but the reduction of female reproduction as well as fruit quality and yield associated with this pleitropic mutant was never eliminated. A spontaneous genic male-sterile (*ms*) from China (Xia et al., 1988) is stable for male sterility with minimal pleitropic effects, if any. The lack of tapetum and abnormal deposition of fluorescing protein precluded meiocyte development after telophase II and caused degeneration of meiocyte in the *ms* sterile anthers (Zhang et al., 1994). The consistent inhibition of ontogeny of meiocytes at telophase II, the absence of pollen production under the varying environmental conditions, and the normal female fertility of the male-sterile genotype make the *ms* mutant useful in hybrid seed production. Murdock (1993) found that, in an autotetraploid, segregation of the trait approached a chromosomal inheritance pattern, suggesting a close linkage to the centromere and/or barriers to crossing over in the autotetraploid. Thus, there is no indication that an *ms* maintainer line would be genotypically unstable due to the occasional production of duplex male-fertile plants (Love et al., 1986). Murdock (1993) and Zhang et al. (1993, 1996) found no indication of linkages with seedling characters *Sp*, *dg*, and *ja*. Other genic male steriles have been reported, but their utility has apparently been limited, and they will not be discussed here.

HYBRID SEED PRODUCTION/MAINTENANCE
OF INBRED LINES

Manual

Hybrid seed can be experimentally produced using the pollination techniques previously described. One individual, if properly trained, can do several hundred pollinations a day. Inbred lines to be hybridized should be planted adjacent to each other in a crossing block. Lines can be selfed for maintenance and hybridized concurrently. A much more efficient procedure is simply to remove the male flower buds from the "female" line each evening during pollination season and allow bees to transfer pollen from the "male" line to the "female" line. The hybrids will be harvested from the "female" line and the male line maintained by harvesting the self-pollinated fruit on the male line. A recessive seedling marker gene in the female will allow identification of any self-pollinations that may occur. Isolation of the female line in another area will maintain the female line.

Use of Male Sterility

If male sterility exists in the female line, there are no functional male flowers to remove by hand. Thus, a substantial savings in labor costs is realized by the use of a gene such as *ms* in watermelon. A procedure has been developed with seedling markers to expeditiously incorporate *ms* into breeding lines and to produce hybrid seed (Zhang et al., 1996; Zhang and Rhodes, 2000).

Tetraploid Plants

Autotetraploid watermelon lines are the maternal parents of triploid watermelon hybrids. Chromosome doubling was first accomplished with the toxic alkaloid colchicine by applying colchicine in lanolin paste to the growing point. More satisfactory tissue culture treatments have been developed (Zhang and Rhodes, 1994, 1995). Dinitroanilines have been used to double chromosome numbers, and their effectiveness has previously been compared with crops other than watermelon (Tosca et al., 1995; Hansen and Andersen, 1996; Zhao and Simmonds, 1995). We compared *in vitro* chromosome doubling effectiveness using colchicine and the dinitroanilines, ethalfluralin (N-ethyl-N-2-methyl-2-propenyl)-2,6-dinitro-4-(trifluoromethyl) benzanine), and oryzalin (3,5-dinitro-N4, N4-dipropylsulfanilamide) and concluded that either ethalfluralin or oryzalin was preferable to colchicine (Li et al., 1999).

Because meiosis is sometimes irregular in autotetraploids, diploids and aneuploids do occur in their offspring (Anghel, 1967; Lower, 1969). The leaves, flowers and pollen grains of tetraploids are morphological distinct from diploids (Zhang et al., 1995) as well as the number of chloroplasts in the guard cells (Compton et al., 1996). From alfalfa studies, it has been suggested that "semihybrid" tetraploid cultivars of alfalfa can be developed using molecular markers to select parents in a tetraploid F_1 population (Brummer et al., 2000). The concept of a very stable "semihybrid" tetraploid is appealing.

Production of Triploid Hybrid Seed

The reader is referred to our previous publication concerning general considerations for hybrid seed production (Rhodes and Zhang, 1999). Plant spacing for seed production is closer than for fruit production. Depending on rainfall and evapotranspiration, drip irrigation every two weeks is desirable for producing high quality seed. Rotation with an appropriate green manure crop and/or timely reduction of nitrogen at fruiting will encourage early fruit set and development instead of continued vine growth.

Triploid seeds are currently produced using two methods. In the United States, de-budding method is used to produce triploid watermelon seed. Almost all of the productions are located in Northern California. The production fields are planted in a ratio of 2 rows of tetraploid female line and 1 row of diploid male line. All the male flower buds are removed from the female tetraploid plants. The female flowers are open-pollinated by bees. The fruit set during de-budding period are marked and harvested for triploid hybrid seed. At least six skilled workers are required per acre of hybrid watermelon seed production for a 2-3 week pollination period. They will insure that pollination proceeds according to plan. These laborers will remove male buds on tetraploid female vines through the pollination season. If a male sterile tetraploid line is available, workers can easily remove the male fertile plants in the tetraploid female row with much less time and efforts. All the fruit set on the male-sterile tetraploid plants can be harvested for hybrid triploid seed. When the marked male-sterile system is used, seed producer can insure that no female off-types exist in the female tetraploid line and the hybrid triploid seed (Zhang and Rhodes, 2000).

Hand-pollination method is used for triploid seed production in Asia and South America where isolation is difficult to get and hand labor is cheaper. Sow homozygous male parent 7-10 days earlier than homozygous, highly fertile female tetraploid parent. The male parent is usually located on the outside of the crossing block. Male parent is carefully checked for its uniformity by experienced field technician before collecting male flowers. Any off-types that can be recognized based on plant morphology and ovary characteristics are removed. Male plants are removed after pollination is complete to insure that only female fruit are harvested. Plant approximately five to ten tetraploid female plants per male plant to insure adequate pollination.

Pollinators

Production of cucurbit seed and fruit, including watermelon, may be limited by pollinators. Habitat fragmentation and destruction, misuse of pesticides, mites, diseases and other insects have taken their toll on the honey bee in the United States. The introduction of more hardy relatives (unrelated to the so-called killer bee) and the revival of the bumblebee are welcome improvements. Though small in numbers, the bumblebee is a far more efficient pollinator of watermelon than the honeybee (Stanghellini et al., 1997).

Seed Transmitted Diseases

At least four diseases can be transmitted on seed: anthracnose, gummy stem blight, fusarium wilt, and bacterial fruit blotch (Zitter, Hopkins and Thomas, 1996). Seed produced in arid locations with drip irrigation will most likely not carry the foliar diseases, but may carry Fusarium wilt. If seed are produced in an area not previously used for cucurbits, the probability of Fusarium wilt infestation is low. Seed infested with bacterial fruit blotch have also been produced in at least one arid location. Treatment of wet extracted seed with 1% HCl has been used against bacterial fruit blotch (Hopkins et al., 1996), but the best security against this bacterial disease is to insure that seed production begins and ends with bacteria-free seed. We believe thoroughly washing the seed with clean water after seed are extracted from fruit and drying the seed quickly on temperature and air-flow controlled dryer are critical to insure seed quality and seed healthy. Seed testing procedures have been standardized for the seed industry (Randwhawa, 1994), and several commercial labs routinely sample large seed lots for the presence of the pathogen. All new accessions to a breeding program should be produced in isolation and thoroughly tested before entry into the breeding program.

CONCLUSIONS

If the genetically engineered watermelon has a place in the sun, it will be after world hysteria over so-called "genetically-modified" organisms has subsided. The most effective techniques of plant breeding lie now in the molecular realm, and their use can mean rapid responses to new strains of pests and new environmental stresses, and designer watermelons to fit the mood and the palette of the consumer. The expansion of research on the triploid watermelon, driven by a consumer market, suggests it will become an even larger part of the market. Modifications and use of the watermelon seed, with its good quality protein and unsaturated oil, will expand. The world seed market has survived bitter experiences with seed-borne diseases and are coordinating and improving their efforts to insure a safe seed supply.

What will happen to the world's germplasm? This is a good question. Far too little is being spent on *in situ* and *ex situ* conservation of the genes necessary for the evolution and survival of many crops, including the watermelon. Perhaps political forces will realize that global cooperation is vital to protect the world's food supply in time to preserve enough of the genetic base essential to improve the crop. We hope so.

Fruit production is limited by flower pollination. Pollination depends on an ample supply of bees. It is critical to encourage the survival of wild pollinators as well as bee-keeping.

REFERENCES

Alstrom, S. 1991. Induction of disease resistance in common bean susceptible to halo blight bacterial pathogen after seed bacterization with rhizophere pseudomonads. J. Gen. Appl. Microbiol. 37:495-501.

Arumuganathan K., and E.D. Earle. 1991. Nuclear DNA content of some important plant species. Plant Mol. Biol. Rep. 9:211-215.

Anghel, I. 1967. A comparative study of 2n, 3n, and 4n forms of *Citrullus vulgaris*. Genetika 5(8):1054-1059.

Brummer, E. Charles, Diane Luth, and Heathcliffe Riday. Breeding for heterosis using traditional and marker-assisted methods. http://genes.alfalfa.ksu.edu/TAG/TAGpapers/brummer.html

Carrasco, A., A.M. Boudet and G. Marigo. 1978. Enhanced resistance of tomato plants to *Fusarium* by controlled stimulation of their natural phenolic production. Physiological Plant Pathology 12:225-232.

Chambliss, O. and C.M. Jones. 1966. Cucurbitacins: Specific insect attractants in Cucurbitaceae. Science 153:1392-1393.

Choi, J.S., K.R. Kang, K.H. Kang, and S.S. Lee. 1992. Selection of cultivars and improvement of cultivation techniques for promoting export of cucumbers. Res. Rpt., Min. Sci. & Technol., Seoul, Republic of Korea. p. 74.

Colson, E. and B. Deverall. 1996. Helping plants fight their own disease battles. Australian Cottongrower 17(5): 76-80.

Compton, M.E., D.J. Gray and G.W. Elmstrom. 1996. Identification of tetraploid regenerants from cotyledons of diploid watermelon cultures *in vitro*. Euphytica. 87:165-172.

Edelstein, M., F. Corbineau, J. Kigel and H. Nerson. 1995. Seed coat structure and oxygen availability control low-temperature germination of melon (*Cucumis melo*) seeds. Physiologia Plantarum 93:451-456.

Gao, X.Y., X.Y. Lin, C.Y. Yang, Y.Y. Wang, B.L. Sun and S.H. Liu. 1983. Study on clonal propagation of seedless watermelon (in Chinese with English summary). J. China Agri. Sci. 2:58-63.

Hammerschmidt, R. and J.A. Smith-Becker. 1999. The role of salicylic acid in disease resistance. In: A.A. Arawal, S. Tuzun and E. Bent, editors: Induced Plant Defenses Against Pathogens and Herbivores: Biochemistry, Ecology and Agriculture. APS Press, Saint Paul. pp. 37-53.

Han, D.Y., D.L. Coplin, W.D. Bauer, and H.A.J. Hoitink. 2000. A rapid bioassay for screening rhizosphere microorganisms for their ability to induce systemic resistance. Phytopathology 90(4): 327-332.

Hashizume, T., I. Shimamoto, Y. Harushima, M. Yui, T. Sato, T. Imai, and M. Hirai. 1996. Construction of a linkage map for watermelon (Citrullus lanatus (Thunb.) Matsum & Nakai) using random amplified polymorphic DNA (RAPD). Euphytica 90: 265-273.

Hawkins, L.K., F. Dane, T.L. Kubisiak, B.B. Rhodes, and R.L. Jarret. 2001. Linkage mapping in a watermelon populations segregating for fusarium wilt resistance. J. Amer. Soc. Hort. Sci. 126: 344-350.

Hawkins, L.K., F. Dane, and T.L. Kubisiak. 2001. Molecular markers associated with morphological traits in watermelon. Hortscience 36: 1318-1322.

Henderson, W.R. and S.F. Jenkins, Jr. 1977. Resistance to anthracnose in diploid and polyploid watermelons. J. Amer. Soc. Hort. Sci. 102(6): 693-695.

Hopkins, D.L., J.D. Cucuzza, and J.C. Watterson. 1996. Wet seed treatments for the control of bacterial fruit blotch of watermelon. Plant Disease 80(5): 529-532.

Itagi, T. 1992. Status of transplant production systems in Japan and new grafting techniques (in Korean) Symp. Protected Hort. Expt. Sta., Rural Development of Admin., Suwon, Republic of Korea. pp. 32-67.

Ito, T. 1992. Present state of transplant production practices in Japanese horticultural industry. In: Kurata K. and Kozai T. (eds.) Transplant Production Systems. Kluwer. Dordrecht, pp. 65-82.

Jarret, R.L., L.C. Merrick, T. Holms, J. Evans, and M.K. Aradhya. 1997. Simple sequence repeats in watermelon (Citrullus lanatus (Thunb.) Matsum. & Nakai). Genome 40: 433-441.

Jeffrey, C. 1975. Further notes on Cucurbitaceae: III. Some African taxa. Kew Bull. 30: 475-493.

Kalloo, G. 1993. Vegetable Breeding Vol. 1. Chap. 6. CRC Press. p. 177.

Kang, K.S., S.S. Choi, and SS. Lee. 1992. Studies on rootstocks for stable production of cucumber. Kor. Soc. Hot. Sci. 19(2): 122-123 (Abstr.).

Kessman, H., T. Staub, C. Hofmann, T. Maetzke, G. Herzog, E. Ward, S. Uknes, and J. Ryals. 1994. Induction of systemic acquired disease resistance in plants by chemicals. Ann. Rev. Phytopath. 32: 439-460.

Knoester, M., C.M.J. Pieterse, and L.C. van Loon. 1999. Systemic resistance in Arabidopsis induced by rhizobacteria requires ethylene-dependent signaling at the site of application. Molecular Plant-Microbe Interactions 12(8): 720-727.

Kobayashi, K. 1991. Development of grafting robot for the fruit-vegetables. Plant Cell Technol. 3(6): 477-482.

Kuc, J. 1999. Specificity and lack of specifity as they relate to plant defence compounds and disease control. In: H. Lyr, P.E. Russell, H.-W. Dehne and H.D. Sisler, editors: Modern Fungicides and Antifungal Compounds. II. 12th International Reinhardsbrunn Symposium, Friedrichoroda, Thuringia, Germany, 24th-29th May. pp. 31-37.

Kurata, K. 1994. Cultivation of grafted vegetables. 2. Development of grafting robots in Japan. HortScience 29: 240-244.

Lee, J. 1994. Cultivation of grafted vegetables. I. Current status, grafting methods, and benefits. HortScience 29: 235-239.

Lee, S.J., D.S. Shin, K.W. Park, and Y.P. Hong. 1996. Detection of genetic diversity using RAPD-PCR and sugar analysis in watermelon [*Citrullus lanatus* (Tunb. Mansf.] germplasm. Theor. and Appl. Genetics 92: 719-725.

Levi, A., C.E. Thomas, A.P. Keinath, and T.C. Wehner. 2000. Estimation of genetic diversity among *Citrullus* accessions using RAPD markers. Acta Horticulturae 510: 385-390.

Levi, A., C.E. Thomas, X.P. Zhang, T. Joobeur, R.A. Dean, T.C. Wehner, and B.R. Carle. 2001. A genetic linkage map for watermelon based on randomly amplified polymorphic DNA (RAPD) markers. J. Amer. Soc. Hort. Sci. 126: 730-737.

Li, Ying, John F. Whitesides, and Bill Rhodes. 1999. *In vitro* generation of tetraploid watermelon with two dinitroanilines and colchicine. Cucurbit Genetics Cooperative 22: 38-40.

Liu, L., J.W. Kloepper and S. Tuzun. 1995. Induction of resistance in cucumber by plant grow-promoting rhizobacteria: Duration of protection and effect of host resistance on protection and root colonization. Phytopathology 85: 1064-1068.

Loon, L.C. van, P.A.H.M. Bakker, and C.M.J. Pieterse. 1998. Induction and expression of PGPR-mediated induced resistance against pathogens. Bulletin OILB/SROP 21(9): 103-110.

Loon, L.C. van, P.A.H.M. Bakker, and C.M.J. Pieterse. 1998. Systemic resistance induced by rhizosphere bacteria. Annual Review of Phytopathology 36: 453-483.

Love, S.L., B.B. Rhodes and P.E. Nugent. 1986. Controlled pollination transfer of a nuclear male-sterile gene from a diploid to a tetraploid watermelon line. Euphytica 35: 633-638.

Lovic, Branko. 2001. Production steps to reduce seed contamination by pathogen of cucurbits. HortSceience 36(3): 479.

Lower, R.L. and K.W. Johnson. 1969. Observations on sterility of induced auto-tetraploid watermelons. J. Amer. Soc. Hort. Sci. 94(4): 367-369.

Malamy, J., J.P. Carr, D.F. Klessig, and I. Raskin. 1990. Salicylic acid: A likely endogenous signal in the resistance response of tobacco to viral infection. Science 250: 1002-1004.

Maynard, D.N. 1989. Triploid watermelon seed orientation affects seed coat adherence on emerged cotyledons. HortScience 24(4): 603-604.

Metraux, J.P., H. Signer, J. Ryals, E. Ward, M. Wyss-Benz, J. Gaudin, K. Raschdorf, E. Schmid, W. Blum, and B. Inverardi. 1990. Increase in salicylic acid at the onset of systematic acquired resistance in cucumber. Science 250: 1004-1006.

Murdock, Brent Alan. 1993. Characterization of male sterility and polyploidy in watermelon. PhD Dissertation, Clemson University, Clemson, SC.

Mussen, Eric C. and Robbin W. Thorp. 1997. Honey bee pollination of cantaloupe, cucumber, and watermelon. University of California Division of Agriculture and Natural Resources Publication 7224. p. 3.

Myer. G. de, K. Capieau, K. Audenaert, A. Buchala, J.P. Metraux, and M. Hofte. 1999. Nanogram amounts of salicylilc acid produced by the rhizobacterium *Pseudomonas aeruginosa* 7NSK2 activate the systemic acquired resistance pathway in bean. Molecular Plant-Microbe Interactions 12(5): 450-458.

Navot, N. and D. Zamir. 1986. Linkage relationships of 19 protein coding genes in watermelon. Theor. Appl. Genet. 72: 274-278.

Navot, N., M. Sarfatti, and D. Zamir. 1990. Linkage relationships of genes affecting bitterness and flesh color in watermelon. J. Hered. 81: 162-165.

Oda, M. 1999. Grafting of vegetables to improve greenhouse production. FFTC Extension Pulletins. December:1-11. http://www.agnet.org/library/article/eb480.html

Oda, M., K. Okada, K. Sasaki, S. Akazawa, and M. Sei. 1997. Growth and yield of eggplants grafted by a newly developed robot. HortScience 32: 848-849.

Ok, S., Chung, Y.S., Um, B.Y., Park, M.S., Bae, J.M., Lee, S.J., and Shim, S.J. (2000). Indentification of expressed sequence tags of watermelon (*Citrulus lanatus*) leaf at the vegetative stage. Plant Cell Reports. 19: 932-937.

Orober, M., J. Siegrist, and H. Buchenauer. 1999. Induction of systemic acquired resistance in cucumber by foliar phosphate application. In: H. Lyr, P.E. Russell, H.-W. Dehne and H.D. Sisler, editors. Modern Fungicides and Antifungal Compounds. II. 12th International Reinhardsbrunn Symposium, Friedrichroda, Thuringia, Germany 24th-29th May 1998. pp. 339-348.

Peer, R., G.J. van, Nieman, and B. Shippers. 1991. Induced resistance and phytoalexin accumulation in biological control of Fusarium wilt of carnation by *Pseudomonas* sp. strain WCS417r. Phytopathology 81: 728-733.

Porter, D.R. 1933. Watermelon breeding. Hilgarda 7: 585-624.

Randhawa, Parm. 1994. An improved protocol for watermelon fruit blotch seed assay. Seed and Plant Lab Notes. 2: 1.

Rhodes, B.B. 1991. Late male fertility in a glabrous, male-sterile (*gms*) watermelon line. Cucurbit Genetics Cooperative 14: 85-89.

Rhodes, B.B., J.N. Zhang, X. Zhang, and W.C. Bridges. 1992. Improved germination rate after 8 years in certain watermelon families. Cucurbit Genetics Cooperative 15:82-83.

Rhodes, B.B. and X. Zhang. 1999. Hybrid seed production in watermelon. J. New Seeds 1(3/4): 69-88.

Rhodes, B.B., X. Zhang, W.V. Baird, and H.T. Knapp. 1999. A tendrilless mutant in watermelon: phenotype and inheritance. Cucurbit Genetics Cooperative 22: 28-30.

Rhodes, B.B. and F. Dane. 1999 Gene list for watermelon (*Citrullus lanatus* L.). Cucurbit Genetics Cooperative 22: 61-77.

Rhodes, B.B., C.O. Huey, Jr., A. Abramovich, and T.B. Platt. 2000. An overview of enhancement of triploid watermelon seed germination. J. New Seeds 2(2): 69-76.

Robinson, R.W. 1999. Hybrid seed production in vegetables. J. New Seeds 1(3/4): 1-47.

Sachs, M. 1977. Priming of watermelon seeds for low-temperature germination. J. Amer. Soc. Hort. Sci. 102(2): 175-178.

Schuster, D.J. and J. E. Polston. 1998. Microbe-induced resistance: a novel strategy for control of insect-transmitted diseases in vegetables. University of Minnesota CICP. National IPM Network.

Siegrist, J., S. Muhlenbeck and H. Buchenauer. 1998. Cultured parsley cells, a model system for the rapid testing of abiotic and natural sustances as inducers of systemic acquired resistance. Physiological and Molecular Plant Pathology 53(4): 223-238.

Singletary, C.C. and E.L. Moore. 1965. Hybrid watermelon seed production. Mississippi Farm Res. 28(6): 5.

Stanghellini, M.S., J.T. Ambrose, and J.R. Schulteis. 1997. The effects of honey bee and bumble bee pollination on fruit set and abortion of cucumber and watermelon. American Bee Journal 137: 386-391.

Thorton, M.L. 1968. Seed dormancy in watermelon *Citrullus vulgaris* Schrad. Proceedings of the Association of Official Seed Analysts 58: 80-84.

Wang, Y.Y., C.L. Li, Z.R. Jiang, and X.Y. Gao. 1980. Clonal propagation of seedless watermelon. Acta Hortic. 7(4): 64 (in Chinese).

Watts, V.M. 1962. A marked male-sterile mutant in watermelon. J. Ameri. Soc. Hort. Sci. 81: 498-505.

Watts, V.M. 1967. Development of disease resistance and seed production in watermelon stocks carrying *msg* gene. J. Ameri. Soc. Hort. Sci. 91: 579-580.

Wees, S.C.M. van, M. Luijendijk, I. Smoorenburg, L.C. van Loon and C.M.J. Pieterse. 1999. Rhizobacteria-mediated systemic resistance (ISR) in Arabidopsis is not associated with a direct effect on expression of known defense-related genes but stimulates the expression of the jasmonate-inducible gene Atvsp upon challenge. Plant Molecular Biology 41(4): 537-549.

Wei, G., J.W. Klepper, and S. Tuzun. 1991. Induction of systemic resistance of cucumber to *Colletotrichum orbiculare* by select strains of plant growth-promoting bacteria. Phytopathology 81: 1508-1512.

Whitaker, T.W., and G.N. Davis. 1962. Cucurbits Botany, Cultivation and Utilization. Interscience Publishers, Inc., New York.

Whitaker, T.W., and W.B. Bemis. 1976. Cucurbits. In Evolution of Crop Plants. Edited by N.W. Simmonds. Longman, London. pp. 64-69.

Widders, I., R. Hammerschmidt, M. Uebersax, and L. Newell. 1992. Evidence for acquired disease resistance to angular leaf spot in field grown pickling cucumbers. 1992 Cucurbit Conference, Raleigh, NC, Sept. 20-23.

Xia, X., Y. Liu, W. Liu, and A. Chen. 1988. Selection of watermelon (*Citrullus vulgaris*) male-sterile line G17AB. J. Shengyang Agr. Univ. 19: 9-13.

Xie, Wenhua. 1985. Cucumber seeds' rest and effect of H_2O_2 on their germination. Chinese Vegetables 2: 1.

Zhang, X., and M. Wang. 1987. Studies of watermelon germplasm resources and breeding. III: Correlation between parents and their F1, phenotypic correlation among characters and path analysis. Acta Univ. Septentrionali Occiden. Agric. 15(1): 82-87.

Zhang, X. and B. Rhodes. 1993. *Male Sterile, ms*, in watermelon not linked to delayed green, *dg* and *I-dg*. Cucurbit Genetics Coop Rpt. 16: 79.

Zhang, X.P., H.T. Skorupska, and B.B. Rhodes. 1994. Cytological expression in the male-sterile *ms* mutant watermelon. J. Heredity 85:279-285.

Zhang, X.P., B.B. Rhodes, H.T. Skorupska, and W.C. Bridges. 1995. Generating tetraploid watermelon using colchicine in vitro. G. Lester and J. Dunlap et al. (eds.) Cucurbitaceae '94. Evaluation and Enhancement of Cucurbit Germplasm, Gateway Publ., Edinburgh, TX, pp. 134-139.

Zhang, X.P., B.B. Rhodes, W.V. Baird, H.T. Skorupska, and W.C. Bridges. 1996. Phenotype, inheritance, and regulation of expression of a new virescent mutant in watermelon: Juvenile albino. J. Amer. Soc. Hort. Sci. 121(4): 609-615.

Zhang, X.P., B.B. Rhodes, W.V. Baird, W.C. Bridges, and H.T. Skorupska. 1996. Development of genic male-sterile watermelon lines with juvenile albino seedling marker. HortScience 31(3): 426-429.

Zhang, X.P., B.B. Rhodes, W.V. Baird, H.T. Skorupska, and W.C. Bridges. 1996. Development of genic male-sterile watermelon lines with delayed-green seedling marker. HortScience 31(1): 123-126.

Zhang, X.P., and B.B. Rhodes. 2000. Method using male sterility and a marker to produce hybrid seeds and plants. U.S. Patent 6,018,101.

Zehnder, G., J. Kloepper, C. Yao, and G. Wei. 1997a. Induction of systemic resistance in cucumber against cucumber beetles by plant growth-promoting rhizobacteria. J. Econ. Entomol. 90: 391-396.

Zehnder, G., J. Kloepper, S. Tuzun, C. Yao, G. Wei, O. Chamblis, and R. Shelby. 1997b. Insect feeding on cucumber modified by rhizobacteria-induced plant resistance. Entomol. Exp. Et Applicata 83: 81-85.

Zitter, T.A., D.L. Hopkins, and C.E. Thomas. 1996. Compendium of Cucurbit Diseases. APS Press. The American Phytopathological Society. St. Paul, MN 55121-2097.

Hybrid Melon Development

A. D. Munshi
J. M. Alvarez

SUMMARY. Melons are an unavoidable item of western dietary. It is
an unique crop where in one hand several pollination control mechanism,
viz., genic male sterility, monoecy and gynoecy are available and had
been exploited successfully for the commercial hybrid seed production,
whereas in other hand most of the varieties had been developed through
conventional breeding by continuous inbreeding and selection without
loosing vigor, which is usually unexpected in cross-pollinated crop. The
crop is highly affected with insect pest and diseases. The studies have
been made on these aspects and the varieties are released for their commercial exploitation. *[Article copies available for a fee from The Haworth Document Delivery Service: 1-800-HAWORTH. E-mail address: <docdelivery@
haworthpress.com> Website: <http://www.HaworthPress.com> © 2004 by The
Haworth Press, Inc. All rights reserved.]*

KEYWORDS. Melon, floral biology, pollination, genetics, hybridization, sex expression, heterosis, male sterility, gynoecy, monoecy, hybrid
seed production

Melon (*Cucumis melo* L., 2n = 2x = 24) is annual climbing, creeping or
trailing vine of length up to 3 m. The dessert types quench thirst and add to the

A. D. Munshi is affiliated with the Division of Vegetable Crops, Indian Agriculture
Research Insitute, New Delhi 110012, India.

J. M. Alvarez is affiliated with the Unidad de Tecnología en Producción Vegetal,
Zaragoza.

[Haworth co-indexing entry note]: "Hybrid Melon Development." Munshi, A. D., and J. M. Alvarez.
Co-published simultaneously in *Journal of New Seeds* (Food Products Press, an imprint of The Haworth
Press, Inc.) Vol. 6, No. 4, 2004, pp. 323-362; and: *Hybrid Vegetable Development* (ed: P. K. Singh, S. K.
Dasgupta, and S. K. Tripathi) Food Products Press, an imprint of The Haworth Press, Inc., 2004, pp. 323-362.
Single or multiple copies of this article are available for a fee from The Haworth Document Delivery Service
[1-800-HAWORTH, 9:00 a.m. - 5:00 p.m. (EST). E-mail address: docdelivery@haworthpress.com].

http://www.haworthpress.com/web/JNS
© 2004 by The Haworth Press, Inc. All rights reserved.
Digital Object Identifier: 10.1300/J153v06n04_02

nutrient content of main diet. Different melon types like netted, salmon flesh cantaloupe, the smooth skinned green fleshed 'Honey Dew', the wrinkled-skinned white fleshed 'Golden Beauty' and several other melons are an unavoidable item of western dietary. Other forms with very different in plant and fruit characters are found in India. In addition, several wild form and non desert types are found in Africa and India. Edible portion of melon contain water (90 percent) and carbohydrate (10 percent). Melons are good sources of vitamin A. The yellow and orange fleshed melon contains more than 350 mg of β-carotene, a precursor of vitamin A. Cantaloupe contains 45 mg and Honey Dew contains 32 mg of vitamin C per 100 g of edible portion. It is an unique crop where in one hand several pollination control mechanism, viz., genic male sterility, monoecy and gynoecy are available and had been exploited successfully for the commercial hybrid seed production, whereas in other hand most of the varieties had been developed through conventional breeding by continuous inbreeding and selection without loosing vigor, which is usually unexpected in cross pollinated crop.

ORIGIN AND TAXONOMY

The place of origin of muskmelon is not known with certainty, but as the wild species of *Cucumis* occur in Africa, it is likely that it originated on that continent more specially in the eastern region of south Sahara desert. Secondary centers of diversity occur in India, China, Persia, and southern Russia. Whitaker and Davis (1962) and Whitaker (1979) thought that central Asia comprising some parts of southern Russia, Iran, Afghanistan, and northwest India may be regarded as a secondary center of muskmelon. According to Gane san (1991), India was one of the secondary centers of origin of melon and subsequent diversification had resulted in several cultivated and wild botanical varieties. Lal and Dhaliwal (1992) outlined the origin and introduction of melon in India. Pitrat et al. (2000) identified the synonymous epithets in different combination and gave comments on in intraspecific classification of cultivars of melon.

The species *Cucumis melo* L. is a large polymorphic, in comprising a larger number of botanical and horticultural varieties. Horticulturally, muskmelon and cantaloupe differ some what in physical characteristics and regional adaptation. Today, cantaloupe simply refers to cultivars that are highly uniform in overall netting and corky tissue on the rind. Internally, the cantaloupe flesh is thick, salmon orange in color with a characteristics flavor and small seed cavity. Muskmelon cultivars on the other hand, have a stronger aroma, skin color varies from salmon orange, green and white, juicier flesh and larger seed cavity. In the USA, most of the cantaloupe cultivars have been developed to meet requirements for packing in crates and long distances transportation.

Naudin (1859), an African botanist, divided this species into several botanical varieties adopting a trinomial classification. Even though this classification is not botanically recognized, because of absence of distinct botanical variation and of free. Inter-crossability, the botanical epithets used by him are still in current usage.

C. melo var. agrestis–Plant with slender vines, small flowers, and inedible fruits.

C. melo var. reticulatus–Netted melons, fruit medium size, surface strongly tested often length wise, rind fleshy with tough skin. Most American cultivars like cantaloupe and Persian melon belong to this group.

C. melo var. cantaloupensis–Cantaloupe melons, fruits warty, scaly and rough with hard skin, warted surface, distinct, netting absent, still cultivated in parts of Europe.

C. melo var. inodorus–Called winter melons of the USA, comprising Honey Dew (green flesh), Casaba and Crenshaw. Fruits with little musky odor, ripening late fruits surface usually smooth, leaves light green or medium green.

C. melo var. flexuosus–Snake or serpent or long melon fruit long and slender 2.5-5.0 cm in diameter, 45-90 cm long. It is same as C. melo var. utilissimus of India, known as kakri, tar, etc., and eaten like salad cucumber.

C. melo var. conomon–Pickling melon, fruit smooth, glabrous to without musky odor

C. melo var. chito–Mango lemon or lemon cucumber, fruit small with size of an ordinary orange; also known as vegetable orange or melon apple, etc.

C. melo var. dudain–Fruits small like chito, surface marbled with rich brown, very fragment grown for ornamental purpose.

C. melo var. momordica–It is known as phoot in north and eastern India. Also called snap melon whose fruit skin burst or cracks on maturity. The rind is yellow, flesh fluffy and sourish.

Most of the central Asian varieties like those of Afghanistan, Uzbekistan, are long duration types grown during dry period April-May to October. These are generally oblong or round, netted or smooth skinned, green, orange or white in flesh color 'Sarda' melon available in north Indian cities during Sept.-Oct. is one of such varieties imported from Afghanistan. Among the leading cantaloupe cultivars, Hales Best, Edisto, Campo, Jacumba, Top Mark, Perlita, Weslan, Planter, Jumbo, Primo, Caravelle, Explorer, Laguna, Mission, etc., can be mentioned. In the persion group, Persian Large and Persian Small are recognized. Among the long duration winter melons, Honey Dew, Casaba, Golden Beauty, Crenshaw and Santa Clans are important. Melon known as Charentais in France is sometimes grown in green houses. Improved varieties of muskmelon developed by different ICAR Institute and State Agricultural Universities in India are Pusa Sharbati (IARI, New Delhi), Pusa Madhuras (IARI, New Delhi), Pusa Rasraj (F_1 hybrid, IARI, New Delhi), Hara Madhu (PAU, Ludhiana), Punjab Sunheri (PAU, Ludianan), Punjab Hybrid (F_1 hybrid, PAU, Ludiana), Arka Rajhans (IIHR, Bangalore), Arka Jeet (IIHR, Bangalore), and Durgapura Madhu (Dept. of Agricultural, Jaipur). Some private seed companies in India are marketing hybrid muskmelon Sona (Cantaloupe, Indo-American Hybrid Seed, Bangalore), Swarna (Cantaloupe, Indo-American Hybrid Seed, Bangalore), MHC-5 (Maharastra Hybrid Seed Co., Jalna), and MHC-6 (Maharastra Hybrid Seed Co., Jalna).

There are several local cultivars of musk melon grown in different regions of India (Nandpuri, 1989; More et al., 1990). Among the fully-netted desert cultivars are Lucknow Safeda, Kanpuria, Jogia, Mathuria, Batti, Allahabad Kajra, Jaunpuri, Bagpat, Mau, and Mahaban in U.P.; Sunkheda and Galteshwar and Panam in Gujarat; Kharri Jalgaon, Kabri Gurbeli and Kavita in Madhya Pradesh; Sharbat-e-Anar, Bhathesa, Papaya, Chiranji and Ladoo in Andhra Pradesh; Kadapa in Karnataka; Batti, Goose and Jam Neel in Maharastra; Kutana and Bagpat in Haryana; Sanganer, Hara Gola, Motta and Tonk melon in Rajasthan; and Haridhari and Musa in Punjab.

BOTANY

It is a polymorphic species where most cultivars are andromonoecious (staminate (male) and perfect flowers (hermaphrodite) on the same plant), but other sex forms (monoecious, gynomonoecious, and hermaphrodite) are also reported. Male flower appear first as a cluster on main and secondary branches but hermaphrodite flowers appear alone on secondary, or higher order, branches on short stout pedicels. Stem is soft-hairy to glabrous, striate or angled, leaves, palmate types, orbicular to ovate to reniform, usually five angled, some time shallowly five to seven, lobed, hairy or some what scabrous, 3-5 inches across. The shape of the ovary varies from ovoid to elongated. After pollination, the ovary wall rapidly expands and develops into pericarp with an exocarp, mesocarp and endocarp. The edible portion is mostly mesocarp. The number of fruits that can reach maturity never exceeds 4 to 5 because of physiological balance. Immature fruits are pubescent, mature fruits are glabrous. The small white to tan seeds are *oblong* to *elliptical*. Melon stems are nearly round as compared to angular in cucumber and the pubescence of melon is not as harsh. Plant are typically trailing vine and tendrils are unbranched and borne single at the nodes. Leaves are usually sub-cordate and less lobed than those of cucumber. The fruits of different cultivars are quite diverse making melon the most variable species within the genus. Some cultivars have fruit with reticulate netting on the rind where as other have smooth rind. Indented vein tracts distinguish in various cultivars but these sutures are lacking in other cultivars. Rind color varies from green some times with white stripes to yellow, tan or white when mature. Flesh color may be orange, pink, green or white. An abscission layer is formed at the attachment of the fruit to the peduncle in some cultivars causing the fruit to separate from the vine at maturity. In other cultivars, peduncle is persistent.

SELFING AND HYBRIDIZATION TECHNIQUE

Muskmelon is andromonoecious (hermaphrodite and staminate flowers) in nature. However, monoecious (pistillate and staminate flowers), and, sometimes, gynomonoecious (monoecious and perfect flowers), are also prevalent. Hybridization of andromonoecious type is a two step process. On the day prior to anthesis, the hermaphrodite flower is emasculated with the help of a fine tipped forceps. The proper size of the bud selected for emasculation decides to a great extent the success of hybridization. Usually buds about to open or due to open the following days are selected. Both pistillate and staminate flowers are covered to prevent insect contamination. Pistillate flower is covered with perforated butter paper bag, while the staminate flower is covered by cotton. Hand pollination is done during the morning (6 a.m.-7 a.m.) by gently rubbing pollen from anther of staminate flower to the stigma of the pistillate flower. After pollination pistillate flower is again covered by butter paper bag to prevent contamination. Muskmelon pollen is sticky so hands and forceps should be washed with alcohol to prevent contamination. Emasculation is not required in monoecious and gynoecious types. In case of selfing pollen from one plant is used to pollination the stigma of the pistillate flowers from the same plant. It has been established that percent selfing is high in early set fruits as compared to late set fruits. The fruits reach maturity at about 48 days after pollination depending on the variety. The efficiency of utilization of the pistillate flowers under hand pollination is doubled in monoecious lines over that in the andromonoecious lines with hermaphrodite flowers (More and Seshadri, 1998). Hand pollination of andromonoecious melon is more difficult than in monoecious cultivars since the anther need to be manually excised from each perfect flower.

FLORAL BIOLOGY AND POLLINATION BEHAVIOR

Most of the muskmelon cultivars are andromonoecious in nature but gynomonoecious and monoecious cultivars have also been reported (Peterson et al., 1983; More et al., 1980). Muskmelon do not suffer from inbreeding depression and in many ways their population structure is more similar to that of inbreeder than outbreeder. This was supported by Lippert and Leg (1972a,b) who suggested that muskmelon can be handled in breeding program in similar way as that of self-pollinated crop.

In muskmelon, the pollen grains adhere together and honey bees are required for pollen transfer thus require cross-pollination. So honey bees are the main pollinating agent. Anthesis in musk melon takes place in the morning around 5.30 a.m. to 6.30 a.m. at a temperature of 22-29°C (Nandpuri and Brar, 1966). Dehiscence take place shortly after anthesis. The stigma becomes receptive from 2 hours before to 2-3 hours after anthesis. Regarding the extent of out crossing, various workers reported varying amounts of out-crossing ranging from 5.40 to 67.80 percent with the average range of 20-30% (Whitaker and Bohn, 1952). It is crossable with long melon and snap melon but not with round melon and cucumber (Dutta and Nath, 1970). Rosa (1927) studied six andromonoecious and one monoecious cultivars and found that andromonoecious lines had 5.4 to 67.8% out crossing while the monoecious lines had 73.2% out crossing. It has been established that success of selfing is high in early set fruit as compared to late set fruit. The varying amount of out crossing could be attributed to genotypic behavior, sexform, isolation distance, wind direction, and bee activity. Fruit set by controlled pollination is slightly higher in the early part of the season than in the later part. Awasthi (1969) observed higher fruit set (98%), under natural pollination as compared to hand pollination (68%). Under open natural system, it is partly cross and partly self-pollinated by bees. Anyway vigoros lines could be obtained after 25 generations of selfing (Robinson et al., 1976). The efficiency of crossing by hand pollination is doubled in monoecious lines over that in the andromonoecious lines with hermaphrodite flowers because emasculation is not needed in the former lines.

GENETICS

Plant breeding is the application of genetic principles to the improvement of plants. A knowledge of the genetics of the various qualitative and quantitative characters of muskmelon is necessary for the improvement of various traits in muskmelon. Robinson et al. (1976) reviewed the genetics of Cucurbitaceae family and out of total 170 genes known for the family, 37 were reported to be of melon (Table 1). Many additional muskmelon genes and alleles have been identified in the past two decades. The Cucurbit Genetic Co-Operative published and updated the gene list of major cucurbitaceous crops including melon. In the recently published gene list, Pitrat (1990, 1994, 1998) reported genes of important traits in melon, i.e., flower, disease resistance, insects resistance, and isoenzymes (Table 2).

GENETICS AND INHERITANCE OF SEX EXPRESSION

Diversity in sex expression within genera is more prevalent among cucurbits than in most plant families (Whitaker, 1931). Most cucurbit species are monoecious but many are dioecious, and a few species bear hermaphroditic flowers. Intermediate sex forms are also frequent in the Cucurbitaceae. Plants may be andromonoecious, with both perfect and staminate flowers, and gynomonoecious, with both hermaphroditic and female flowers (Roy and Saran, 1989). Six different sex forms have been reported in *C. melo*, i.e., monoecious, andromonoecious, gynomonoecious, hermaphrodite, trimonoecious, and gynoecious (Poole and Grimball, 1939). Sex expression seems to be controlled by two genes A and G, whose interaction is assumed to determine the inheritance of the various sex forms. The genotypes suggested by Poole and Grimball (1939) for the different sex forms of melon are:

Monoecious: $a^+/-$, $g^+/-$

Andromonoecious: a/a, $g^+/-$

Gynomonoecious or gynoecious: $a^+/-$, g/g

Hermaphrodite: a/a, g/g

In cucumber, the development of true gynoecious breeding lines has led to the exploitation of heterosis and development of hybrids (More and Munger, 1986). In melon, most cultivars are andromonoecious and the utilization of gynoecious lines in muskmelon hybrid seed production was earlier suggested by Anais (1971). The use of gynoecious lines as seed parents in melon hybrid seed production would eliminate hand pollination, chemical treatments, and manual rouging, which are necessary when andromonoecious, gynomonoecious, and monoecious lines are used as seed parents; but stable gynoecious breeding lines have only become available in the early 80s (Peterson et al., 1983).

Poole and Grimball (1939) considered gynoecious forms to be a transitory expression caused by environmental fluctuations; however, Kubicki (1966) suggested that gynoecious genotypes were controlled by genetic factors different from the above, and according to Frankel and Galun (1977), gynoecious sex expression may be modified by minor genes. A muskmelon gynoecious germplasm with acceptable fruit quality, 'Wisconsin-998' ('WI-998'), was released by Peterson et al. in 1983. The population of 'WI-998' segregated for gynoecious and gynomonoecious expression, and at some stage, homozygous, gynoecious plants could be identified in this population. Kenigsbuch and Cohen (1990) studied the genetic control of gynoecy in 'WI-998' and concluded

TABLE 1. Mode of inheritance of various qualitative and quantitative traits in muskmelon

Character/Attribute	Mode of inheritance
A. QUALITATIVE	
Rind color	Yellow monogenic and partially dominant over cream (Bains, 1960; Bains and Kang, 1963; Sandhu, 1990); dark green dominant over green, greenish grey dominant over the light green, stripped and mottled pericarp dominant over white (Dutta, 1967); green monogenically dominant yellow (Chadha et al., 1972).
Flesh color	Orange monogenically dominant over light green (Bains and Kang, 1963; Sandhu, 1990). White monogenically dominant over green (Chadha et al., 1972); complex inheritance with some maternal effects (Ramaswamy et al., 1977).
Flesh thickness	Thinner partially dominant over thicker and monogenic (Bains, 1960; Bains and Kang, 1963).
Flesh taste	Monogenic with partial dominance (Sandhu, 1990).
Juiciness	Less juiciness monogenically dominant over juiciness (Chadha et al., 1972).
Fruit slipping	Slipping monogenically dominant over non-slipping (Sandhu, 1990).
Rind cracking	Digenic with both complementary and supplementary gene action (Sandhu, 1990).
Ribbing	Pronounced and regular ribbing dominant over non-ribbing (Dutta, 1967).
Sutures	Suture less monogenically dominant over sutured (Bains and Kang, 1963; Ramaswamy et al., 1977; Sandhu, 1990); absence of sutures digenically dominant with supplementary relationship (Chadha et al., 1972).
Netting	Digenic with netting dominant over smoothness, 'N' gene controls netting, 'n' controls smoothness, 'S' gene suppresses netting when present in homozygous condition and 'N' is epistatic over 'S' and so was 'n' over 's' (Ramaswamy et al., 1977); netting monogenic and partially dominant over smoothness (Sandhu, 1990).
Resistant to fruit fly	Susceptibility digenic with complementary gene action (Sambandam and Chelliah, 1972).
Resistance to red pumpkin beetle	Resistance monogenically dominant over susceptibility (Vashistha and Choudhury, 1974).
Resistance to powdery mildew	Resistance dominant over susceptibility (Nath et al., 1973); resistance recessive to susceptibility (Choudhury and Sivakami, 1972).

Character/Attribute	Mode of inheritance
B. QUANTITATIVE	
Earliness	Dominance variance more than the additive variance, over dominance also present (Chadha et al., 1972; Munshi and Verma, 1998, 1999); earliness partially dominant over lateness (Chadha, 1973; Singh et al., 1976), both additive and dominance genetic variance significant, dominance variance exceeded additive variance (Singh et al., 1989); dominance and additive gene effects in some crosses while additive, dominance as well as epistatis in others (Sandhu, 1990).
Rinds thickness	Additive, dominance and epistasis (Sandhu, 1990).
Flesh thickness	Partial dominance along with adequate additive genetic variance (Chadha et al., 1972; Munshi and Verma, 1998, 1999); dominance gene effects (Kalloo and Dixit, 1983); additive, dominance and epistatis (Sandhu, 1990).
T.S.S.	Partially dominant along with adequate additive genetic variance (Chadha et al., 1972, Munshi and Verma, 1998, 1999); dominance gene effect (Kallo and Dixit, 1983); more than seven groups of genes with over dominance (Swamy and Dutta, 1985); additive, dominance and epistatis (Sadhu, 1990).
Number of fruits per vine	Partial dominance along with adequate additive genetic variance (Chadha et al., 1972); dominance gene effects (Kalloo and Dixit, 1983); additive, dominance and epistatis (Sandhu, 1990).
Fruit weight	Thirty-nine genes with geometrically cumulative effects, small fruited-ness partially dominant over larger fruited-ness (Sambandam and Chelliah, 1972); partial dominance with additive genetic variance (Munshi and Verma, 1998, 1999).
Total yield	Dominant variance more than additive variance over dominance also present (Chadha et al., 1975; Munshi and Verma, 1998, 1999); dominance (Kalloo and Dixit, 1983); additive gene effects (Sandhu, 1990).
Vine length	Both additive and dominant genetic variances significant, dominance variances higher than additive variance (Singh et al., 1989).

that there is a gene that interacts with a and g to produce stable gynoecious plants, they symbolized this gene as m, and suggested that gynoecious stable plants are genotypes a^+a^+ggmm. The genotype a^+aggmm is expressed also as a gynoecious sex type, but produced hermaphrodite and gynoecious segregated upon selfing.

Another gene that has been claimed to influence sex expression in melon is the abrachiate mutant ab (Foster and Bond, 1967) that interacts with a and g (e.g., $ababaag^+g$ plants produce only staminate flowers).

TABLE 2. *Cucumis melo* Gene Index (adapted from Pitrat, 1998)

Preferred symbol	Synonym	Character
a	(M)	*andromonoecious.* Mostly staminate, fewer perfect flowers; interact with g
ab		*abrachiate.* Lacking lateral branches; ab a plants produced only staminate flowers
Ac		*Alternaria cucumerina* resistance (in MR-1)
Af*		*Aulacophora foveicollis* resistance. Resistance to the red pumpkin beetle
Ag		*Aphis gossypii* tolerance. Freedom from leaf curling following aphid infestation
Ala		*Acute leaf apex.* Dominant over obtuse apex, linked with lobed leaf. (Ala in Maine Rock, ala in PV Green)
Al-1	Al_1	*Abscisión layer-1.* One of two dominant genes for abscisión layer formation. See Al-2. (Al-1 Al-2 in C68, al-1 al-2 in Pearl)
Al-2	Al_2	*Abscisión layer-2.* One of two dominant genes for abscisión layer formation. See Al-1.
Ap-1^1	APS-1^1	*Acid phosphatase-1^1.* One of two codominant alleles, each regulating one band. The heterozygote has two bands. See Ap-1^2
Ap-1^2	APS-1^2	*Acid phosphatase-1^2.* One of two codominant alleles, each regulating one band. The heterozygote has two bands. See Ap-1^1
bd		*Brittle dwarf.* Rosette growth with thick leaf. Male fertile, female sterile (in TAM-Perlita45).
Bi		*Bitter.* Bitter seedling (common in honeydew or in Charentais types, while most American cantaloupes are Bi).
Bif-1	Bif	*Bitter fruit-1.* Bitterness of tender fruit in wild melon. Relations with Bi are unknown.
Bif-2		*Bitter fruit-2.* One of two complementary independent genes for bitter taste in young fruit. Bif-2 Bif-2 Bif-3 Bif-3 are bitter. (Relations with Bi and Bif-1 are unknown.)
Bif-3		*Bitter fruit-3.* One of two complementary independent genes for bitter taste in young fruit. Bif-2 Bif-2 Bif-3 Bif-3 are bitter. (Relations with Bi and Bif-1 are unknown.)
cab-1		*cucurbit aphid borne* yellows virus resistance-1. One of two complementary independent genes for resistance to this virus: cab-1 cab-1 cab-2 cab-2 plants are resistant (in PI 124112).
cab-2		*cucurbit aphid borne* yellows virus resistance-1. One of two complementary independent genes for resistance to this virus: cab-1 cab-1 cab-2 cab-2 plants are resistant (in PI 124112).

Preferred symbol	Synonym	Character
cb	cbl	*cucumber beetle resistance.* Interacts with Bi, the non-bitter bibi cbcb being the more resistant (in C922-174-B).
cf		*cochleare folium.* Spoon-shaped leaf with upward curling of the leaf margins (spontaneous mutant in Galia).
cl		*curled leaf.* Elongated leaves that curl upward and inward. Usually male and female sterile.
dc-1		*Dacus cucurbitae-1 resistance.* One of two complementary recessive genes for resistance to the melon fruit fly. See dc-2.
dc-2		*Dacus cucurbitae-2 resistance.* One of two complementary recessive genes for resistance to the melon fruit fly. See dc-1.
dl		*dissected leaf* (in USSR 4). Highly indented leaves.
dl^v	cl	*dissected leaf Velich.* First described as *cut leaf* in Cantaloup de Bellegarde. Allelic to dl.
dl-2		*dissected leaf-2.* First described as 'hojas hendidas.'
dlet	dl	*delayed lethal.* Reduced growth, necrotic lesions on leaves and premature death.
f		*flava.* Chlorophyl deficient mutant. Growth rate reduced (in K 2005).
fas		*fasciated* stem (in Vilmorin 104).
fe		*Fe* (iron) inefficient mutant. Chlorotic leaves with green veins. Turns green when adding Fe in the nutrient solution.
Fn		*Flaccida necrosis.* Semi-dominant gene for wilting and necrosis with F pathotype of Zucchini Yellow Mosaic Virus (Fn in Doublon, fn in Védrantais).
Fom-1	Fom$_1$	*Fusarium oxysporum melonis* resistance. Resistance to races 0 and 2 of Fusarium wilt.
Fom-2	Fom$_2$	*Fusarium oxysporum melonis* resistance. Resistance to races 0 and 1 of Fusarium wilt.
Fom-3		*Fusarium oxysporum melonis* resistance. Same phenotype as Fom-1 but segregates independently from Fom-1 (Fom-3 in Perlita FR, fom-3 in Charentais T).
g		*gynomonoecious.* Mostly pistillate, fewer perfect flowers; a g plants produced perfect flowers exclusively.
gf		*green flesh* color. Recessive to salmon.
gl		*glabrous.* Trichomes lacking.
gp		*green petals.* Corolla leaf-like in color and venation.
Gs		*Gelatinous sheath* around the seeds. Dominant to absence of gelatinous sheath.

TABLE 2 (continued)

Preferred symbol	Synonym	Character
gyc		*greenish yellow* corolla.
gy	n, M	*gynoecious*. Interacts with a and g to produce stable gynoecious plants (A- gg gygy in WI 998).
h		*halo cotyledons*. Yellow cotyledons, later becoming green.
jf		*juicy flesh*. Segregates discretely in a monogenic ratio in segregating generations.
L		*Lobed leaf*. Dominant on non-lobed. Linked with *Acute leaf* apex.
lmi		*long mainstem internode*. Affects internode length of the main stem but not of the lateral ones.
Mc		*Mycospharella citrullina* resistance. High degree of resistance to gummy stem blight.
MC-2	(MC-i)	*Mycospaerella citrullina* resistance. High degree of resistance to gummy stem blight.
Mca		*Macrocalyx*. Large, leaf like structure of the sepals in staminate and hermaphrodite flowers (Mca in makuwa, mca in Annamalai).
Me		*Mealy* flesh texture. Dominant to crisp flesh (Me in C. callosus, me in makuwa).
ms-1	(ms^1)	*male sterile-1*. Indehiscent anthers with empty pollen walls in tetrads.
ms-2	(ms^2)	*male sterile-2*. Anthers indehiscent containing mostly empty pollen walls, growth rate reduced.
ms-3	ms-L	*Male sterile-3*. Waxy and translucent indehiscent anthers, containing two types of empty pollen sacs.
ms-4		*male sterile-4*. Small indehiscent anthers. First male flowers abort at bud stage (in Bulgaria 7).
ms-5		*male sterile-5*. Small indehiscent anthers. Empty pollen (in Jivaro, Fox).
Mt		*Mottled rind* pattern. Dominant to uniform color. Epistatic with Y (not expressed in Y-) and st (Mt- stst and Mt- StSt mottled; mtmt stst striped, mtmtstst uniform).
Mu		*Musky* flavor (olfactory). Dominant on mild flavor (Mu in *C. melo callosus*, mu in makuwa or Annamalai).
n		*nectar less*. Nectaries lacking in all flowers
Nm		*Necrosis* with *Morocco* strains of Watermelon Mosaic Virus.
nsv		Melon *necrotic spot virus* resistance.

Preferred symbol	Synonym	Character
O		*Oval fruit* shape. Dominant to round.
P		*pentamerous*. Five carpels and stamens; recessive to trimerous.
Pa		*Pale* green foliage. PaPa plants are white (lethal); Papa are yellow.
Pc-1		*Pseudoperonospora cubensis* resístanse. One of two complementary incompletely dominant genes for downy mildew resistance (in PI 124111). See Pc-2.
Pc-2		*Pseudoperonospora cubensis* resístanse. One of two complementary incompletely dominant genes for downy mildew resistance (in PI 124111). See Pc-1.
Pc-3		*Pseudoperonospora cubensis* resístanse. Partial resistance to downy mildew (in PI 414723).
Pc-4		*Pseudoperonospora cubensis* resístanse. One of two complementary genes for downy mildew resistance in PI 124112. Interacts with Pc-1 or Pc-2.
Pgd-1^1	6-PGDH-2^1 Pgd-2^1	*Phosphoglucodehydrogenase-1^1*. One of two codominant alleles that regulates 6-Phosphoglucodehydrogenase, each regulates one band. The heterozygote has one intermediate band. See Pgd-1^2.
Pgd-1^2	6-PGDH-2^2 Pgd-2^2	*Phosphoglucodehydrogenase-1^2*. One of two codominant alleles that regulates 6-Phosphoglucodehydrogenase, each regulates one band. The heterozygote has one intermediate band. See Pgd-1^1.
Pgi-1^1	PGI-1^1	*Phosphoglucoisomerase-1^1*. One of two dominant alleles, each regulating two bands. The heterozygote has three bands. See Pgi-1^2.
Pgi-1^2	PGI-1^2	*Phosphoglucoisomerase-1^2.* One of two dominant alleles, each regulating two bands. The heterozygote has three bands. See Pgi-1^1.
Pgi-2^1	PGI-2^1	*Phosphoglucoisomerase-2^1*. One of two dominant alleles, each regulating two bands. The heterozygote has three bands. See Pgi-2^2.
Pgi-2^2	PGI-2^2	*Phosphoglucoisomerase-2^2*. One of two dominant alleles, each regulating two bands. The heterozygote has three bands. See Pgi-2^1.
Pgm-1^1	PGM-2^1 Pgm-2^1	*Phosphoglucomutase-1^1*. One of two codominant alleles, each regulating two bands. The heterozygote has three bands. See Pgm-1^2.
Pgm-1^2	PGM-2^2	*Phosphoglucomutase-1^2*. One of two codominant alleles, each regulating two bands. The heterozygote has three bands. See Pgm-1^1.

TABLE 2 (continued)

Preferred symbol	Synonym	Character
Pm-1	(Pm[1]) Pm-A?	*Powdery mildew* resistance. Resistance to race 1 of *Sphaerotheca fuligenea*.
Pm-2	(Pm[2]) Pm-C?	*Powdery mildew* resistance-2, Interacts with Pm-1. Resistance to race 2 of Sphaerotheca fusca.
Pm-3	(Pm[3])	*Powdery mildew* resistance-3. Resistance to race 1 of *S. fusca*. Derived from PI 124111.
Pm-4	(Pm[4])	*Powdery mildew* resistance-4. Resistance to *S. fusca*. Derived from PI 124112.
Pm-5	(Pm[5])	*Powdery mildew* resistance-5. Resistance to *S. fusca*. Derived from PI 124112.
Pm-E		*Powdery mildew* resistance-E. Interacts with Pm-C in PMR 5 for *Erysiphe cichoracearum* resistance.
Pm-F		*Powdery mildew* resistance-F. Interacts with Pm-G in PI 124112 for *Erysiphe cichoracearum* resistance.
Pm-G		*Powdery mildew* resistance-G. Interacts with Pm-F in PI 124112 for *Erysiphe cichoracearum* resistance.
Pm-H		*Powdery mildew* resistance-H. Resistance to *Erysiphe cichoracearum* and susceptibility to *S. fusca* (in Nantais Oblong).
Pm-w	Pm-B?	*Powdery mildew* resistance in WMR 29. Resistance to *S. fusca* race 2.
Pm-x		*Powdery mildew* resistance in PI 414723. Resistance to *S. fusca*.
Prv[1]	Wmv	*Papaya Ringspot Virus* resistance. Resistance to W strain of PRSV. Dominant to Prv[2].
Prv[2]		*Papaya Ringspot Virus* resistance. Allele at the same locus as Prv1 but different reaction with some strains of the virus. Recessive to Prv[1].
Px-1[1]	PRX-1[1]	*Peroxidase-1[1]*. One of two codominant alleles, each regulating a cluster of four adjacent peroxidase bands. The heterozygote has five bands. See Px-1[2].
Px-1[2]	PRX-1[2]	*Peroxidase-1[2]*. One of two codominant alleles, each regulating a cluster of four adjacent peroxidase bands. The heterozygote has five bands. See Px-1[1].
Px-2[1]	Px2A Prx2	*Peroxidase-2[1]*. One of two codominant alleles, each regulating a cluster of three adjacent peroxidase bands. The heterozygote Px-21 Px-22 has 4 bands. See Px-2[2].
Px-2[2]	Px2B Prx2	*Peroxidase 2[2]*. One of two codominant alleles, each regulating a cluster of three adjacent peroxidase bands. See Px-2[1].

Preferred symbol	Synonym	Character
r		*red stem.* Red pigment under epidermis of stems, especially at nodes.
ri		*ridge.* Ridged fruit surface, recessive to ridge less.
s		*sutures.* Presence of vein tracts ('Sutures'); recessive to rib less.
Sfl	S	*Subtended floral leaf.* The floral leaf bearing the hermaphrodite flowers is sessile, small and encloses the flower.
si-1	B	*short internode-1.* Extremely compact plant habit (bush type) (in UC Topmark bush).
si-2		*short internode-2.* Short internodes from 'birdnest' melon (in Persia 202).
si-3		*short internode-3.* Short internodes in 'Maindwarf' melon.
Skdh-1^1		*Shikimate dehydrogenase-1.* One of two codominant alleles, each regulating one band (see Skdh-1^2).
Skdh-1^2		*Shikimate dehydrogenase-1.* One of two codomiant alleles, each regulating one band (see Skdh-1^1).
So		*Sour taste.* Dominant to sweet.
sp		*spherical fruit* shape. Recessive to obtuse; dominance incomplete.
st		*striped epicarp.* Recessive to nonstriped fruit.
v		*virescent.* Chlorotic foliage.
v-2		*virescent-2.*
v-3		*virescent-3.* White cotyledons which turn green, light green young leaves which are normal when they are older.
Vat		*Virus aphid transmission* resistance. Resistance to the transmission of viruses by *Aphis gossypii* (in PI 161375).
w		*white* color of mature fruit. Recessive to dark green fruit skin.
wf		*white flesh.* Recessive to orange. Wf epistatic to Gfgf.
wi		*White* color immature fruit. Dominant to green.
Wmr		*Watermelon Mosaic Virus 2* resistance (in PI 414723).
Wt		*White testa.* Dominant to yellow or tan seed coat color.
Y		*Yellow* epicarp. Dominant to white fruit skin.
yg		*yellow green* leaves. Reduced chlorophyll content.
yg^w	lg	*yellow green Weslaco.* First described as *light green* in a cross Dulce × TAM-Uvalde. Allelic to yg.
yv		*yellow virescent.* Pale cotyledons; yellow green young leaves and tendrils; bright and yellow petals and yellow stigma; etiolated; older leaves becoming green.

TABLE 2 (continued)

Preferred symbol	Synonym	Character
yv-2		*yellow virescence-2*. Young leaves yellow green, old leaves normal green.
Zym-1	Zym	*Zucchini Yellow Mosaic* Virus resistance. Resistance to pathotype 0 of this virus (in PI 414723).
Zym-2		*Zucchini Yellow Mosaic* Virus resistance. One of three complementary genes (see Zym-1 and Zym-2) for resistance to this virus (in PI 414723).
Zym-3		*Zucchini Yellow Mosaic* Virus resistance. One of three complementary genes (see Zym-1 and Zym-2) for resistance to this virus (in PI 414723).

GENETICS OF OTHER CHARACTERS OF MUSKMELON

Skin Color

Chadha et al. (1972) studied the inheritance of green color by crossing Hara Madhu (light yellow skin) with New Melon (green skin). The green skin color of F_1 and phenotypic segregation in F_2 of 3 green:1 yellow skin indicated the monogenic inheritance of skin color with the dominance of green skin color over yellow. Chauhan (1970) has reported a single incompletely dominant gene for yellow or green rind color.

Fruit Shape

Ramaswamy et al. (1977) studied oblong × flat, flat × round, and oblong × round crosses and concluded that fruit shape in muskmelon was not inherited in a simple way, and several genes may be involved. Shinde (1979) reported the possibility of a single gene dominance of oblong fruit shape over round fruit shape. However, in flat × round and flat × oblong crosses he did not find any differences in segregation. Robinson et al. (1976) assigned gene symbol 0 for oval fruit shape which was dominant over round and associated with a gene a (which was assigned for andromonoecious sex form). More and Seshadri (1998) proposed to assign gene symbol 0-1 for oval fruit shape of (Shinde, 1979) instead of the symbol (0) for Robinson et al. (1976), because Shinde (1979) did not find any linkage between sex form and fruit shape, although Risser (1984) found a correlation between gene a and fruit shape, she attributed it to a pleiotropic effect of the gene and not to linkage.

Fruit Stripe

Ramaswamy et al. (1977) reported that inheritance of stripe was the same in netted × smooth, netted × netted, and smooth × smooth crosses. In cross between striped and non striped parents, F_1 fruits were not striped. The F_2 plant produced striped and non-striped fruits in the ratio of 1:3. In the backcross of F_1 with stripped parents, the striped and none striped fruits segregated in the ratio of 1:1, while back crossing with the non-striped parents produced all non-striped fruits. They concluded that striping was under control of one single recessive gene. However, Chadha et al. (1972) reported that striping is controlled by two genes with complimentary gene action.

According to Ramaswamy et al. (1977) a single recessive gene st govern fruit stripe, while dominant allele st determine non striped fruits. Robinson et al. (1976) based on the study of Hagiwara and Kamimura (1936) has also assigned gene symbol St for striped epicarp recessive to the non-striped fruit.

Netting

Ramaswamy et al. (1977) studied netting in crosses between (Edisto, Pusa Sharbati) and smooth (SI 445, Durgapura Madhu, Lucknow Safeda, Persiiski 5) parents. All the F_1 plants produced netted fruit. In the F_2 population of SI-445 × Edisto, Durgapura Madhu × Pusa Sharbati, and Pusa Sharbati × Lucknow Safeda, the ratio of netted to smooth fruits showed a ratio of 14:2. The backcross ratio of (F_1 × smooth) showed a good fit for 3:1 ratio in all the three crosses. A good fit for 14:2 ratio was obtained when netted parent was used as female on male parent. These results indicated the assumption of two genes linked to each other to govern the manifestation of netting. The gene N is assigned for netting and its allele n for smoothness. The gene S is assigned a role of suppressing netting when present in homozygous conditions. N is epistatic over S and s. In the absence of N, SS caused smoothness, where as Ss causes varying degree of netting. The gene n is epistatic over S and so a double recessive genotype will be smooth. The gene symbol proposed by Ramaswamy et al. (1977) based on a very comprehensive study, i.e., netted (NNss) and smooth (nnSS or nnss) are the only acceptable ones.

Flesh Color

Ramaswamy et al. (1977) crossed green flesh colored fruit (Durgapura Madhu) with white (Lucknow Safeda) flesh fruit and obtained all white flesh fruits in F_1 while in F_2 white and green flesh fruit segregated in the ratio of 3:1. A backcross with Durgapura Madhu (green flesh) gave the monogenic ratio of 1:1. In other word presence of green pigment (chlorophyll) is conditioned by recessive gene and its dominant allele prevents the production of chlorophyll.

Chadha et al. (1972) studied the inheritance of white vs. green flesh in a cross of Hara Madhu (green flesh) × Japanese Cantaloupe (white flesh) and on the basis of F_1, F_2 and backcross generation data they revealed the monogenic inheritance of flesh color with dominance of white over green flesh. On the basis of these studies More and Seshadri (1998) proposed the symbol gf.1 which was distinguish from gf gene symbol assigned by Robinson et al. (1976) to the green flesh, which is recessive to salmon color flesh as reported by Huges (1948). Ramaswamy et al. (1977) also studied inheritance of orange × green, green × orange and orange × white flesh colour and did not come to any conclusion be cause of the complex inheritance of these crosses.

Juicy Flesh

Chadha et al. (1972) studied the inheritance of juiciness by crossing Hara Madhu (more juicy) with New Melon (less juicy) and Japanese Cantaloupe (less juicy). In both the crosses, the F_1 resembled the less juicy parent and the F_2 segregation gave ratio of 3 less juicy to 1 juicy. This indicated the recessive ness of juiciness. Backcrosses generation also supported the findings. On the basis of the study More and Seshadri proposed gene symbol jf for juicy flesh as suggested by Robinson et al. (1976) as Chadha et al. (1972) did not propose any gene symbol for juicy and less juicy condition. Inheritance of fruit shape, weight, rind, and flesh color, flesh thickness and T.S.S. was also reported by Ma et al. (1993).

Relevant information with respect to various qualitative and quantitative traits of muskmelon are given in Tables 1 and 2.

Breeding Objectives

For development of hybrids and open pollinated varieties in muskmelon, the major characters of interest, their inheritance pattern and sensitivity to environment is of utmost consideration. In muskmelon earliness in terms of total early yield (80 days) or higher proportion of total yield as early yield, are more important to fetch high market price and also to help utilising yielding potential of the crop before the onset of monsoon specially in North India. Similarly quality in terms of T.S.S. (pre-dominantly sweetness) and flesh thickness (or proportion) are also important in this dessert fruit crop. While T.S.S. is more sensitive to environment, flesh thickness (fruit diameter-cavity diameter) should be more than cavity and is more easily fixable also. Fruit shape, flesh color and fruit skin are also major considerations. Round fruit with orange juicy, thick flesh and rough, tough, netted skin will be an attraction in market and also help in long distance transportation. Slight deviation in shape, e.g., flattish round to oval or flesh colour orange to green juicy flesh with assured sweetness may also be acceptable in the market. Nevertheless it is worthy to point out that many of these fruit characteristics depend very much on the market the fruits will be sold. Short vining plant type will accommodate more plants to get more yield per unit area and stress resistant genotypes, particularly biotic stress resistant types will assure and ensure sustainable yields under adverse conditions. High total yields realizable through more number of medium sized (600-800 g) fruits are more desirable for packing, transport, as well as consumer preference rather than big sized fruits. Most of the earliness and quality attributes are polygenic in inheritance with environmental sensitivity and are measured in terms of higher uniformity in the hybrids as compared with that in the open pollinated varieties. Keeping in view the above facts, the breeding objectives of muskmelon are as follow:

1. Early maturity.
2. Medium vine length.
3. High female to male ratio.
4. High T.S.S. The sugar content in muskmelon is a premium attribute. It should vary from 11 to 13% or higher but not less than 10%.
5. Higher early and total marketable yield.
6. Resistance to common diseases like powdery mildew, downy mildew, and virus.
7. Resistance to important insect pest like red pumpkin beetle, fruit fly, and aphids.

Breeding Methods

Usually inbreeding depression has not been reported in melons, therefore, melon can be handled as self pollinated crop for breeding purposes through judicious application of selfing and selection for desirable traits.

Mass Selection

The muskmelon varieties Arka Jeet, Arka Rajhans, Hara Madhu were developed by single plant selection and later maintained by mass selection.

Pedigree Method

This method is applicable to develop genotypes of crossing parental lines having complementary traits followed by selection of desired types in the selfed/inbred generations up to 5-6 generations till homozygosity is attained. The varieties Pusa Sharbati and Punjab Sunheri were developed through pedigree method.

Polyploid Breeding

Attempts have been made to produce triploid seedless muskmelon (Batra, 1953). Triploids are produced by crossing tetraploids with diploids. Tetraploids are synthesised using colchicine (0.4 percent emulsion) applied twice at 4 days intervals at shoot apex at cotyledonary stage. The quality of tetraploid strain was superior to that of diploids. However, for other characters such as productivity, fruit size, fertility, and number of seeds, the tetraploid strains were inferior to the diploids. This indicates that there is not much scope for improvement through polyploidy breeding. However, regeneration spontaneous tetraploid through tissue culture was reported by Adelberg et al. (1993).

Heterosis Breeding

Muskmelon is a predominantly cross-pollinated crop and several pollination control mechanism, viz., genic male sterility, monoecy, gynoecy, use of growth regulators and larger number of seeds per fruit can be successfully exploited for commercial seed production by economizing the cost of hybrid seed production. Like most of the vegetable hybrids, early maturity, higher yield, better quality and uniformity of fruits are the major advantage of muskmelon hybrids. Heterosis breeding in melon was discussed in detail by Dhaliwal (1997) and Sandha and Lal (1999).

Manifestation of Heterosis

First report of heterosis is muskmelon was reported by Munger (1942) when the observed that hybrid produced from two different parents produced 30% higher yield and are more uniform than their parents. Foster (1967) reported that a group of hybrids with one common parents out yielded the commercial parent by 84%. Heterosis for earliness, T.S.S., early and total yield and netting intensity was reported by Bohn and Davis (1957), Lippert and Leg (1972a,b), Nandpuri et al. (1979), Pandey and Kalloo (1976), Chadha and Nandpuri (1980), Mishra and Seshadri (1985), More and Seshadri (1998), and Munshi and Verma (1997). Despite lack of inbreeding depression in melon heterosis for earliness, fruit size, fruit weight, flesh thickness and soluble solid had been reported by McCreight et al. (1993), Lee et al. (1996), Dhaliwal (1996), and Liou et al. (1995). Sharma (1975) observed heterosis percentage of 10.9, 46.9, 142.6, and 135.0 for days for first fruit harvest, fruit weight, yield per plant and early yield, respectively, over better parent. Dixit and Kalloo (1983) observed higher percentage of heterosis over better parent for number of fruits per plant (54.3%) in the cross combinations of Punjab Sunheri × Sel 1, cross combination Pusa Sharbati × Sarda melon for yield (46.76) in Arka Jeet × Pusa Sharbati and T.S.S. (26%) in Arka Jeet × Sarda

Melon. Mishra and Seshadri (1985) crossed 2 genic male sterile like with 32 andromonoecious varieties and reported heterosis over better parent for cross combination ms-1 × Harela (31.3%) for yield, ms-2 × Pusa Madhuras (13.5%) for T.S.S. and ms-1 × Persiiski-5 for earliness (19.0%) and early yield (43.37%). More and Seshadri (1980) suggested the possibility of using a monoecious line as female parents for production of F_1 hybrids. They observed 50 hybrids and found that M_1 × Mix Top Mark, M1 × Golden perfection, M_2 × Yauco Treat, M2 × PMR-6 and M2 × Top Mark superior over better parents for earliness, yield, and quality attributes. Monoecious like M1, M2, M3, and M4 were developed by More et al. (1980) from monoecious (M1) segregate isolated by Sharma (1975). A F_1 hybrid M3 × Durga Pura Madhu was found to be promising with respect to earliness and yield (Kesavan et al., 1987) and was release for commercially cultivars as M_3 hybrid (Pusa Rasraj) in 1990. Monoecious character was controlled by a single pair of recessive genes. Munshi and Verma (1997) observed appreciable heterosis over better parent and top parent for most of the characters studied, i.e., earliness, number of fruits per plant, fruit weight and yield per plant. However, they did not find any significant heterosis with respect to T.S.S. and fruit flesh thickness. Pusa Madhuras × Ravi, Pusa Sharbati × Pusa Madhuras, and Pusa Madhuras × Hara Madhu were observed as three most promising F_1 hybrids. The best performing hybrid combination was Pusa Madhuras × Ravi which recorded 11.71% heterosis for number of fruits per plant and 28.15% heterosis for yield per plant over top parent Pusa Madhuras a widely accepted commercial variety. Resistance against certain specific diseases can also be transmitted to the hybrids. Foster (1960) transmitted crown blight resistance to F_1 hybrids. There are many instances where resistance had been reported to be dominant (Dhillon and Wehner, 1991), which might be utilized in the breeding programme. Kesavan and More (1987, 1991) utilized FM1, FM2 and FM5 three monoecious line (introduced from Cornell) in hybrid program and identified four promising hybrids resistance to powdery mildew along with desirable horticultural characters. According to Andrus and Bohn (1967), high genotypes × environmental interaction was responsible for lack of wider adaptability of the open pollinated varieties and under such situation hybrids would be advantageous as they were higher adaptability and less prone to ecological change.

Muskmelon, being a dessert fruit, quality of fruits will be a deciding factor in heterosis breeding beside the yield. Quality character (especially T.S.S.) is highly influenced by environmental factors, hence, it is not possible to fix it with in arithmetical limits even in pure varieties. Since its inheritance and expression is very complex, it is necessary to find out whether F_1 hybrids can prove superior to the pure bred varieties and heterozygosity *per se* can bring about the genetic balance necessary in the manifestation of this quality attributes. Besides, growing season in some parts of north India is limited and hence earliness is much sought after here to avoid pre-monsoon showers which deteriorates the fruit quality and invites many diseases (More, 1998).

METHODS OF HYBRID SEED PRODUCTION

Emasculation and Pollination

Since muskmelon fruits contain large number of seeds, hybrid seeds can be manually produced by emasculation and pollination. According to Munger (1942), 3000 viable seeds are required to produce seedlings for one acre. These seeds can be obtained from at the most 10 fruits, as even small size fruit yields more than 300 seeds. Only ten successful pollination are required to produce hybrid seeds sufficient for an acre. Hybrid seed can also be produced by using one andromonoecious line having small number of male flower. In such case, male flowers were removed before anthesis at two or three days interval (Gableman 1974). However, total elimination of male flowers was difficult to achieve and this resulted in contamination of hybrid seed.

Monoecy

The andromonoecious nature of muskmelon involves tedious emasculation of the perfect flowers of the female line for hybrid seed production. Though mostly it is andromonoecious, some genotypes show monoecious sex expression. Monoecious lines reduced the extent of self-pollination in female parents, when seed production is done under open pollination with adjacent rows of male and female parents (Foster, 1968). Since the use of monoecy exclude emasculation, it can reduce the time required for hand pollination by 50% and enhance fruit set fruit set by 40-70% as compared to 5-10% the andromonoecious parent (More et al., 1982). The problem in using monoecy in hybrid seed production is undesirable linkage between genes controlling monoecious sex expression on and fruit shape (Risser 1984 and Kubicki 1992), consequently F_1 combinations with round fruits cannot be easily obtained. The problem was evident in Pusa Rasraj (M3 hybrid), which was not acceptable commercially because of its undesirable fruit shape and poor external appearance (Sandha and Lal, 1999). Some workers have also assessed the possibility of eliminating emasculation by using seedling marker. Foster (1968) reported that by using a glabrous marker gene, field crossing could produce 30-35% hybrid seed using andromonoecious lines and 60% by using monoecious lines as seed parent. He also advocated that by over planting and rogueing, a satisfactory field stand of F_1 hybrids could be established. Similarly other recessive seedling mutants reported by Pitrat (1990) can also be used in hybrid seed production by using the mutant as seed parent. Commercial hybrid production using monoecy under open field condition is handicapped on account of larger mixture of selfed or sib seed in hybrid progeny (Seshadri et al., 1983 and More et al., 1991a).

Male Sterility

First recessive male sterile gene *ms-1*, in muskmelon was reported by Bohn and Whitaker (1949). Since then at least four additional male sterile recessive alleles, viz., *ms-2* (Bohn and Principe, 1962), ms-3 (McCreight and Elmstrom 1984), ms-4 (Pitrat 1990), and ms-5 (Lecouviour et al., 1990) had been identified. The phenotype of ms-4 plant is different from ms-1, ms-2, ms-3, and ms-5 (McCreight, 1993). Sterility in all these male sterile mutants is monogenic recessive. Male sterile line ms-1 identified by Bohn and Whitaker (1949) had proved boon to the hybrid seed industry. The line showed good nicking ability when used in a series of hybrid combination (Nandpuri et al., 1974a,b). Punjab Hybrid a F_1 between MS-1 × Hara Madhu was released in Punjab state and subsequently at national level in 1984. This hybrid still occupies a larger percentage of muskmelon growing area (Sandha and Lal, 1999). Genic male sterility in muskmelon is maintained under heterozygous (Msms) condition in isolation by back crossing with the recessive (msms) parent every year. For commercial seed production heterozygous seed stack (Msms) is grown which segregates into 50% heterozygous male fertile (Msms) and 50% homozygous male sterile line plant (msms). The former is removed from the population before flowering where as the latter is kept for hybrid seed production. For maintenance of sterility, a male sterile line is sown in an isolated field. At flowering, each plant is examined and marked as male sterile or male fertile. An adequate supply of bees is provided. After the fruit set, all male fertile plants are rouged before the fruit become mature and seed is harvested only from male sterile plants. These seeds, which will segregates as male sterile and male fertile plant in 1:1 ratio serves as a stock seed to repeat the cycle. For hybrid seed production, the same seed is planted at the ratio of 3:1 (female:male). Male fertile plants are identified and rouged at the appearance of first male flower. Hybrid seed is harvested from the fruits which develop on male sterile plants (Dhaliwal, 1989; Kumar and Dhaliwal, 1991). To get maximum hybrid seed, six seedlings per hill are planted so that sufficient population is maintained after rouging the fertile plant (Dhillon, 1994). Commercial exploitation of genic male sterility is handicapped by identification of male fertile plants before flowering. Male sterile line ms-3 had been found superior to ms-1 and ms-2 due to the fact that male sterile plants are very easy to identify and possess superior horticultural traits. In male sterile line ms-5, male flower buds abort pre-maturely, hence male sterile plants can easily be identified. Use of markers gene to simply the procedure of identification and hybrid seed production was reported by Foster (1968) by utilizing glabrous seedling markers which was controlled by single recessive gene. This could eliminate the tedium method of identification of male sterile plant and keep down the cost of hybrid seed production. Efforts were made by Mishra (1981) to incorporate male sterility and genetic markers in monoecious lines to facilitate the identification and rouging those seedlings on germination. It was expected that male sterility coupled with marker in monoecious sex form world promote greater cross-pollination for the production of hybrid seed. But 100% open-pollinated pure seed could not be produced on field scale even with complex female parent monoecious + genic male sterility + marker gene, because male sterility was being simply inherited recessive character, could not maintained under homozygous condition. The poor linkage of genetic marker with ms line was suggested by Sandha and Lal (1999). Functional male sterility induced in muskmelon was induced by applying two sprays of 0.3 percent FW 450 (sodium alpha, beta dichloroisobutyrate) by Pundir and Singh (1965). However, this method had not been exploited commercially for the production of F$_1$ hybrid.

Gynoecy

Even monoecious sex form and isolation of male sterile stock (Gohn and Principe, 1964; Lo Zanov, 1969; McCreight and Elmstrom, 1983; Seshadri et al., 1983) has not simplified the technique of hybrid seed production. The muskmelon hybrid seed is 12-30 times costlier than that of open-pollinated cultivars (McCreight and Elmstrom, 1984). The gynoecious lines having natural bias of producing only female flowers ensure 100% hybrid seed production, thus having potential to lower the lost of hybrid seed considerably. By using gynoecy in heterosis breeding, the tedious emasculation and of identification and rouging of male fertile from the mixed population is avoided. Probably, for this reason, Frankel and Galun (1977) and Loy et al. (1979) advocated the use of gynoecious lines in hybrid seed production. The exact genetic make up of gynoecism in muskmelon is not yet understood due to influence of temperature and photoperiod for its expression. This breeds true and produce all pistillate flowers under short day and low temperature conditions, but when grown at high temperature and long day it becomes gynomonoecious (Kubicki, 1969). Wisconsin 998 was the first gynoecious line in muskmelon developed by Peterson et al. (1983). Wisconsin 998, when used in hybrids seed production, was exhibited good combining ability for yield and earliness (Lal and Dhaliwal, 1993 and Dhaliwal and Lal, 1996). A cross between WI 998 × Punjab Sunheri (MHL-10) was found promising and was released form commercial cultivation in Punjab during 1995. It was not only early and high yielding but also had good shipping quality (Lal and Dhaliwal, 1996). Procedure for commercial hybrid seed production of MHL-10 was discussed in detail by Lal (1995). More et al. (1987) had developed three true breeding gynoecious line, viz., 86-104, 105, and 118 and by subsequent selfing and selection for good horticultural characters and stability for gynoecy, finally seven lines namely GH 3-2, 5E-1, 7-7, 4D, 5D, 6C-4, and 6E-7 were selected (More et al., 1991). Several cross combination were made by the above lines (Anonymous 1996, 1997, 1999) but none of them were found to be commercially acceptable, i.e., not desirable from quality point of view having low T.S.S. and undesirable fruit shape. Prospects and practicability of genoecy in heterosis breeding had been described by More and Seshadri (1989), Seshadri and More (1993), and More et al. (1987, 1991). Additional information on problems and prospects of heterosis breeding in muskmelon was discussed by Seshadri et al. (1983) and later on appraised by More et al. (1991).

Unlike cucumber gynoecious lines of muskmelon could not be maintained by GA, however, perfect flower induction was reported with the application of 5-methyl-7-chloro-4-ethoxy carboxy-methoxy 2,1,3, benzothio-diazole (MCEB) at the fourth true leaf stage and following hand pollination, selfed seeds were produced (Byers et al., 1972). Use of $AgNO_3$ @ 100-200 ppm (Owen et al., 1980) induced production of 12 percent flowers on the first 20 nodes, which can be used to self or sib for maintenance of gynoecious lines. Use of Ethral and Alar (Rudich et al. 1970) and silver thiosulphate (More and Seshadri, 1987) had also been suggested for inducing maleness. The possibilities of using gynoecious lines zh 18-4-2, zh v 5-4, and zh 29 and androecious A37 and AD, for breeding melon was discussed by Ivonov (1995).

The utilization of gynoecious lines for the production of hybrid suffers with various types of limitation and therefore further need to search the possibility to utilize the existing gynoecious lines for the production of hybrid.

APPLICATION OF PLANT GROWTH REGULATOR

Foliar spray of Ethral @ 50 to 100 ppm at the two to three true leaf stage accelerates female flowers at early nodes and staminate or mixed forms of staminate and female flower at a later stage in most of the cucurbitaceous crop. Thus, the Ethral sprayed plant behaves as a female line in early stage. In muskmelon also application of Ethral at 2-3 true leaf stage induced temporary suppression of male flower and produced pistillate flower at early node thus the Ethral sprayed plant behave as female line at an early stage (Kalloo, 1974; Kalloo et al., 1982; and Byers et al., 1972). Korzeniewska et al. (1995) advocated the use of Ethephon (300-350 ppm) for F_1 hybrid seed production with monoecious lines. Foliar application of telephone induced temporary gynoecious stage (averaging 7-19 days, during which development of male bud was inhibited and bees could be used to pollinate flowers). The method of hybrid seed production was the selection of any two good combination on the basis of heterosis percentage out of which female line should be monoecious and planting them in the ratio at 4:2 (monoecious:pollen parent), where formers was sprayed with growth regulatory. The monoecious line would produce female flowers at early nodes and at the same time, the corresponding pollen parents produced male flowers which hybridise the female flower of the female line. Pollinator rows would be destroyed after fruit set. The first few fruits produced in the female lines would be F_1 hybrid. However, according to More (1974) and More and Seshadri (1975, 1998) even phenotypically, a temporary change of sequence of flowers attempted in monoecious seed parents through exogenous application of 2-chloro-ethyl-phosphonic acid had shown the distinct possibility of inducing perfect and female flowers at earlier nodes (five to six) in andromonoecious and monoecious lines. Its practical use on larger scale hybrid seed production under open field was yet to be worked out.

RESISTANCE BREEDING

Disease Resistance

Resistance to Fungi

Powdery mildew: Powdery mildew is a very important world wide disease in melons, the fungus *Sphaerotheca fusca* (Fr.) *Blumer* (= *S. fuliginea* Poll.) is the main agent inciting the disease (Sitterly, 1978).

Genetics of melon resistance to *S. fusca* has been studied for a long time, since 1938 when Jagger et al. found that the resistant line 'PMR 45' developed the disease. Two physiological races of the fungus were then defined. Thomas in 1978 described a third race in USA that was able to develop on PMR 5. Race 3 was also isolated in India (Kaur and Jhooty, 1986) and Israel (Cohen et al., 1996).

Molot and Lecoq (1986) reported that a *S. fusca* isolate found in Montfavet (France) could not incite disease on 'Edisto 47', line that in USA was susceptible to race 2. These results were supported by McCreight et al. (1987) results that concluded that race 2 isolates from France had a different virulence pattern than those from USA. Mohamed et al. (1995) classified them as race 2 "France" and race 2 "USA."

A fifth race (race 0) was described by Bertrand (1991) who found several isolates that could not infect some lines that were considered to be susceptible to all races. Finally in 1996 Bardin described two new races, race 4 and race 5.

All the physiological races described up to the present are summarized in Table 3.

The study of the resistance genetics is not yet finished, few allelism tests have been done because of the difficulties in maintaining large isolate collections. In most of the studied cases the inheritance of the resistance is monogenic and dominant (Pitrat et al., 1998).

In 'PMR 45' and 'PMR 5' resistance is due to one dominant gene, but in other cases the genetics of the resistance is more complex, being controlled by several dominant and recessive genes, and also some modifier genes can influence the expression of the resistance (Kenigsbuch and Cohen, 1992). There are also some differences among the different authors that could be explained because the techniques and isolates used are different.

Fusarium wilt: The plant wilt incited by *Fusarium oxysporum* f. sp. *melonis* is actually present in all production areas and represents a big threat for the crop because it reduces the production, and sometimes also causes the death of the plants, making necessary to leave the infected plots. Several resistance genes have been identified in different materials, but here, like in powdery mildew, several physiological races have been found according with its different virulence pattern on a set of differential hosts (Table 4).

Downy mildew: Pseudoperonospora cubensis (Berk. And Curt.) Rostw, the causative agent of downy mildew of melon has a wide distribution on production areas and often causes important damages. Its distribution area comprises mainly tropical and subtropical regions because its development is enhanced

TABLE 3. Reaction of melon differential hosts when inoculated with different physiological races of *S. fusca* (S = susceptible, R = resistant, H = heterogeneous reaction, nt = not tested).

Differential hosts	S. fusca Races						
	0	1	2 France	2 USA	3	4	5
Iran H	S	S	S	nt	nt	S	S
Védrantais	R	S	S	S	S	S	S
PMR 45	R	R	S	S	S	S	S
WMR 29	R	R	R	H	nt	S	S
Edisto 47	R	R	R	S	R	R	S
PMR 5	R	R	R	R	S	R	R
PI 124 112	R	R	R	R	R	R	R
PI 414723	R	R	R	S	nt	R	R

TABLE 4. Resistance genes, origin of the resistance, and reaction to the different physiological races of *F. oxysporum* f. sp. *melonis* (S = susceptible, R = resistant, r = partially resistant).

Differential hosts	Charentais T	Doublon	CM. 17187	Ogon 9
Resistance genes Physiological races	---	Fom 1	Fom 2	Oligogenic and recessive
0	S	R	R	r
1	S	S	R	r
2	S	R	S	r
1-2	S	S	S	r

by high humidity, after rains or sprinkle irrigation, and temperatures between 20-25°C.

In melon several resistance sources are known. Partial resistance was described in accessions from India (Ivanoff, 1944). Othser accessions were also described with different levels of resistance, 'PI 12411', 'PI 124112', 'PI 414723', and 'Phoot' (Pitrat et al., 1998). There seems to be general agreement that 'PI 124111 ('MR-1') and 'PI 124112' have the highest level of resistance.

Two incompletely dominant complementary genes (*Pc-1*, and *Pc-2*) have been described in 'MR-1' (Thomas et al., 1988). Two complementary genes (*Pc-4*, and *Pc-1* or *Pc-2* present in 'MR-1') control resistance in 'PI 124111' (Kenigsbuch and Cohen, 1992). Using 'Phoot' as a source of resistance, two dominant genes were observed in some crosses, but two recessive genes were observed in other (Samkuwar and More, 1993).

The conclusion is that the highest level of resistance is under di- or oligogenic control and is partially dominant.

Resistance to Viruses

Cucumber Mosaic Virus (CMV)

CMV causes very important losses in melon, it induces plant stunting and mosaic, size reduction, and distortion of leaves. Symptoms on fruits are discolorations and, occasionally, deformations.

CMV is one of the most common viruses infecting cucurbits world wide, especially under temperate climatic conditions. In tropical and subtropical regions CMV is less frequent in cultivated cucurbits (Quiot et al., 1983).

Various levels of resistance to CMV are present in some melon germplasm like 'Freeman's Cucumber' or 'PI 161375' in which the resistance seems to be oligogenic recessive (Lecoq et al., 1998).

Resistance to CMV transmission by one of its major vectors, *Aphis gossypii*, has been introduced into commercial melon cultivars. This resistance is also efficient against potyviruses, and to a lesser extent against CABYV (Lecoq et al., 1998).

Potyviruses

Several distinct members of the Potyvirus genus infected cultivated cucurbits. Three of them, papaya ring spot virus (PRSV), watermelon mosaic virus 2 (WMV 2), and zucchini yellow mosaic virus (ZYMV) have a world wide distribution, and cause important yield reductions in melon.

PRSV (formerly watermelon mosaic virus 1, WMV 1) causes several mosaic and deformations on leaves and fruits of melon. Some strains can induce systemic necrotic spots and systemic necrosis in some melon cultivars.

PRSV epidemics are common in tropical or subtropical areas, but they are only occasionally observed in temperate regions.

Resistance genes have been found in PI 180280 (*Prv¹*), PI 180283 (*Prv²*), or PI 124112 (monogenic dominant) (Pitrat and Lecoq, 1983). Several other cucurbit potyviruses (WMV-M, ZYFV and some other only partially characterized potyviruses from Africa) are biologically related to PRSV, they share similar host plant resistance genes (Quiot-Douine et al., 1990). They are, however, now considered as distinct entities.

Watermelon mosaic virus 2 (WMV 2) has been reported for more than 40 years in different parts of the world. WMR 2 and PRSV were once thought to be different strains of the same virus, now they are recognized as different entities based on serological, molecular and biological properties. It may induce very severe vein banding, green mosaic, deformation and plant stunting in melon. It is prevalent in Mediterranean and temperate regions while is only rarely detected in tropical areas (Lecoq et al., 1998).

A good level of resistance controlled by a single dominant gene (*Wmr*) has been found in PI 414723 (Gilbert et al., 1994).

A good level of resistance has been obtained using the coat protein mediated resistance approach (Fuchs et al., 1997).

Zucchini yellow mosaic virus (ZYMV) was first described in Italy in 1973 by Lisa et al. (1981). The same year Lecoq et al. (1981) described a new virus called muskmelon yellow stunt virus (MYSV) in France. These two viruses were found to be serologically indistinguishable (Lecoq et al., 1984). As many other similarities between them were found, the two viruses were considered to be the same and the name zucchini yellow mosaic virus proposed (Lisa and

Lecoq, 1984). Since then, ZYMV has spread all over the world and became a major threat to all cucurbit crops. It induces on melon, mosaic, yellowing, stunting fruit, and seed deformation. These symptoms are generally very severe but may vary according to the virus isolate or the plant cultivar.

Two pathotypes NF and F were defined in ZYMV according with their ability (pathotype F) or inability (pathotype NF) of inducing a wilting reaction in *C. melo* cv. 'Doublon'. This reaction is controlled by a partially dominant gene named Flaccida necrosis and symbolized by *Fn* (Risser et al., 1981). The resistance detected in *C. melo* 'PI 414723' characterized by no symptoms when plants are inoculated with ZYMV pathotype 0, and systemic chloro-necrotic spotting when inoculated with pathotype 1, was controlled by one dominant allele *Zym* (Pitrat and Lecoq, 1984).

Cucumis melo 'PI 414723' is also resistant to the transmission of ZYMV by *Aphis gossypii*, because this line carries the *Vat* gene conferring resistance to virus transmission by this aphid.

Melon Necrotic Spot Virus (MNSV)

MNSV is a carmovirus that is transmitted by the soil fungus *Olpidium bornovanus*. Alternatively, it can be transmitted mechanically and possibly by some chewing insects. MNSV was found in the USA, Europe, the Mediterranean Basin, and Japan, and has a host range mostly limited to cucurbits.

Typical symptoms are systemic necrotic spots, streaks on petioles and stems that may eventually lead to plant necrosis. MNSV has been shown to be seed transmitted, the virus in carried on the seed (probably in the seed coat or papery layers), released in the soil during germination, acquired and then introduced in the plant by the zoospores of the fungus (Campbell et al., 1996).

One single recessive gene, *nsv*, has been reported as controlling resistance to MNSV in melon. This gene is present in cvs. 'Gulfstream', 'PMR-5', and 'Planters Jumbo' (Coudriet et al., 1981), and in 'PI 161375' (Maestro, 1992). This is the only resistance found in melon. Homozygous lines for *nsv* did not react producing necrotic spots after mechanical inoculation of MNSV (Coudriet et al., 1981).

Insect Resistance

Aphis gossypii Glover

The cotton aphid, *Aphis gossypii* Glover, is the only aphid species able to colonize melon. It is a very important pest that causes losses all over the world. It may produce leaf rolling, stop of growing, and sometimes plant death (Pitrat and Lecoq, 1982). Nevertheless it may cause much more important damages acting as virus vector, in fact it is a very efficient vector for melon virus transmitted in a non-persistent manner (Labonne et al., 1982; Fereres et al., 1992).

Many studies have been done aimed to find genetic resistances to *Aphis gossypii* in melon. In 1944, Ivanoff did find resistance to this aphid in four melon genotypes.

At the beginning of the 70s, a research team from California started looking for resistance to insects in melon. They evaluated in open field and greenhouses all melon materials reported as resistant to *A. gossypii* by other authors, and detected a good level of resistance within line 'LJ 90234', derived from the India accession 'PI 175111' (Kishaba at al., 1971). The resistance mechanisms in this line are, non preference, antibiosis, and tolerance (Bohn et al., 1972). The tolerance seems to be controlled by one dominant gene named *Ag* (Bohn et al., 1973). The genetic control of antibiosis is complex and influenced by the environmental conditions (Kishaba et al., 1976).

Later on some works describing some resistant commercial varieties were published (Shinoda and Tanaka, 1987), or on the identification of new resistance sources (Bohn et al., 1996).

In 1979, Lecoq et al. reported a resistance to *A. gossypii* virus transmission identified within the Corean accession 'PI 161375'. Later they found that this line was also resistant to *A. gossypii* colonization, and these two resistances were both controlled by the same dominant gene named *Vat* (*virus aphid transmission*) (Pitrat and Lecoq, 1980,1982). These authors also stated that this gene and the one controlling resistance by antibiosis in 'LJ 90234' (Kishaba et al., 1976) are the same. According with Lecoq et al. (1980), the *A. gossypii* virus transmission resistance controlled by *Vat* is specific for the aphid, but non-specific for the virus.

Others

A high degree of resistance to red pumpkin beetle (*Aulacophovus foveicollis*) was observed in varieties Casaba and PI 70683 (Vashishta and Choudhury 1971). Resistance was inversely related to their moisture and nitrogen content. Fruit fly (*Dacus cucurbitae*) is a serious pest of cucurbits. *Cucumis callosus* had been reported to be resistant to this pest (Sambandam and Chelliah 1969; Chelliah and Sambandam 1972). Pareek and Kavadia (1995) screened 17 muskmelon cultivars for resistance against fruit fly (*Dacus cucurbitae*) and found the cultivars were either susceptible (51-75% fruit damage) or highly susceptible (76-100%). None of the cultivars were found to be resistant under Jaipur (Rajasthan) condition. They (1993) also reported that Durgapura Madhu was the most susceptible variety against red pumpkin beetle (*Aulacophora faveicollis*) followed by Hara Madhu, No. 45, Punjab Hybrid and Hara Gola. C. *callosus* a non commercial species was cross compatible with cultivated species of melon and resistant to caterpillars of *Margaronia indica*. Durgapura Madhu, Sel-1, Kocha. 4, Punjab Sunheri, Planter's Jumbo, Pusa Sharbati, and Racoid Mill showed lower infestation at mites (*Tetranychus cinnabarinus*) (Dhuria and Sukhija, 1986). Morales and Bastidas (1997) reported Edisto, Galleon, Laredo, Durango, Concorde, Amarelo, Honey Dew, and Noy Amid cultivars of Cantaloupe were most resistant to whitefly (*Bemisia tabaci*).

REFERENCES

Anaïs G. 1971. Nouvelles orientations dans la selection du melon (*Cucumis melo* L.). Utilization de la gynoecie. Modification de l'expression du sexe par traitements chimiques (giberelline-ethrel) et par greffage. Ann. Amelior. Plantes, 21, 55-65.

Anonymous 2001. *Proc. of XIXth, Group Meeting of A.I.C.V.I.P.*, I.I.V.R. Varanasi 15-18 Jan, 2001.

Anonymous, 1996. Annual scientific Report 1995-96. Division of Vegetable Crops, IARI, New Delhi-12.

Anonymous, 1997. Annual Scientific Report 1996-97. Division of Vegetable Crops, IARI, New Delhi-12.

Anonymous, 1999. Annual Scientific Report 1998-99. Division of Vegetable Crops, IARI, New Delhi-12.

Bains, M.S. and Kang, U. 1963. Inheritance of some flower and fruit character in muskmelon. *Indian J. Genet., Pl. Breed*, 23: 101-106.

Bardin, M. 1996. Diversité phenotypique et genetique des oïdiums des Cucurbitacées, *Sphaerotheca fuliginea et Erysiphe cichoracearum*. PhD These, Université Claude Bernard-Lyon (France), 161 pp.

Batra, S. 1952. Induced tetraploidy in muskmelon. *J. Hered.*, 43: 141-148.

Bertrand, F. 1991. Les oïdiums des Cucurbitacées. Maintien en culture pure, étude de leur variabilité et de la sensibilité chez le melon. PhD These, Université de Paris-Sud (France), 259 pp.

Bohn, G.W. and Principe J.A. 1964. A second male sterile gene in muskmelon. *J. Hered.*, 55: 211-215.

Bohn, G.W. and Whitaker, T.A. 1949. A gene for male sterility in muskmelon. *Proc. Amer. Hort. Sci.*, 53: 309-314.

Bohn, G.W., Kishaba, A.N., and Toba, H.H. 1972. Mechanism of resistance to melon aphid in a muskmelon line. *HortScience*, 5: 281-282.

Bohn, G.W., Kishaba, A.N., Principe, J.A., and Toba, H.H. 1973. Tolerance to melon aphid in *Cucumis melo* L. *J. Amer. Soc. Hort. Sci.*, 98: 37-40.

Bohn, G.W., Kishaba, A.N., and McCreight, J.D. 1996. A survey of tolerance to *Aphis gossypii* Glover in part of the world collection of *Cucumis melo* L. In: Proc. of the VIth EUCARPIA Meeting on Cucurbit Genetics and Breeding, 28-30 May 1996. Málaga (Spain): 334.

Campbell, R.N., Wipf-Scheibel, C. and Lecoq, H. 1996. Vector assisted seed transmission of melon necrotic spot virus in melon. *Phytopathology*, 86: 1294-1298.

Chadha, M.L. and Nandpuri, K.S. 1980. Hybrid vigour studies in muskmelon. *Indian J. Hort.*, 37: 55-61.

Chadha, M.L. and Lal, T. (1993). Improvement of cucurbits. In: K.L. Chadha and G. Kalloo (eds.) *Advances in Horticulture vol. 5-Vegetable Crops*: Part I, pp. 137-197. Malhotra Publ. House, New Delhi, India.

Chadha, M.L., Nandpuri, K.S. and Singh, S. (1972). Inheritance of quantitative characters in muskmelon. *Indian J. Hort.*, 29: 174-178.

Chauhan, S.V.S. 1970. Studies on inheritance of some fruit characters in *Cucumis melo* L. *J. Agro. Sci. Res.*, 10: 41-45.

Choudhury, B. and Sivakami, N. 1972. Screening muskmelon (*Cucumis melo* L.) for breeding resistant to powdery mildew. *Third Inst. Symp. Subtrop. Trop. Horticul.* 2: 10.

Cohen Y., Katzir N., Schreiber S. and Greenberg R. 1996. Occurrence of *Sphaerotheca fuliginea* race 3 on cucurbits in Israel. *Plant Dis.*, 80: 344.

Coudriet, D.L., Kishaba, A.N. and Bohn, G.W. 1981. Inheritance of resistance to muskmelon necrotic spot virus in a melon aphid resistant breeding line of muskmelon. *J. Amer. Soc. Hort. Sci.*, 106: 789-791.

Dhaliwal, M.S. 1989. Techniques of developing Punjab Hybrid muskmelon seed. *Seeds and Farms*, 15: 25.

Dhaliwal, M.S. 1997. Heterosis breeding in muskmelon: A review. *Agril. Reviews.* 18(1): 35-42.

Dhillon, N.P.S. 1994. F1 hybrid seed production in muskmelon (*Cucumis melo* L.). Management of male sterile population. *Seed Sci. and Technol.*, 22: 601-605.

Dhillon, N.P.S. and Wehner, T.C. 1991. Host plant resistance in insects in cucurbits-germplasm resources. genetics and breeding. *Tropical Pest Management*, 37: 421-428.

Dhiman, J.S., Lal, T., Dhaliwal, M.S. and Lal, T. 1997. Downy mildew resistance in snapmelon and its exploitation for muskemelon improvement. *Pl. Disease Res.* 12(1): 88-90.

Dixit, J. and Kalloo. 1983. Heterosis in muskmelon (*Cucumis melo* L.). *Haryana Agric. Univ., J. Res.*, 13: 549-553.

Dutta, O.P. and Nath, P. 1970. Cross ability in melon. *Indian J. Hort.*, 27: 60-67.

Esquinas-Alcozar, J.T. and Bullick, P.J. 1983. *Genetic resources of cucurbitaceae*. A *Global Report*. International Board, for plant genetic resources, Rome.

Fereres, A., Blua, M.J. and Perring, T.M. 1992. Retention and transmission characteristics of zucchini yellow mosaic virus by *Aphis gossypii* and *Myzus persicae*. *J. Econ. Entomol.*, 85: 759-765.

Foster R.E., and Bond W.T. 1967. Abrachiate, an androecious mutant muskmelon. *J. Hered.*, 58: 13-14.

Foster, R.E. 1968. F1 hybrid muskmelon. V. Monoecism and male sterility in commercial seed production. *J. Hered.*, 59: 205-207.

Foster, R.E. 1960. Breeding for disease resistance and variety. In: *Cantaloupe Res in Arizona*. Arizona Univ. Rpt. 195: 1-4.

Foster, R.E. 1967. F1 hybrid muskmelon. I. Superior performance of selected hybrids. *Proc. Amer Soc. Hort. Sci.*, 91: 205-207.

Frankel R. and Galun E. 1977. Pollination mechanisms, reproduction and plant breeding. Monogr. Theor. Appl. Genet., vol 8. Springer, Berlin Heidelberg New York. 281 pp.

Fuchs, M., McFerson, J., Tricoli, D., McMaster, J., Deng, R., Boeshore, M., Reynolds, J., Russell, P., Quemada, H. and Gonsalves, D. 1997. Cantaloupe line CZW-30 containing coat protein genes of cucumber mosaic virus, zucchini yellow mosaic virus, and watermelon mosaic virus 2 is resistant to these aphid-borne viruses in the field. Mol. Breeding, 3: 279-290.

Gaableman, W.H. 1974. F_2 hybrids in vegetable production. In: Proc. XIXth Int. Hong Cong. I (II): 419-426.

Ganesan, J. 1991. Botanical nomenclature of indian melons (Cucumis melo L.). Plant Breeding News Letter, 1: 3-4, 2.

Gilbert, R.Z., Kyle, M.M., Munger, H.M. and Gray, S.M. 1994. Inheritance of resistance to watermelon mosaic virus in Cucumis melo L. HortScience, 29: 107-110.

Ivanoff, S.S. 1944. Resistance of cantaloupes to downy mildew and the melon aphid. J. of Heredity, 35: 34-39.

Ivanov, D. 1995. Possibilities of using different sexual types in breeding melon. Rasteniev "dni-Nauki. 32(1-2): 164-165.

Jagger, I.C., Whitaker. T.W., Porter, D.R. 1938. A new biotic form of powdery mildew on muskmelon in the Imperial Valley of California. Plant Dis Rept., 51: 1079-1080.

Kaur, J. and Jhooty, J.S. 1986. Presence of race 3 of Sphaerotheca fuliginea in Punjab. Indian Phytopathol., 39: 297-299.

Kenigsbuch, D., and Cohen, Y. 1990. The inheritance of gynoecy in muskmelon. Genome, 33: 317-320.

Kenigsbuch, D. and Cohen, Y. 1992. Inheritance of resistance to downy mildew in Cucumis melo PI 124112 and commonality of resistance genes with PI 124111F. Plant Dis. 76 615-617.

Kenigsbuch D. and Cohen Y. 1992. Inheritance of resistance to powdery mildew in a gynoecious muskmelon. Plant Dis.. 76: 626-629.

Kesavan. P.K. and More. T.A. 1987. Powdery mildew resistance in F1 hybrid in muskmelon. In. National Symp. on Heterosis. Exploitation, Accomplishment and prospects Parbhani Maharastra, October 15-17. MAU, Parbhani (Maharastra) p. 54 (Abstr).

Kesavan, P.K. and More, T.A. 1991. Use of monoecious lines in heterosis breeding muskmelon (Cucumis melo L.). Veg Sci., 18: 59-64.

Kesavan. P.K., More, T.A. and Sharma, J. 1987. Use of monoecious sex form in muskmelon. In. Nat. Symp. Heterosis Exploitation, Accomplishment and Prospects, Parbhani, Oct. 15-17, p. 54.

Kishaba, A.N., Bohn, G.W., and Toba, H.H. 1971. Resistance to Aphis gossypii in muskmelon. J. Econ. Entomology, 64: 935-937.

Kishaba, A.N., Bohn, G.W. and Toba, H.H. 1976. Genetic aspects of antibiosis to Aphis gossypii in Cucumis melo from India. J. Amer. Soc. Hort. Sci., 101: 557-561.

Kroon, G.H.. Custers, J.B.M., Kho. Y.O., Dennijs, A.P.M. and Varekamp, H.Q. 1979. Inter specific hybribization in Cucumis (L.). I. Need for genetic variation, Biossystematic relations and possibilities to overcome crossability Barriers. Euphytica. 28: 723-728.

Kubicki, B. 1966. Genetic basis for obtaining gynoecious muskmelon lines and the possibility of their use for hybrid seed production. *Genet. Polon*, 7: 27-29.

Kubicki, B. 1969. Sex determination in muskmelon (*Cucumis melo* L.). *Genet. Polon.* 10: 145-146.

Kumar, J.C. and Dhaliwal, M.S. 1991. Techniques of developing hybrids in vegetable crops. Agro-Botanical Publishers, Bikaner, India.

Labonne, G., Quiot, J.B. and Monestiez, P. 1982. Contribution of different aphid species to the spread of cucumber mosaic virus (CMV) in a muskmelon plot. *Agronomie*, 2: 797-804.

Lal, T. and Dhaliwal, M.S. (1992). Genetic diversity in curcurbits. Muskmelon. *Indian Horticulture*, 37 (2): 55-56.

Lal, T. and Dhaliwal, M.S. 1993. Performance of selected muskmelon hybrids over environment. In: *Symp. on Heterosis Breeding in Crop Plants: Theory and Application*, Ludhiana, India, pp. 34-35.

Lal, T. and Dhaliwal, M.S. 1996. Evaluation of muskmelon hybrids over environment. *Pb. Veg. Grower*, 31: 10-13.

Lecoq, H., Cohen, S., Pitrat, M. and Labonne, G. 1979. Resistance to cucumber mosaic virus transmission by aphids in *Cucumis melo* L. *Phytopathology*, 69: 1223.

Lecoq, H., Labonne, G. and Pitrat, M. 1980. Specificity of resistance to virus transmission by aphids in *Cucumis melo. Ann. de Phytopathologie*, 12: 139-144.

Lecoq, H., Pitrat, M. and Clement, M. 1981. Identification et caractérisation d'un potyvirus provoquant la maladie du rabougrissement jaune du melon. *Agronomie*, 1: 827-834.

Lecoq, H., Lisa, V. and Dellavalle, G., 1984. Serological identity of muskmelon yellow stunt and zucchini yellow mosaic viruses. *Plant Disease*, 75: 208-211.

Lecoq, H., Wisler, G. and Pitrat, M. 1998. Cucurbit viruses: The classics and the emerging. In: Cucurbitaceae '98. Evaluation and enhancement of cucurbit germplasms. J.D. McCreight (ed.): ASHS Press, Alexandria, VA, 126-142.

Lecouviour, M., Pitrat, M. and Risser, G. 1990. A fifth gene for male sterility in muskmelon. *Cucurbit Genet Coop. Reptr.* 13: 10-13.

Leppik, E. 1966. Searching gene centers of the genus cucumis through host-parasite relationship. *Euphytica*, 15: 323-328.

Lippert, L.F. and Legg, P.D. 1972a. Appearance and quality characters i muskmelon fruits evaluated by ten cultivar diallel cross. *Amer. Soc. Hort. Sci.*, 97: 84-87.

Lippert, L.F. and Legg, P.D. 1972b. Diallel analysis for yield and maturity characteristics in muskmelon. *Amer. Soc. Hort. Sci.*, 53: 87-90.

Lisa, V., Boccardo, G., D'Agostino, G., Della Valle, G. and D'Aquilino, M. 1981. Characterization of a potyvirus that causes zucchini yellow mosaic. *Phytopathology*, 71: 667-672.

Lisa, V. and Lecoq, H. 1984. Zucchini yellow mosaic virus. CMI/AAB Descriptions of plant viruses, no. 282, 4 pp.

Lozanov, P. 1969. Functional male sterility in melons (*Cucumis melo* L.). *Genet. Selek.*, 2: 195-203.

Ma, Z., Ma, Z. and Zhu, D. 1995. Heredity of main economic characteristics of the *Cucumis melo* L. *Acta Hort.*, 402: 66-71.

Maestro, C. 1992. Résistance du melon aux virus. Interaction avec les pucerons vecteurs. Analyse génétique sur des lignées haplodiploides. PhD Thesis. Université d'Aix-Marseille (France), 134 pp.

Magdum, M.B. 1982. Studies on sex forms in muskmelon (*Cucumis melo* L.). PhD Thesis, P.G. School, IARI, New Delhi.

Magdum, M.B. and Seshadri, V.S. 1983. Inheritance of sex forms in muskmelon. In: *Proc. 15th Intern. Cong. Genet. Part II. p.* 562 (Abst.). Oxford and IBH Pub. Co New Delhi.

Magdum, M.B., Shinde, N.N. and Seshadri. V.S. 1982. Androecious sex forms in *muskmelon. Cucurbits Genet. Coop Reptr.* 5: 24-25.

McCreight, J.D. and Elstorm, G.W. 1984. A third muskmelon male sterile gene. *Hort Sci.,* 19: 268-270.

McCreight, J.D. and Elmstrom, G.W. 1983. A third male sterile gene in muskmelon. *Cucurbita Genet Coop Reptr.,* 6: 46.

McCreight, J.D., Pitrat, M., Thomas, C.E., Kishaba, A.N. and Bohn, G.W. 1987. Powdery mildew resistance genes in muskmelon. *J. Amer. Soc. Hort. Sci.,* 112: 156-160.

Mishra, J.P. and Seshadri. V.S. 1985 Male sterility in muskmelon II. Studies on heterosis. *Genet Agr.,* 39: 367-376.

Mohamed, Y.F., Bardin, M., Nicot, P.C. and Pitrat. M. 1995. Causal agents of powdery mildew of cucurbits in Sudan. *Plant Dis.,* 79: 634-636.

Molot, P.M. and Lecoq, H. 1986. Les oïdiums des Cucurbitacées. I. Donées bibliographiques. Travaux préliminaires. *Agronomie,* 6: 355-362.

Morales, P.A. and Bastidas, Y.R. 1997. Evaluation of the resistance of eight cultivars of melon (*Cucumis melo* L.) to attract by the whitefly *Bemisia tabaci* in the Los Perozosarea. Estado Falcon, Venezuela. *Boletinde-Entomologia-Venezolana.* 12:141-149.

More, T.A., and Munger, H.M. 1986. Gynoecious sex expression and stability in cucumber (*Cucumis sativus* L.) *Euphytica,* 35: 899-903.

More, T.A. and Seshadri, V.S. 1975. Response of different sex forms in muskmelon (*Cucmis melo* L.) to 2-chloro ethyl phosphonic acid. *Veg. Sci.,* 2: 37-44.

More, T.A. and Seshadri. V.S. 1980. Studies on heterosis in muskmelon (*Cucumis melo* L.). *Veg. Sci.,* 7: 108-117.

More, T.A. and Seshadri, V.S. 1998. Improvement and cultivation of muskmelon, *cucumber and water melon.* In: *Curcurbits.* Nayar N.M. and More, T.A. (eds.). Oxford and IBH Publishing Co. Pvt. Ltd., New Delhi, pp. 169-186.

More, T.A. and Seshadri, V.S. 1998. Sex expression and sex modification. In: *Cucurbits.* Nayar, N.M. and More, T.A. (eds.). Oxford and IBH Publishing Co. Pvt. Ltd. New Delhi, pp. 39-66.

More, T.A. and Varma, A. 1991. Breeding for virus resistance in muskmelon (*Cucumis melo* L.) Abstr. II. 612. In: Gold Jub. Symp. on Genetics Research and Education. Current Trends and Next Fifty Years. New Delhi, Feb. 12-15, Oxford and IBH Pub. Co., New Delhi.

More, T.A., Seshadri, V.S. and Sharma, J.C. 1980. Monoecious sex expression in muskmelon (*Cucumis melo* L.). *Cucurbit Genet. Coop. Reptr.* 3: 32-33.

More, T.A., Seshadri. V.S. and Sharma. J.C. 1980. Monoecious sex expression in muskmelon. *Cucurbit Genet Coop. Reptr.* 3: 32-33.

More, T.A., Varma, A. Seshadri, V.S., Somkumar, R.G. and Rajamony, L. 1993. Breeding and development of cucumber green mottled mosaic virus (CGMMV) resistant lines in melon (*Cucumis melo* L.). *Cucurbit Genet. Coop. Reptr.* 16: 44-46.

More, T.A. and Seshadri, V.S. 1987. Maintenance of gynoecious muskmelon with silver thiosulphate. *Veg. Sci.*, 14: 138-142.

Munger, H.M. 1942. The possible utilization of first generation muskmelon hybrids and improved methods of hybridization. *Proc. Amer. Soc. Hort. Sci.*, 40: 405-410.

Munshi, A.D. and Verma, V.K. 1997. Studies on heterosis in muskmelon (*Cucumis melo* L.). *Veg. Sci.*, 24: 103-106.

Munshi, A.D. and Verma, V.K. 1998. A note on gene action in muskmelon (*Cucumis melo* L.). *Veg. Sci.*, 25: 93-94.

Munshi, A.D. and Verma, V.K. 1999. Combining ability in muskmelon (*Cucumis melo* L.). *Indian J. Agric. Sci.*, 69(3): 214-216.

Munshi, A.D. and Verma, V.K. 2000. Field screening of muskmelon genotypes against Cucumber Green Mottled Mosaic Virus. *Indian J. Pl. Genet. Resources*, 13(1): 72-74.

Nandpuri, K.S. and Brar, S. 1966. Studies on floral biology in muskmelon (*Cucumis melo* L.). *J. Res. Ludhiana*, 3: 395-399.

Nandpuri, K.S., Singh, S. and Lal, T. 1974. Study on the comparative performance of Fl hybrids and their parents in muskmelon. *Punjab Agric Univ. J. Res.*, 11: 230-238.

Nandpuri, K.S., Singh, S. and Lal, T. 1982. Punjab hybrid–A new variety of muskmelon. *Prog. Fmg. Feb. Issue*, pp. 3-4.

Nandpuri, K.S., Singh, S. and Lal, T. 1974. Combining ability studies in muskmelon (*Cucumis melo* L.) *Punjab Agric. Univ. J. Res.*, 11: 225-229.

Naudin, C. 1959. Essays d'uhe monographic des especies etdes varietes dugehere (cucumis). *Ann. Sci. Nat.*, 11: 5-87.

Ogawa, R., Sugahara, S., Kasuya, M. Sakamori, M., Aoyagi, M. Sakurai, Y. and Takase, N. 1994. New breeding lines with ressitance to *Fusarium* wilt and powdery mildw in melted melon and studies on F_1 characters. *Research Bulletin of the Aichi Ken Agricultural Research Centre*, 26: 147-156.

Pandey, S.C. and Kalloo, G.S. 1976. Line × tester analysis for the study of heterosis and combining ability in muskmelon. Presented in *Advances in Plant Science Symp.* B.C.K.V., Kalyani, West Bengal.

Pareek, B.L. and Kavidas, V.S. 1993. Screening of muskmelon varieties against red pumpkin beetle (*Aulacophora foveicollis*). *Indian J. Entomol.*, 55(3): 245-251.

Pareek, B.L. and Kavidas, V.S. 1995. Screening of muskmelon varieties against fruit fly (*Dacus cucurbitae*) under field condition. *Indian J. Entoml.*, 57 (4): 417-420.

Peterson, C.E., Owens, K.W. and Rowe, P.R. 1983. Wisconsin 998 muskmelon germplasm. *HortScience*, 18: 116.

Pitrat, M. and Lecoq, H. 1980. Inheritance of resistance to cucumber mosaic virus transmission by *Aphis gossypii* in *Cucumis melo*. *Phytopathology*, 70: 958-961.

Pitrat, M. and Lecoq, H. 1982. Relations génétiques entre les resistances par non-acceptation et par antibiose du melon à *Aphis gossypii*. Recherche de liaisons avec d'autres gènes. *Agronomie*, 2: 503-506.

Pitrat, M. and Lecoq, H. 1983. Two alleles for Watermelon Mosaic Virus 1 resistance in melon. *Cucurbit Genet. Coop. Rpt.* 16:40-41.

Pitrat, M. and Lecoq H. 1984. Inheritance of zucchini yellow mosaic virus resistance in *Cucumis melo* L. *Euphytica*, 33: 57-61.

Pitrat, M. 1998. Gene list for melon. *Cucurbit Genet. Coop. Rpt.*, 21: 69-81.

Pitrat. M. (1994). Gene list for *Cucumis melo* L. *Cucurbit Genet Coop. Reptr.* 17: 148.

Pitrat, M. 1990. Gene list for *Cucumis melo* L. *Cucurbit Genet. Coop Reptr.* 13: 58.

Pitrat, M., Dogimont, C. and Bardin, M. 1998. Resistance to fungal diseases of foliage in melon. In: Cucurbitaceae '98. Evaluation and Enhancement of Cucurbit Germplasm (J.D. McCreight, ed.) ASHS Press. California: 167-173.

Pitrat, M., Halelt, P. and Hammer, K. 2000. Some comments on intraspecific classification of cultivars of melon. *Proc. of 7th EUCARPIA Meeting on Curcurbits Genetic and Breeding*. Eds. N. Katzir and H.S. Paris. *Acta Hort.*, 510: 29-36.

Poole, C.F., and Grimball, P.C. 1939. Inheritance of sex forms in *Cucumis melo* L. *J. Hered.* 30: 21-25.

Quiot, J.B., Labonne, G. and Quiot-Douine, L. 1993. The comparative ecology of cucumber mosaic virus in Mediterranean and tropical regions. In: Plant Virus Epidemiology, R.T. Plumb and J.M. Tresh (eds.). Blackwell Scientific Publications. Oxford. UK: 177-183.

Quiot-Douine, L., Lecoq, H.. Quiot, J.B., Pitrat, M. and Labonne, G. 1990. Serological and biological variability of virus isolates related to strains of papaya ring spot virus. *Phytopathology*, 80: 256-263.

Rajamony, L., More, T.A. and Seshadri, V.S. 1990. Inheritance of resistance to Cucumber Green Mottled Mosaic Virus in muskmelon (*Cucumis melo* L.). *Euphytica*, 47: 93-97.

Rajamony, L., More, T.A., Seshadri, V.S. and Varma, A. 1987. Resistance to cucumber green mottled mosaic virus (CGMMV) in muskmelon. *Cucurbit Genet. Coop Reptr.*, 10: 58-59.

Ramaswamy, B., Seshadri, V.S. and Sharma, J.C. 1977. Inheritance of some fruit characters in muskmelon. *Sci. Hort.*, 6: 107-120.

Risser, G., Banihashemi, Z. and Davis, D.W. 1976. A proposal nomenclature of *Fusarium oxysporum* f.sp. *melonis* races and resistance genes in *cucumis melo*. *Phytopathology*, 66: 1105-1106

Risser, G. and Rode, J.C. 1979. Silver nitrate induction of staminate flowers in gynoecious plants of muskmelon (*Cucumis melo* L.). *Ann. Amel. Plantes*, 29: 349-352.

Risser, G., Pitrat, M., Lecoq, H. and Rode, J.C. 1981. Sensibilité variétale du melon (*Cucumis melo* L.) au virus du rabougrissement jaune du melon (MYSV) et a sa transmisión par *Aphis gossypii*, Hérédité de la réaction de flétrissement. *Agronomie*, 1: 835-838.

Risser, G. 1984. Correlation between sex expression and fruit shape in muskmelon (*Cucumis melo* L.). Proc. IIIrd Eucarpia Meeting on Breeding Cucumbers and Melons. Plovdiv (Bulgaria), 100-103.

Robinson, R.W. and Decker-Walters, D. 1999. Curcurbits. CAB International, Walling ford, Oxon, OX108DE, U.K.

Robinson, R.W., Munger, H.M., Whitaker, T.W. and Bohn, G.W. 1976. Genes of Cucurbitaceae. *Hort. Sci.*, 11: 554-568.

Rosa, J.T. 1928. Results of inbreeding melons. *Proc. Amer. Soc. Hort. Sci.*, 24: 79-84.

Rosa, J.T. 1928. The inheritance of flower type in *Cucumis* and *Citrullus*. *Hilgardia*, 3: 233-250.

Rowe, P.R. 1969. The genetics of sex expression and fruit shape, staminate flower induction and F1 hybrid feasibility of gynoecious muskmelon. PhD Thesis State Univ. East Lansing, MI.

Roy, R.P., and Saran, S. 1989. Sex expression in the *Cucurbitaceae*. In: Bates D.M., Robinson R.W., and Jeffrey C. (eds.). Biology and Utilization of the Cucurbitaceae. Comstock Publ. Assoc., Cornell University Press, Ithaca, pp. 251-268.

Sambandam. C.N. and Chelliah, S. 1972. *Cucumis callosus* (Rottl) Cong. a valuable material for resistance breeding in muskmelon. *Third Intr. Symp. Hort.* Bangalore. Abst. p. 7.

Sandhu, H.S. 1990. Genetic studies in muskmelon. M.Sc. thesis, PAU, Ludhiana.

Seshadri, V.S. (1986). Cucurbits. In: Vegetable Crops in India (Eds. Bose, T.K. and Som, M.G.). Naya Prakash, Calcutta. pp. 91-164.

Seshadri, V.S. Mishra, J.P., Sharma, J.C. and More, T.A. 1983. Heterosis breeding in melons-problems and prospects. *South Indian Hort. Commem*. Special Issue. 74-82.

Sharma, J.C. 1975. Genetical studies in muskmelon (*Cucumis melo* L.). PhD Thesis. P.G. School, IARI, New Delhi.

Shinde, N.N. 1979. Studies on sex forms in muskmelons (*Cucumis melo* L.). PhD Thesis, P.G. School, IARI, New Delhi.

Shinde, N.N. and Seshadri, V.S. 1981. Association of sex forms with fruit shape in muskmelon (*Cucumis melo* L.). *Cucurbit Genet. Coop. Reptr.*, 4: 26-28.

Shinde, N.N. and Seshadri, V.S. 1983. Inheritance of sex forms in muskmelon. *In XV Intern. Congress of Genetics*, New Delhi. Oxford and IBH Publ. Co., New Delhi.

Shinoda, T., and Tanaka, K. 1987. Resistance of melon *Cucumis melo* L. to the melon aphid *Aphis gossypii* Glover. I. Differences in population growth of melon aphid on melon cultivars. Bull. Nat. Res. Instit. Veg. Ornam. Plants & Tea. Japan, Ser. A, no. 1: 157-164.

Singh, D., Nandpuri, K.S. and Sharma, B.R. 1976. Inheritance of some economic quantitative characters in an inter varietal cross of muskmelon (*Cucumis melo* L.), *Punjab Agric. Univ. J. Res.*, 13: 172-176.

Singh, M.J., Randhawa, K.S. and Lal, T. 1989. Genetic analysis for maturity and plant characteristics in muskmelon. *Veg. Sci.*, 16: 181-184.

Sitterly, W.R. 1978. The powdery mildews of cucurbits. In: The Powdery Mildews (D.M. Spencer, ed.), Academic Press, London.

Somkumar, R.G. and More, T.A. 1993. Downy mildew (*Pseudoperonospora cubensis*) resistance in melon (*Cucumis melo* L.). *Cucurbit Genet. Coop Reptr.*, 16: 40-41.

Thomas, C.E. 1978. A new biological race of powdery mildew of cantaloupes. *Plant Dis. Rept.*, 62: 223.

Vashistha, R.N. and Choudhury, B. 1972. Studies on growth and yield potential of muskmelon cultivars resistant to red pumpkin beetle. *Haryana J. Hort Sci.*, 1: 55-61.

Thomas, C.E., Cohen, Y., McCreight, J.D., Jourdain, E.L. and Cohen, S. 1988. Inheritance of resistance to downy mildew in *Cucumis melo*. *Plant Dis.*, 72: 33-35.

Whitaker, T.W. 1931. Sex ratio and sex expression in the cultivated cucurbits. *Amer. J. Bot.*, 18: 359-366.

Whitaker, T.W. 1979. The breeding of vegetable crop. Highlights of the past seventy five years. *Hort Science*, 14: 359-363.

Whitaker, T.W. and Bohn, G.W. 1952. Natural cross pollination in muskmelon. *Proc. Amer. Soc. Hort. Sci.*, 60: 391-396.

Whitaker, T.W. and Davis, G.N. 1962. Cucurbits Cultivation and Utilization. Leonard Hill, London.

Bottlegourd Breeding

Sheo Pujan Singh

SUMMARY. Bottlegourd is a cultivated annual monoecious species with five wild perennial dioecious species belongs to family Cucurbitaceae. It is grown for its tender fruits, basically used as vegetable. The tender edible fruits are also prepared into sweets, pickles, and other delicious preparations. In many cases of ailments, it is a preferred vegetable because of its cooling effects and easy digestibility. Various plant parts have medicinal value too. It is known as poor man's vegetable in India, and is now attaining fast popularity among health conscious urban elite, which has encouraged round the year cultivation of this potentially important vegetable, except in very cool regions during winter. Considerable amount of heterosis exists in this crop and is being exploited commercially due to easy hybrid seed production because of its monoecious nature. *[Article copies available for a fee from The Haworth Document Delivery Service: 1-800-HAWORTH. E-mail address: <docdelivery@ haworthpress.com> Website: <http://www.HaworthPress.com> © 2004 by The Haworth Press, Inc. All rights reserved.]*

KEYWORDS. Bottlegourd, crop biology, sex forms, heterosis, inbred lines, seed production, hybrid seed production

INTRODUCTION

Bottlegourd [*Lagenaria siceraria* (Molina) Standle; 2n = 22] synonymously called white flowered gourd or calabash gourd is a member of family

Sheo Pujan Singh is affiliated with the Department of Vegetable Crops, Narendra Deva University of Agriculture & Technology, Faizabad (U.P.) 224229, India.

[Haworth co-indexing entry note]: "Bottlegourd Breeding." Singh, Sheo Pujan. Co-published simultaneously in *Journal of New Seeds* (Food Products Press, an imprint of The Haworth Press, Inc.) Vol. 6, No. 4, 2004, pp. 363-375; and: *Hybrid Vegetable Development* (ed: P. K. Singh, S. K. Dasgupta, and S. K. Tripathi) Food Products Press, an imprint of The Haworth Press, Inc., 2004, pp. 363-375. Single or multiple copies of this article are available for a fee from The Haworth Document Delivery Service [1-800-HAWORTH, 9:00 a.m. - 5:00 p.m. (EST). E-mail address: docdelivery@haworthpress.com].

Digital Object Identifier: 10.1300/J153v06n04_03

Cucurbitaceae. It is a cultivated annual monoecious species with five wild perennial dioecious species. The wild species are confined to Africa and Madagascar (Willis, 1966). Bottlegourd is considered to be one of the most ancient crops cultivated by man in the tropics with an archeological evidence of human utilization at least 15,000 years in the New World and 12,000 years in the old world (Richardson, 1972). Whitaker (1971) was of the view that bottlegourd was indigenous to tropical Africa (south of Equator) and has diffused to the New World by trans-oceanic drift or human transport. Heiser (1980) supporting the view of Whitaker and concluded Africa as place of origin of bottlegourd. Though he further conceded that there lacks a decisive evidence to distinguish between Africa and America as the original home of this species.

Bottlegourd is grown for its tender fruits, basically used as vegetable. The tender edible fruits are also prepared into sweets, pickles, and other delicious preparations. Hundred gram of tender fruit contains 96 g water, 0.2 g protein, 0.1 g fat, 2.5 g carbohydrate, 0.6 g fiber, 0.5 g minerals, 12 k-cal, 20 mg calcium, 10 mg phosphorus, 0.7 mg iron, 0.3 mg thiamine, 0.01 mg riboflavin, and 0.2 mg niacin (Gopalan et al., 1982). The fruits make delicious supplement to the human diet but the contents are considered of little nutritive value. However, in many cases of ailments, it is a preferred vegetable because of its cooling effect to the stomach and easy digestibility. The dry fruit shells are used to make musical instrument and utensils. Various plant parts have medicinal value too. This species also holds promise for its yet unexploited possible uses of oil and protein contents of seed (Jacks, 1986). The seed kernels contain 45% oil, and about 35% protein.

The crop thrives well in hot humid weather conditions, but it also grows well and continues to produce good harvest under frost free low temperature conditions if the plants have attained sufficient vegetative growth before the onset of cool weather, as is common in northern Indian plains where August-September sown crops of bottlegourd produce remunerative off season produce during cool months of November, December, and January. Bottlegourd, generally known as poor man's vegetable in India, is attaining fast popularity among the health conscious urban elite, which has encouraged round the year cultivation of this potentially important vegetable in almost all part of the country, except in very cool regions during winter.

CROP BIOLOGY

Vegetative Habit

Bottlegourd is an annual viny pubescent herb (Figure 1) with stout five angled stem. The stem is profusely branched. Prostrate or climbing branches of a

FIGURE 1. A bottlegourd plant spread on the ground.

single plant can cover varying space depending upon weather conditions, soil nutrient, available moisture, and area provided for its growth. The plants have well-developed branched root system. The nodes also throw out roots when come in contact with the moist soil. The leaves are commonly simple, reniform, wavy with entire margin. They may have 3-5 shallow to deep lobes. The apex of the leaves may be pointed or blunt. Some genotypes have pinnatified leaves also. Bifid tendrils are borne in the axils of leaves.

Sex Forms

Commonly bottlegourd is monoecious in nature where solitary male and female flowers (Figure 2A) are found separately in the leaf axils of the same plant. A stable andromonoecious sex form bearing hermaphrodite flower (Figure 2B) and male flower in the same plant have also been isolated and reported by Singh et al. (1996). This isolate has been named as Andromon-6. Male flowers are borne on longer pedicels than female and hermaphrodite flowers. Both male and female flowers have generally large and white showy corolla with five petals. Male flowers have larger petals than female flowers. In male flowers stamens are apparently 3, two are 2-celled and one is 1-celled. Female flowers have inferior prominent ovary, which may be round, ovate, long or cylindrical. Ovary is tricarpelary, syncarpous, and unilocular. Style is short and stout. There are three stigmatic lobes. Ovules are many, generally 400-700 in number. Placentation is parietal.

Fruit Shape and Size

Although fruit is essentially a berry, it is called a pepo because of its hard and tough rind at maturity. The term 'gourd' refers to the hard tough rind. The

FIGURE 2. (A) Male (left) and female (right) flower buds an hour before anthesis. (B) Hermaphrodite flower of andromonoecious line-Andromon-6.

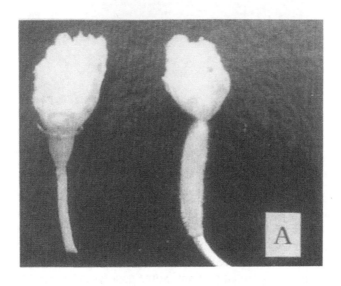

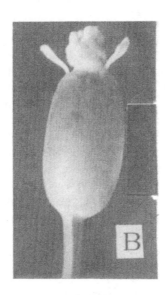

fruits are indehiscent. A great variability is encountered in fruit shapes. They may be long, cylindrical, curved, necked, oblong, round, flat-round, conical, pear-shaped, club-shaped, etc. But bottlegourds are broadly classified as long (Figure 3) and round (Figure 4) types. Variability in fruit characteristics exhibited by bottlegourd has been described by Sirohi and Sivakami (1991). There also occurs a great range of variation in size/weight of fruits. In cultivated types, long fruits commonly vary from 40 cm to 100 cm in length at maturity, while, round type may vary from 10-40 cm in diameter. The fruit weight at

FIGURE 3. A long fruited bottlegourd variety.

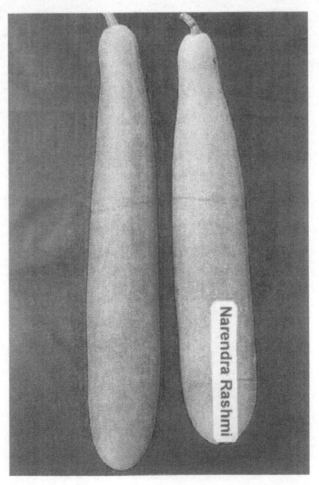

maturity varies from 1-10 kg. However, at edible green stage commonly the fruit weight varies from 0.5 kg to 2.0 kg. The fruits may be sweet or bitter. Bitter types are not edible, although they have medicinal importance. At edible stage the fruits are pale green, green or dark green. Some genotypes are striped too (Figure 5). At maturity the fruit turn into cream or creamish-brown in color.

FIGURE 4. A round fruited bottlegourd variety.

FIGURE 5. A long fruited bottlegourd variety with striped skin.

Seed

Numerous seeds are found in mature fruits. The number of seeds per fruit varies from 400-700. Seed weight varies from 10 to 15 g per 100 seeds. The seeds are light brown to dark brown in color. They are oblong, flattened, and distinctly margined. Furrows and ridges are also present on seeds.

Flowering and Fruit Set

In the main summer season crop flowering starts at about 40 days after sowing. Generally male flowers appear first at about 10th node and female flowers follow the sequence and appear at 8th to 15th node. Thereafter the flush of male and female flowers continue to occur in the plants for next 50-80 days till the latter decay and die. The ratio of male:female flower may vary from 5:1 to 15:1 in the common types. Sex ratios are highly sensitive to environment and nitrogen application. Long days, high temperature, and high nitrogen rates promote greater number of male flowers. In cool weather, numerous female flowers may be borne in bottlegourd plants (Figure 6) than male flowers, but all the female flowers do not turn into fruits. In general, higher proportions of female flowers are borne in a plant even during summer but only a few turns into fruits. First green edible fruits are available in 55-65 days after sowing in early varieties.

HETEROSIS

Green fruit yield in open-pollinated varieties of bottlegourd varies from 250 to 400 q/ha. The corresponding figures for hybrids are from 400 to 700 q/ha. Fruit yield, desirable attractive fruits shape, earliness, and resistance against common diseases and pests appear to be the major components required for heterosis breeding program of bottlegourd, apart from other attributes. Rajendran (1961) recorded 266.51 percent heterosis in the yield of the best hybrid combination of bottlegourd over the better parent. Roy (1964) reported 81.42 and 78.60 percent heterosis in bottlegourd over their respective better parents in the yield of two best hybrids. Choudhury and Singh (1971a) developed two high yielding bottlegourd hybrids which gave 75 percent and 106 percent higher yield over their respective better parent. In a genetical study of two crosses, Sharma et al. (1983) observed heterosis over better parent for fruit number, fruit weight and total yield in both the crosses. Pal et al. (1984) observed more rapid germination (2-4 vs. 5-6 days) and earlier fruit maturity (10-11 days) in F_1 hybrids over the better parent. The female flowers were borne at lower nodes and hybrids had 17.28 percent higher flesh thickness. It was also noted that hybrids gave 20% higher early yield than parents

FIGURE 6. A branch bearing one developed fruit and nine female flowers on consecutive nodes during winter.

and had a longer harvesting period (65-71 days) than the latter (55-65 days). Sirohi et al. (1985) reported heterosis of 0.44 to 2.90 per cent for fruit length, 1.38 to 64.71 percent for number of fruit per plant, 1.13 to 22.19 percent for fruit weight and 0.90 to 76.46 percent for fruit yield per plant. The best F_1 hybrid S-46 × S-47, which gave 115 percent higher yield over the commercial cultivar PSPL and 32.70 percent higher yield over the best parental line S-48, was suggested for commercial cultivation.

Janakiram and Sirohi (1989) studied heterosis for 9 yield components in 45 F_1 hybrids, namely S-46 × S-54, S-10 × S-52-7, and S-54 × S-52-7 showed 84.5, 80, and 80 percent heterosis, respectively, for yield over the best parental line, S-41. The higher yields in these crosses were attributed to increased number of fruits/plant, fruit weight, and fruit size. The cross S-46 × S-54 gave 148.97 percent higher yield than the commercial cultivar Pusa Summer Prolific Round. In a diallele set involving nine parents of Bottlegourd, Maurya et al. (1993) studied the performance of 36 F_1 hybrids, excluding reciprocals, under low temperature condition during the winter season. They noted that the highest yielding cross took only 83.33 days for its first picking compared to 111.33 days needed by commercial cultivar PSPL. Heterosis of the highest yielding cross over the top parent for yield per plant was 80.51 percent. Sharma et al. (1993) derived information on heterosis from data on yield components in the parents and F_1 hybrids from 11 lines × 3 tester crosses. The cross Summer Long Green Selection-2 × Faizabadi long had the highest heterosis over the control cultivar PSPL for number of fruits (106.63 percent) and total yield per plant (110.33 percent). Hissar Local Sel-2 × PSPL had the highest heterosis (22.93 percent) over the control for fruit length. Singh et al. (1998) studied the performance of 28 F_1 hybrids along with check (excluding reciprocals). The crosses showing significant heterosis over the better parent were ARBGH-7 × Pusa Naveen for yield per vine, per hectare and no. of fruits per vine ARBGH-7 × LC 2-1 for fruit weight and days to first female flowering, PBOG-61 × NDBG-56 for fruit length, girth and vine length and PSPL × ARBGH-7 for days to first fruit harvest. Cross ARBGH-7 × LC 2-1 showed significant increase over the commercial checks for the yield and yield related characters.

Kumar et al. (2001) conducted experiments at Narendra Deva University of Agriculture & Technology, Faizabad, India, to study the environmental effect on the standard heterosis of Bottlegourd with the objectives: (i) to explore the appropriate seasons for exploitation of standard heterosis and (ii) to identify season specific best hybrids. They conducted the experiments on 10 diverse parents and their 45 F_1 hybrids in three seasons, viz., September sown winter crop, November sown early summer crop, and March sown main summer crop. Pusa Naveen an elite variety was used as standard check. The results revealed that though the yield levels were low during off season crops of winter and early summer seasons as compared to main summer season, the degrees of standard heterosis were higher during winter season (up to 61.65%) and early summer season (up to 71.93%) as compared to main summer season crop (only up to 18.14%). Hence, they concluded that standard heterosis in bottlegourd could better be exploited in off-season crops than in main season crop.

POLLINATION CONTROL MECHANISMS

Dehiscence of anthers in bottlegourd takes place between 11 a.m. to 2.00 p.m., whereas anthesis of both staminate and pistillate flowers takes place between 5.0 p.m. and 8.0 p.m. as reported by Nandapuri and Singh (1967). Staminate and pistillate flowers open more or less at the same time, floral biology of bottlegourd was also studied by Joshi and Gour (1971). They reported that both male and female flowers generally open between 5.00 to 7.00 p.m., except when the minimum temperature is low when flower opening is delayed. The stigma remained receptive 6 hours before and 30 hour after the flower opened. Fruit formation could be recognized only after 24 hours of pollination. Genetic male sterility controlled by *ms1* gene has been reported by Dutta (1983), which can be used for hybrid seed production.

Bottlegourd being monoecious in nature, controlled pollination is necessary for pure line breeding and maintenance of germplasm, a part from crop improvement through hybridization programme, and hybrid seed production through hand pollination; controlled pollination becomes imperative because of monoecious nature of the crop. In cases where absolute purity is necessary, male and female flowers, which are likely to open in the offing evening, must be completely wrapped in the morning before 10 a.m. with strips of bandage cotton. Controlled pollinations are made in the afternoon when the dehiscence begins. Female flowers are once again wrapped with the help of cotton strips to protect them from undesirable pollen transfer through insects or wind. However, in cases where absolute purity is not so necessary, controlled pollinations can be attempted directly without covering of male or female flowers. However, precaution must be observed to complete the pollination work while anthers and stigmas of male and female flowers, respectively, are still completely covered by their petals and the petals have not yet started loosening out for anthesis. This practice reduces the labor cost and also saves time of covering the flowers buds.

Due care should be taken to remove the anthers well before dehiscence from the hermaphrodite flowers are of trimonoecious or andromonoecious lines, if such flowers to be used in the hybridization work.

HYBRID SEED PRODUCTION

Bottlegourd is endowed with following unique combination of desirable attributes that make it suitable for heterosis breeding program.

  i. It is generally monoecious in nature, which makes hand emasculation easier in female parents by merely pinching off the male flower buds prior to dehiscence and anthesis,
 ii. both male and female flowers are large due to which controlled hand pollination becomes an easier task,

iii. the fruit set success in the hand-pollinated female flowers is very high, provided chosen flowers are at the proper position in the plant,
iv. it is a cross-pollinated crop where high degree of economic heterosis is reported for desirable traits such as yield, earliness, number of fruits, etc.
v. the crop suffers only negligible inbreeding depression in the process of development of uniform inbred lines,
vi. good parents produce several hundred seeds per fruit which weigh nearly 100 g, and
vii. due to wider spacing requirements 2.5 kg to 3.0 kg seeds are enough for planting one hectare land area which can be produced only by 25 to 30 good fruits.

Thus hybrid seed production in bottlegourd is economically feasible. However, very little information is available regarding the standardized method of hybrid seed production in this crop. Choudhury and Singh (1971b) proposed a method of hybrid seed production in which male buds are pinched off when male and female parents are grown in adjacent rows. Thus all the fruit set in female parent would be through cross-pollination by insects. The precaution should be made that no single male bud is left in the female parent, otherwise it could promote selfing or sibbing within the female parent. Hybrid seed production can also be done by direct and reciprocal controlled hand pollination on both the parents. This procedure doubles the amount of hybrid seed produced within a given area, as compared to the method suggested by Choudhury and Singh (1971b). The use of genic male sterile line as proposed by Dutta (1983) in hybrid seed production needs further investigation. Similarly, search of dominant genetic markers to facilitate the identity of true to the type F_1 hybrid plants before the onset of flowering is also an important aspect to be worked out to strengthen the hybrid seed production in this crop.

MAINTENANCE OF INBREDS

Since there is no much vigor loss due to inbreeding, uniform inbreds of bottlegourd could be developed with comparable vigor of open pollinated varieties. They can be maintained through open pollination by growing them at proper isolation distance. A distance of 1000 m is quite appropriate. Alternatively, the inbreds can also be maintained by hand pollination along with other genetic stocks. Seed production principles such as land requirement, field inspection, etc., should be carefully observed. Care must be taken for rouging out the undesirable off types at right stages of plant growth, viz., (i) before flowering, (ii) at the time of flowering, and (iii) at the time of fruit set and maturity to maintain true-to type plant characteristics. To bring about improvement in heterosis level in the F_1 hybrids, progeny breeding, recurrent, and reciprocal recurrent selections may also be practiced to improve the inbreds, as suggested by Sharma et al. (1983). The seed production of open pollinated varieties is similar to that of maintenance of inbreds.

CONCLUSIONS

Bottlegourd is an important vegetable crop of several countries of the tropical world, but unfortunately the crop remains an under exploited vegetable (Indira and Peter, 1988). Due to non-availability of improved cultivars, its cultivation has been largely dependent on local land races. However, over the last two decades emphasis has been laid for crop improvement in this crop, at least in India, owing to which a sizeable number of improved open-pollinated varieties as well as hybrids have been released for cultivation by State Agricultural Universities, Indian Council of Agricultural Research Institutes, and private seed companies in India.

Emphasis in future bottlegourd breeding should be laid to develop varieties resistant to major diseases and pests apart from breeding for earliness, high yield, uniform attractive fruit shape, and large fruiting period. As the consumer preference is increasing for the use of bottlegourd fruits, an effort will also be required to develop varieties with varied qualities for export purposes. Future thrust on bottlegourd breeding should also be given to develop season and region specific improved varieties. Marker based selection in the segregating generations derived after hybridization would be another desirable facet of bottlegourd breeding. The concerted efforts for crop improvement, and production technology would probably very soon give bottlegourd its due place among vegetables which it deserves.

REFERENCES

Choudhury, B. and Singh, B. (1971a). Two high yielding bottlegourd hybrids. *Indian J. Hort.*, 18: 15-32.

Choudhury, B. and Singh, B. (1971b). Pusa Meghadoot and Pusa Manjari, two high yielding Bottlegourd hybrids. *Ind. Hort.* 16: 15-16.

Dutta, O.P. (1983). Male sterility in okra (*Abelmoschus esculentus* (L.) Moench) and bottlegourd (*Lagenaria siceraria* (Mol.) Standl.) and its utilization in hybrid seed production. *Thesis Abstr.* 9: 341.

Gopalan, C.; Rama Sastri, B.U. and Balasubramanian, S.C. (1982). Nutritive Value of Indian Foods. Indian council of Medical Research. National Institute of Nutrition. Hyderabad.

Heiser, C.B. (1980). The Gourd Book. University of Oklahoma Press, Norman.

Indira, P. and Peter, K.V. (1988). Under Exploited Tropical Vegetables pp. 68. Directorate of Extension, Kerala Agricultural University, Trichur. India.

Jacks, T.J. (1986). Cucurbit Seed Protein and Oil. In Plant Protein: Applications, Biological Effects and Chemistry. Ed. Robert L. Ory. U.S. Department of Agriculture/ American Chemical Society. Washington DC.

Janakiram, T. and Sirohi, P.S. (1989). Heterosis studies in round fruited bottlegourd. *Madras Agric J.* 76 (6): 339-342.

Joshi, D.P. and Gour, S.K.S. (1971). Floral biological studies of *Lagenaria siceraria* Standl. (Bottlegourd). *J. Res. Punjab Agric. Univ.* 8: 420-426.

Kumar, Randhir; Singh, S.P.; Singh, S.B. and Singh, N.K. (2001). Environmental effect on the standard heterosis of Bottlegourd. In Poster Abstracts of 'Diamond Jubilee Symposium on Hundred Years of Post-Mendelian Genetics and Plant Breeding-Retrospects and Prospects.' November 6-9, 2001. New Delhi, organized by Indian Society of Genetics and Plant Breeding pp. 156-157.

Maurya, I.B., Singh, S.P. and Singh, N.K. (1993). Heterosis and combining ability in bottlegourd [*Lagenaria siceraria* (Molina) Standl.]. *Veg. Sci.*, 20: 77-81.

Nandapuri, K.S. and Singh, J. (1967). Studies on floral biology of bottlegourd [*Lagenaria siceraria* (Mol.) Standl.]. *J. Res. Punjab Agric. Univ.* 4: 54.

Pal, A.B., Srivanandappa, D.T. and Vani, A. (1984). Manifestation of heterosis in Bottlegourd [*Lagenaria siceraria* (Molina) Standl.]. *South Ind. Hort.*, 32: 32-38.

Rajendran, R. (1961). Studies on the exploitation of hybrid vigour in bottlegourd (*Lagenaria siceraria*). M.Sc. Thesis, I.A.R.I., New Delhi, India.

Richardson, J.B. (1972). The Pre-Columbian distribution of bottlegourd (*Lagenaria siceraria*). A re-evaluation. *Econ. Bot.* 26: 265-273.

Roy, H.K. (1964). Studies on exploitation of hybrid vigour in Bottlegourd. M.Sc. Thesis, I.A.R.I., New Delhi, India.

Sharma, B.R., Singh, J., Singh, S. and Singh, D. (1983). Genetical studies in bottlegourd. *Veg. Sci.*, 10: 102-111.

Sharma, N.K., Dhankar, B.S. and Tewatia, A.S. (1993). Line × tester analysis for combining ability studies in bottlegourd. *Haryana J. Hort. Sci.*, 22: 324-327.

Singh, S.P., Maurya I.B. and Singh, N.K. (1996). Occurrence of andromonoecious form in Bottlegourd (*Lagenaria siceraria*) exhibiting monogenic recessive inheritance. *Curr. Sci.* 70: 458-459.

Singh, P.K., Kumar J.C. and Sharma J.R. (1998). Heterosis studies in long fruited bottlegourd. *Veg. Sci.* 25 (1): 55-57.

Sirohi, P.S. and Sivakami, N. (1991). Genetic variability in cucurbits Bottlegourd. *Ind. Hort.* 36: cover page II & pp. 44-45.

Sirohi, P.S., Sivakami, N. and Choudhury, B. (1985). Heterosis in long-fruited bottlegourd. *Ann. Agric. Res.*, 6: 210-214.

Whitaker, T.W. (1971). Men across the sea. J.C. Kelley, C.W. Pennington and R.L. Rands (Eds.), University of Texas Press, pp. 320-327.

Willis, J.G. (1966). Dictionary of Flowering Plants and Ferns (7th Ed.), Cambridge University Press.

Hybrid Cucumber

U. K. Kohli
Amit Vikram

SUMMARY. Cucumber is a member of Cucurbitaceae comprising 90 genera and 750 species. It is consumed as salad or in pickled form. Cucumber is a low energy and high water content vegetable and is mainly used as refreshing condiment Considerable heterosis has been manifested in cucumber for various traits such as number of fruits, early and high yield. However, hybrid seed production requires development of superior lines for production of good quality hybrids. Three major genes Acr/acr, M/m, and A/a besides environmental factors and modifying genes mainly control various sex types. Of the various sex forms, gynoecious and monoecious are important from hybrid production point of view. *[Article copies available for a fee from The Haworth Document Delivery Service: 1-800-HAWORTH. E-mail address: <docdelivery@haworthpress.com> Website: <http://www.HaworthPress.com> © 2004 by The Haworth Press, Inc. All rights reserved.]*

KEYWORDS. Cucumber, heterosis, floral biology, sex expression, maintenance of gynoecious lines, hybrid seed production

Cucumber (*Cucumis sativus* L.) is a member of Cucurbitaceae comprising 90 genera and 750 species. It is consumed as salad or in pickled form. The cu-

U. K. Kohli and Amit Vikram are affiliated with the Department of Vegetable Crops. Dr. YS Parmar University of Horticulture and Forestry, Nauni, Solan 173230 (H.P.), India.

[Haworth co-indexing entry note]: "Hybrid Cucumber." Kohli, U. K., and Amit Vikram. Co-published simultaneously in *Journal of New Seeds* (Food Products Press, an imprint of The Haworth Press, Inc.) Vol. 6, No. 4, 2004, pp. 377-382; and: *Hybrid Vegetable Development* (ed: P. K. Singh, S. K. Dasgupta, and S. K. Tripathi) Food Products Press, an imprint of The Haworth Press, Inc., 2004, pp. 377-382. Single or multiple copies of this article are available for a fee from The Haworth Document Delivery Service [1-800-HAWORTH, 9:00 a.m. - 5:00 p.m. (EST). E-mail address: docdelivery@haworthpress.com].

http://www.haworthpress.com/web/JNS
© 2004 by The Haworth Press, Inc. All rights reserved.
Digital Object Identifier: 10.1300/J153v06n04_04

cumber is reported to be indigenous to India (de Candolle, 1886). The chief evidence for this suggestion is the occurrence of *Cucumis hardwickii* Royle, a cucumber-like plant, in the foot-hills of the Himalayas in India. *C. hardwickii* is similar to *C. sativus* except that the exterior of the fruit is smooth and flesh is extremely bitter. Since, *C. hardwickii* crosses freely with *C. sativus* (Deakin et al., 1971), this has led to the conclusion that *C. hardwickii* is either a feral or progenitor form of the cultivated cucumber (de Candolle, 1886).

Cucumber cultivars are currently classified as either pickling or slicing types. The fruits are eaten at immature stage. Cucumber is a low energy and high water content vegetable and is mainly used as refreshing condiment. In 100 grams fresh edible portion, the fruit contains 96 g water, 0.6 g protein, 2.2 g carbohydrates, 0.1 g fat, 45 IU vitamin A, 0.03 mg vitamin B, 0.02 mg vitamin B_2, 0.3 mg niacin, 12 mg vitamin C, 12 mg calcium, 0.3 mg iron, 15 mg magnesium, and 24 mg phosphorus (Esquinas-Alcazar and Gullick, 1983).

Considerable heterosis has been manifested in cucumber for various traits such as number of fruits, early and high yield. However, hybrid seed production requires development of superior lines for production of good quality hybrids.

HETEROSIS

Hayes and Jones (1916) were the first investigators to report heterosis in cucumber. They reported 24-39 percent yield increase in F_1 over the highest yielding parent. However, heterosis for number of fruits per plant was reported to be 6-27 percent. Solanki et al. (1982) have reported a heterosis of 120.23% for fruit yield plant. Nishi (1967) reported that cucumber hybrids were grown only on 18% of the farms in the chief producing districts of Japan in 1940, but 95% in 1965. The cost of cucumber hybrid seed was about thrice compared to open pollinated cucumber seed. Takahashi (1987) noted that 100% of the cucumber hybrids grown in Japan for fresh market were hybrids.

FLORAL BIOLOGY

Cucumber is an annual and day neutral plant. Within *C. sativus*, staminate, pistillate, and hermaphroditic flowers occur in various arrangements. Cucumbers are primarily monoecious, i.e., bearing separate staminate and pistillate flowers on the same plant. Male flowers occur in clusters with each flower on a slender stem and having three stamens.

Female flowers are borne singly and have an inferior distinguishable large ovary. The ovary has three carpels and several rows of ovules. The ovary is having a short, thick style with three stigma lobes. Flowers are having yellow, wrinkled petals. Nectar is produced by male as well as female flowers. Bees visit the flowers for nectar. Female flowers produce a higher volume of nectar than male flowers.

Cucumbers also bear many other flower types ranging from staminate, pistillate, and hermaphroditic flowers in various arrangements. Normally, the ratio of male to female flowers is ten to one in monoecious cultivars. Therefore, the new cultivars of cucumber having gynoecious character and hybrids based on gynoecious lines have been developed which have substantially higher number of pistillate flowers and, therefore, fruit prolifically.

POLLINATION

Honeybees are the major pollinating agents of cucumber. Ideal conditions for honeybee pollination include clear skies and low wind speed. Under low humidity also bee activity is more. The optimum time for pollination is between 9:00 a.m. to 12:00 noon. Cucumber pollen is sticky and not actively colleted by bees. To ensure adequate pollination, bees should visit the flowers at least 12 to 15 times for normal fruit development. To increase fruit set, the pollination can be enhanced by supplementing the natural bee population with additional beehives in the field. At least two beehives should be placed per acre. Each bee hive should contain around 15,000 bees. Shemetkov (1957) reported that a cucumber flower should be visited 8 to 10 times for satisfactory fruit set but the number of seeds and fruit weight continued to increase with the increasing number of visits. Anderson (1941) stated that nubbins, balls, and crooks were the result of poor pollination. The gynoecious cultivars or hybrids should be protected from pollination, because due to fertilization, their fruits may become misshapen.

Cucumber is primarily an insect-pollinated crop. Therefore, while selfing or crossing bees should be excluded to prevent pollen contamination. For the purpose of hybrid seed production gynoecious lines are used which eliminate the need for manual crossing. Since gynoecious lines bear 100 percent female flowers, therefore, any seed obtained from them is F_1 hybrid. During seed production, however, sufficient isolation distance should be kept between two cultivars or hybrids to prevent contamination.

If hybrid seed production is to be carried out by hand pollination, the flower to be pollinated should be covered before anthesis to prevent insect pollination (Robinson, 1999). Male flowers can also be covered a day before anthesis. The hand-pollinated flowers are bagged and tagged to protect from insect pollination and easy identification.

SEX EXPRESSION

Sex inheritance is important in cucumber hybrid seed production. There are a number of sex types in cucumber occurring as a result of differences in frequency and distribution of staminate, pistillate, and hermaphrodite flowers. The main sex types occurring in cucumber are as follows:

- Monoecious plants: Staminate and pistillate flowers on the same plant.
- Androecious plants: Staminate flowers only.
- Gynoecious plants: Pistillate flowers only.
- Hermaphroditic plants: Hermaphrodite flowers only.
- Androecious plants: Staminate and hermaphrodite flowers on the same plant.

Three major genes Acr/acr, M/m, and A/a besides environmental factors and modifying genes are mainly controlling various sex types (Tatlioglu, 1993). Of the various sex forms, gynoecious and monoecious are important from hybrid production point of view.

Combinations of genotypes at these loci lead to the development of basic sex types, i.e., androecious, monoecious, hermaphroditic, gynoecious. The sex expression is reported to be affected by number of other major loci. In addition, photoperiod and temperature are two environmental factors specifically affecting sex expression. Almost all cucumbers are day neutral. However, cucumbers have been reported to respond to varying photoperiodic regimes by showing variations in flower number and type (Lower and Edwards, 1986). Generally, shorter days as well as lower temperatures are reported to increase female tendency. Male sterility in cucumbers is of little use because of the availability of genetic control over staminate versus pistillate flowering. Cytoplasmic male sterility has not been reported in cucumber (Lower and Edwards, 1986).

HYBRID SEED PRODUCTION

Hybrids of cucumber are produced mainly by crossing gynoecious lines with monoecious lines. Though, other systems of producing gynoecious hybrid seed such as gynoecious × gynoecious have been proposed but gynoecious × monoecious hybrids are still the most widely grown (Robinson, 1999).

Most of the commercial hybrids based on gynoecious cucumber lines are a blend of gynoecious hybrid and monoecious seed. About 10% blending is of a monoecious genotype seed with gynoecious hybrid seed has been advocated by Peterson and DeZeew (1967). In addition, homozygous gynoecious hybrid cucumber seed has been produced by crossing two gynoecious lines after one parent has been treated with a growth regulator to induce male flowers (Robinson, 1999).

Hybrid seed homozygous for gynoecious sex expression could also be produced by crossing gynoecious and hermaphroditic lines (Kubicki, 1970).

MAINTENANCE OF GYNOECIOUS LINES

The commercial production of gynoecious cucumber seed was made possible only when it was discovered that gynoecious inbreds could self reproduce if a growth regulator is applied to induce male flower formation (Robinson, 1999).

Peterson and Anhder (1960) for the first time discovered the effect of gibberellic acid on promotion of male flower formation in cucumber. However, due to erratic male flower induction by use of gibberellic acid, application of silver compound such as silver nitrate is done to induce male flowers. Silver ions inhibit ethylene action and thus promote male flower formation in gynoecious cucumber plants (Beyer, 1976). However, due to phytotoxic effects such as burning of plants, silver thiosulphate is now widely used by seed producers for the maintenance of gynoecious cucumber lines. It induces male flowering of cucumber plants over a longer period and is less phytotoxic compared to silver nitrate.

REFERENCES

Anderson, W.S. 1941. Growing cucumbers for pickling in Mississippi. Miss. Agr. Expt. Sta. Bull. 355: 17 pp.

Beyer, E. Jr. 1976. Silver ion: A potent antiethylene agent in cucumber and tomato. HortSci. 11: 195-196.

De Candolle, A. 1886. Origin of Cultivated Plants. Kegan. Paul. Trench & Co., London.

Deakin, J.R., Bohn, G.W. and Whitaker, T.W. 1971. Interspecific hybridisation in Cucumis. Econ. Bot. 25: 195-211.

Esquinas-Alcazar, J.T. and Gullick, P.J. 1983. Genetic resources of Cucurbitacae. International Board for Plant Genetic Resources, Rome.

Hayes, H.K. and Jones, D.F. 1916. First generation crosses in cucumbers. Conn. Agr. Expt. Sta. Ann. Rept.: 319-322.

Kubicki, D. 1970. Cucumber hybrid seed production based on gynoecious lines multiplied with the aid of complimentary hermaphroditic lines. *Genet. Palanica* 11: 181-186.

Lower, R.L. and Edwards, M.D. 1986. Cucumber Breeding. In: *Breeding Vegetable Crops*. AVI Publishing Co., Connecticut, USA. 173-207 pp.

Nishi. 1967. F_1 seed production in Japan. *Proc. XVII International Hort. Congress* 3: 231-257.

Peterson, C.E. and Anhder, L.D. 1960. Induction of staminate flowers on gynoecious cucumbers with gibberllin A_3. *Science* 131: 1673-1674.

Peterson, C.E. and Dezeew, D.J. 1967. The hybrid pickling cucumber. *Spartan Dawn. Quart. Bull. Mich. Agr. Exp. Sta.* 46: 267-273.

Robinson, R.W. 1999. Rationale and methods for producing hybrid cucurbit seed. *Journal of New Seeds.* 1: 1-47.

Shemetkov, M.F. 1957. The use of bees for pollinating cucumbers in hot houses and forcing beds. *Biul. Nauch. Tekh. Inf. Inst. Pchelovod.* 2: 21-24

Solanki, S.S., Seth, J.N. and Lal, S.D. 1982. Heterosis and inbreeding depression in cucumber (*Cucumis sativus* L.)V. *Progr. Hort.* 14: 136-140.

Takahashi, O. 1987. Utilization and seed production of hybrid vegetable varieties in Japan. In: *Hybrid Seed Production of Selected Cereal Oil and Vegetable Crops*. Eds. W.P. Feistritzer and A.G. Kelly. FAO Plant Production and Protection Paper 82: 313-328.

Tatlioglu, T. 1993. Cucumber: *Cucumis sativus* L. In: *Genetic Improvement of Vegetable Crops*. Eds. G. Kalloo and B.O. Bergh. Pergamon Press, USA. 197-234 pp.

Mechanisms for Hybrid Development
in Vegetables

Sanjeet Kumar
P. K. Singh

SUMMARY. Availability of cost effective mechanism/method to produce large-scale F_1 seeds utilizing selected parental lines is an important factor, which ultimately determines the commercial viability of the hybrid varieties. In vegetables, although experimental crosses (few seeds for research purpose) can be developed through manual emasculation (in case of hermaphrodite crops) followed by manual pollination of emasculated flowers or pistilate flowers (in case of monoecious crops with separate staminate and pistilate flowers) seed production of commercial hybrids (large quantity of seeds for cultivation) based on such methods is economically feasible only in tomato, eggplant, sweet pepper, cucurbits and few other vegetables, in which a large number of F_1 seeds are obtained from one manually pollinated crossed fruits. Ever since (since 1930s) the discovery of male sterility (in onion) and self-incompatibility (in cabbage) mechanisms and their proposed utilization in hybrid seed production, several mechanisms and methods have been evolved for the development of experimental and commercial hybrids. This chapter describes genetic (inherited) and non-genetic mechanisms utilized for hybrid development in selected vegetable crops with special reference to male

---

Sanjeet Kumar is affiliated with the Indian Institute of Vegetable Research, 1, Gandhinagar (Naria), P.B. 5002, Varanasi 221005, India.

P. K. Singh is affiliated with the Sungro Seeds Limited, Shalimar Bagh, New Delhi 110088, India.

[Haworth co-indexing entry note]: "Mechanisms for Hybrid Development in Vegetables." Kumar. Sanjeet, and P. K. Singh. Co-published simultaneously in *Journal of New Seeds* (Food Products Press, an imprint of The Haworth Press, Inc.) Vol. 6, No. 4, 2004, pp. 383-409; and: *Hybrid Vegetable Development* (ed: P. K. Singh, S. K. Dasgupta, and S. K. Tripathi) Food Products Press, an imprint of The Haworth Press, Inc., 2004, pp. 383-409. Single or multiple copies of this article are available for a fee from The Haworth Document Delivery Service [1-800-HAWORTH, 9:00 a.m. - 5:00 p.m. (EST). E-mail address: docdelivery@haworthpress.com].

sterility, self-incompatibility, gynoecism, auxotrophy, use of sex regulators, and chemical hybridizing agents. *[Article copies available for a fee from The Haworth Document Delivery Service: 1-800-HAWORTH. E-mail address: <docdelivery@haworthpress.com> Website: <http://www.HaworthPress. com> © 2004 by The Haworth Press, Inc. All rights reserved.]*

KEYWORDS. Apomixis, auxtrophy, CHA, gynoecism, hybrids, male sterility, self-incompatibility, sex regulation, vegetables

INTRODUCTION

The successful demonstration of maize hybrids in 1920s promoted studies to examine possible exploitation of heterosis in the form of F_1 varieties in several crop species. Among the vegetables, first F_1 hybrid of eggplant was released during 1924 in Japan (Nishi, 1967). Subsequently, hybrids of watermelon (1930), cucumber (1933), radish (1935), tomato (1940), and cabbage (1942) were developed (Liedle and Anderson, 1993). Seeds of most of these hybrids were produced through natural crossing exploiting competitive fertilization between self and cross pollen. Thus only 40-80% of the seeds were actual hybrids (Liedle and Anderson, 1993), which were far below than the current acceptable level for contamination in hybrid seeds. Therefore, search for methods to produce pure hybrid seeds at commercial scale was realized. Pearson (1933) using self-incompatibility mechanism in cabbage and Jones and Clarke (1943) using cytoplasmic male sterility mechanism in onion, proposed the methods to produce large scale pure hybrid seeds. Now a day, F_1 hybrid breeding method is commonly utilized to exploit heterosis in several economically important vegetables including tomato, eggplant, hot and sweet peppers, onion, cabbage, cauliflower, other cole crops, radish, carrot, melons, etc. Vegetable breeders prefer to select hybrid breeding because it is comparatively easy to incorporate resistant genes for biotic and abiotic stresses in F_1 hybrid and right of the bred variety is protected in terms of parental lines. Moreover, despite high cost of hybrid seeds, there has been increasing concern of the farmers on the cultivation of hybrids. This is because under optimum crop production and protection management, crop raised from the seeds of F_1 hybrid has several advantages like better yield, adaptability, uniformity and reactions to certain stresses in comparison to crop raised from the seeds of improved pure line or population.

EXPERIMENTAL CROSSES VS. COMMERCIAL HYBRIDS

In seed propagated crops including vegetables, development of F_1 hybrids involves three broad steps: (i) development, maintenance, and multiplication of parental lines; (ii) assessment of combining ability between the parental lines; and (iii) hybrid seed production utilizing selected parental combination(s). Albeit success of the former two steps is essentially required for developing heterotic hybrids, it is the success of the third step, i.e., large-scale hybrid seed production, which finally determines the fate of developed hybrids in terms of their commercial viability. The failure to produce economic hybrid seeds of otherwise very heterotic cross combination for yield and quality may result in commercially non-viable hybrids will remain experimental crosses.

MECHANISMS FOR HYBRID DEVELOPMENT

Although experimental crosses (few seeds for research purpose) can be developed through manual emasculation (in case of hermaphrodite vegetables) followed by manual pollination of emasculated flowers or pistillate flowers (in case of monoecious vegetables), seed production of commercial hybrids (large quantity of seeds for cultivation) based on such mechanisms and methods is economically feasible only in vegetables like tomato, eggplant, many cucurbits, in which a large number of F_1 seeds are obtained from one manually pollinated crossed fruits. Nevertheless, in these vegetables also, cost of F_1 seed production can be brought down, if practically applicable mechanism(s) to avoid selfing and maximize out crossing is resorted in the hybrid seed production field. For example, in tomato, sharp reduction in labor expenditure of hybrid seed production can be achieved by the elimination of manual emasculation process, as it represents about 40% of the total expenditure (Yordanov, 1983). Likewise, expenditure on manual pollination can be saved during pepper hybrid seed production, if considerable amount of natural cross pollination takes place on the plants of female parent (Kumar et al., 2002). Several mechanisms and methods have been evolved for the development of hybrids in vegetable crops, however, only selected once are utilized to develop commercial hybrids of specific vegetable (Table 1). The commercial utilization many of these mechanisms are not feasible, therefore, utilization of such mechanisms has been restricted to develop only experimental crosses (Bassett, 1986; Kalloo and Bergh, 1993; Table 2).

In this chapter, attempts have been made to describe genetic (inherited) or non-genetic mechanisms, viz., male sterility, self-incompatibility, gynoecism, auxotrophy, use of sex regulators, and chemical hybridizing agents, based on their relative importance in hybrid development of selected vegetable crops.

MALE STERILITY

Onion crop provides one of the rare examples of very early recognition of male sterility (Jones and Emsweller, 1936), its inheritance and use in hybrid

TABLE 1. The most commonly utilized mechanisms/methods for developing commercial hybrids in vegetables

Mechanism	Commercially exploited in:
Hand emasculation + HP	Tomato, eggplant, sweet pepper, okra, hot pepper
Pinching of staminate flowers + HP	Cucurbits (bitter gourd, bottlegourd, etc.)
Male sterility + HP	Tomato, hot pepper, sweet pepper
Male sterility + NP	Onion, cabbage, cauliflower, carrot, radish, muskmelon, hot pepper
Self-incompatibility + NP	Most of the cole vegetables like broccoli, cabbage, etc.
Gynoecism + NP	Cucumber, muskmelon
Pinching of staminate flowers* + NP	Cucurbits including bitter gourd, summer squash, etc.
PGR and pinching of staminate flowers* + NP	Summer squash, winter squash, etc.

HP = hand pollination; NP = natural pollination; PGR = plant growth regulator.
* Genotypes with increased proportion of pistillate flowers are desirable for hybrid development.

seed production (Jones and Clarke, 1943). Since then male sterility has been reported in several vegetables. These male sterile plants were either isolated in natural populations or were artificially induced through mutagenesis (Kaul, 1988). In recent past, male sterility systems have been also developed through genetic engineering (Williams et al., 1997; Kumar et al., 2000) and protoplast fusion (Pelletier et al., 1995).

Types of Male Sterility

Kaul (1988) classified male sterility in two major groups, viz., genetic (spontaneous or induced) and non-genetic (induced) male sterility. On phenotypic basis, genetic male sterility has been classified in three classes, i. e., sporogenous, structural, and functional. Similarly, non-genetic male sterility has been classified as chemical, physiological, and ecological male sterility. Further, on genotypic basis genetic male sterility was grouped as genic, cytoplasmic and gene-cytoplasmic male sterility (Kaul, 1988). Based on the location of gene(s) controlling genetic male sterility, spontaneously isolated, artificially induced through mutagenesis, artificially incorporated through protoplast fusion or genetically engineered male sterility systems can be classified as (i) genic male sterility (gms; more precisely nuclear male sterility) and (ii) cytoplasmic male sterility (cms; more precisely cytoplasmic-nuclear male sterility).

Nuclear or Genic Male Sterility (gms)

As the name suggests, nuclear male sterility (earlier termed as gms) is controlled by the gene(s) from the nuclear compartment. Most of the naturally oc-

TABLE 2. Commercially unexploited mechanisms for the development of hybrids in vegetables

Mechanism	Vegetable	Remark(s)	Reference
Nuclear male sterility	Tomato	Monogenic recessive mutant was utilized to develop cost effective experimental crosses.	Sawhney, 1997; Kumar et al., 2001
Auxotrophy	Tomato	A very attractive and feasible model was proposed utilizing monogenic recessive nutritional mutants, e.g., thiamin dependent.	Barabas, 1991
Incongruity	Tomato	Models to transfer barrier (incompatibility between pollen and stigma) genes and corresponding penetration genes were demonstrated.	Hogenboom et al., 1978
Synthetic seeds	Celery & lettuce	Celery and lettuce hybrids were successfully multiplied (*in vitro*) through embryoids.	Sakamoto et al., 1991
Nuclear male sterility	Watermelon	The utilization of monogenic recessive mutant was proposed.	Zhang et al., 1994
Nuclear male sterility	Cabbage	Proposed feasible use of a dominant male sterile mutant to produce hybrid seeds; multiplication of male sterile line has been proposed with the aid of tissue culture.	Fang et al., 1997
Cytoplasmic ms + Self-incompatibility	Radish	The combined use of both these mechanisms to enhance efficiency of hybrid seed production has been proposed.	Cho et al., 1985
Functional male sterility	Eggplant	A monogenic recessive mutant was identified and proposed for commercial utilization.	Phatak & Jaworski, 1989
Nuclear male sterility	Bottlegourd	Male sterile plants were identified and characterized and utilized to develop experimental crosses.	Dutta, 1983
Nuclear male sterility	Okra	Male sterile plants were identified and characterized and utilized to develop experimental crosses.	Dutta, 1983
Gynoecism	Bitter gourd	Gynoecious plants were identified and proposed for utilization after genetic characterization.	Ram et al., 2002
Chemical hybridizing agents	Several vegetables	Experimental crosses were developed and proposed for commercial utilization.	McRae, 1985
Transgenic male sterility	Several vegetables	Few of them are at the edge of commercial utilization.	Williams et al., 1997

curring or induced male sterile mutants are recessive in nature with few exceptions in cole vegetables (e.g., cabbage and broccoli) and genetically transformed male sterile lines (Kaul, 1988; Williams et al., 1997). Certain mutants, which although produce functional pollen, pollen fail to self fertilize, either due to non-dehiscence of pollen or their special flower morphology, e.g., positional sterility in tomato (Atanassova, 1999) and functional male sterility in eggplant (Phatak and Jaworski, 1989). The occurrence of predominantly recessive male sterility clearly indicates that gms is the result of mutation in any gene(s) controlling microsporogenesis (pollen development process), stamen development or microgametogenesis (male gamete development process).

EGMS Line: Certain gms lines are conditional mutants, meaning thereby in a particular environment male sterile mutant plants turn into male fertile. After determination of critical environment (usually temperature or photoperiod) for sterility and fertility expression, such GMS mutants are classified under environmental sensitive genic male sterile (egms) lines. In vegetable crops, mostly temperature sensitive egms lines have been reported (Table 3). From practical application viewpoint, it is necessary to identify critical temperature or photoperiod for the fertility/sterility expression in temperature and photoperiod sensitive genetic male sterility, respectively.

Development of Genic Male Sterility

The monogenic recessive gene controlling male sterility is transferred in the desirable genotype through backcrossing program. After the identification of recessive male sterile mutant plant, the first step is to cross it with plant of the same variety. In the F_2 and subsequent generations, male sterile plants are pollinated with the bulk pollen from all the male fertile segregants and seeds from only male sterile plants are harvested to raise next generation. Thus after

TABLE 3. Environmental sensitive male sterile mutants in vegetables

Vegetable	Mutant	Reference
Cabbage	TGMS, PGMS	Rundfeldt, 1961
Brussels sprout	TGMS	Nieuwhof, 1968
Broccoli	TGMS	Dickson, 1970
Pepper	TGMS, TCMS	Daskalov, 1972; Shifriss, 1997*
Carrot	TGMS	Kaul, 1988
Tomato	TGMS	Rick, 1948; Sawhney, 1983

TGMS–Thermosensitive genic male sterility
PGMS–Photoperiod sensitive genic male sterility
*TCMS–Thermosensitive cytoplasmic male sterility

four generations, homozygous male fertile (*MsMs*) plants are eliminated from the population and only heterozygous male fertile (*Msms*) and male sterile (*msms*) plants remain in the population. Thereafter, gms plants (*msms*) are maintained by back crossing it with heterozygous isogenic line (*Msms*) for male sterility.

Utilization of Genic Male Sterility

Since gms is maintained through backcrossing, in hybrid seed production field, 50% male fertile segregants (*Msms*) need to be identified and removed before they shed pollen. In some gms lines, *ms* genes are tightly linked with the recessive phenotypic marker genes (Table 4). Such marker genes, especially which expresses at seedling stage, are good proposition for the identification of sterile/fertile plants at seedling stage. Hybrid seed production using EGMS line is more attractive because of the ease in seed multiplication of male sterile line. Seeds of EGMS line can be multiplied in an environment where it expresses male fertility trait while hybrid seeds can be produced in other environment, where it expresses male sterility. Because of more tedious maintenance process and non-availability of suitable marker gene among the vegetable crops, utilization of gms is restricted only in few vegetables (Table 4). The identification of fertilizing cytoplasm for specific nuclear male sterile gene (Horner and Palmer, 1995), is an interesting research area, which upon success, may provide opportunity for most efficient utilization of gms lines, like cms line.

Cytoplasmic Male Sterility (cms)

The expression of male sterility in cms plants is the result of incompatibility between recessive nuclear gene (called maintainer gene; *rf*) and male sterile specific cytoplasmic genome. Now it is well documented that specific open reading frame (ORFs) in mitochondrial genome (mt-genome) are responsible for the expression of male sterile trait (Kumar et al., 2000). Once dominant restorer (*Rf*) gene (located in nuclear genome) responsible for pollen fertility of a cytoplasmic male sterile line is identified, it is commonly known as cytoplasmic-genic male sterility (cms). Therefore, those cytoplasmic male sterile lines for which *Rf* gene(s) have been identified are widely known as genic-cytoplasmic male sterility (g-cms) and more often treated as a separate class of male sterility system. However, both cms and g-cms can be described under common head (i.e., cms) because of the fact that in both these systems, expression of male sterility is due to the defect in the cytoplasm (mt-genome). Based on mode of action of the pollen fertility restorer (*Rf*) and mainainer (*rf*) alleles, cms are of two types, viz., (i) gametophytic and (ii) sporophytic. In gameto-

TABLE 4. Nuclear and cytoplasmic male sterility (gms and cms) in selected vegetable species

Vegetable spp.	Salient features of gms	Salient features of cms
Tomato (*Lycopersicon esculentum*)[†]	More than 55 recessive genes have been reported (Kaul, 1988; Georgiev, 1991); *sl-2*, *ms-13*, and *ms-15* are temperature sensitive (Sawhney, 1997); *ps-2* gene has been exploited at commercial scale (Atanassova, 1999); YAC clone containing *ms-14* gene has been cloned (Gorman et al., 1996).	Sterile cytoplasm has been derived from the distinct species through protoplast fusion (Melchers et al., 1992; Petrova et al., 1999); restorer gene (Rf) is not available (Petrova et al., 1999).
Eggplant (*Solanum melongena*)	Monogenic recessive gene has been reported (Kaul, 1988); monogenic recessive functional sterility available (Phatak and Jaworski, 1989).	Not reported
Pepper (*Capsicum spp.*)[†]*	More than 12 recessive genes have been reported (Shifriss, 1997); MS-12 (*ms-509/ms-10*) and *ms-3* genes are commercially utilized in India and Hungry, respectively (Kumar et al., 2000). The *ms-10* gene is linked with taller plant height, erect growth and dark purple anther (Dash et al., 2001).	First reported in a Indian accession (Peterson, 1958); most of the cms lines are temperature sensitive (Shifriss, 1997); occurrence of Rf allele is common in small fruited (usually hot pepper) and rf in large fruited (usually sweet pepper) lines (Shifriss, 1997); RAPD markers linked to Rf gene have been identified (Boaxi et al., 2000).
Cole vegetables (*Brassica oleracea*)		
Cauliflower (var. *botrytis*)*	Both recessive and dominant genes have been reported (Kaul, 1988; Kumar et al., 2000).	In cole vegetables, sterile cytoplasm derived from *B. nigra* (Pearson, 1972) and *Raphanus sativus*, Ogura type (Ogura, 1968); problem of seedling yellowing (at low temperature) associated with Ogura based cms lines of broccoli, cauliflower, cabbage. Brussels sprout has been solved using protoplast fusion. Cybrid cms lines of cabbage and cauliflower are being utilized by seed companies in France (Pelletier et al., 1995); protoplast fusion has been utilized to transfer Ogura cytoplasm from brocolli into cabbage (Sigareva and Earle, 1997).
Cabbage (var. *capitata*)*	Both recessive (Nieuwhof, 1961) and dominant (Fang et al., 1997) genes have been reported; RAPD marker linked to dominant gene has been identified (Wang et al., 1998) and its use in hybrid seed production has been proposed.	
Brussels sprout (var. *germmifera*)*	Recessive male sterile mutant has been reported (Nieuwhof, 1968; Kaul, 1988).	
Broccoli (var. *italica*)*	Six recessive non-allelic genes have been reported (Dickson, 1970); linkage of ms gene with bright green hypocotyle (Sampson, 1966).	

Crop	Genic male sterility	Cytoplasmic male sterility
Watermelon (*Citrullus lanatus*)[†]	Recessive mutants have been reported (Watts, 1962; Kaul, 1988; Zhang et al., 1990), linkage of ms gene with delayed-green (dg) seedling marker gene (Zhang et al., 1996).	Not reported
Muskmelon (*Cucumis melo*)[†]	Five recessive non-allelic genes have been reported (McCreight, 1993); ms-1 is commercially utilized in India (Kumar et al., 2000).	Not reported
Cucumber (*Cucumis sativus*)	Monogenic recessive gene has been reported (Barnes, 1961; Kaul, 1988); limited scope of utilization because of the availability of gynoceous lines (Kumar et al., 2000).	Not reported
Summer squash (*Cucurbita pepo*)	Monogenic recessive gene has been reported (Eisa and Munger, 1968; Kaul, 1988); very limited scope of utilization because of the availability of sex regulating mechanism using certain chemicals.	Not reported
Onion (*Allium cepa*)*	Monogenic recessive gene has been reported (Kaul, 1988).	Two types of sterile cytoplasms, viz., S (Jones and Clarke, 1943) and T (Berninger, 1965) have been reported; S-cytoplasm is most widely exploited (Pelletier et al., 1995).
Carrot (*Daucus carota*)*	Recessive male sterile genes have been reported (Welch and Grimball, 1947; Kaul, 1988); not utilized at commercial scale because of the availability of cms lines (Singh, 2002).	Two types (petaloid and brown anther) of male sterile lines are available (Welch and Grimball, 1947; Morelock, 1974); genetics of fertility restoration is complex (Peterson and Simon, 1986) because structural variants of mt DNA are numerous (Ranfort, 1995).
Radish (*Raphanus sativus*)*	Three recessive mutants have been reported (Tokumasu, 1951; Kaul, 1988);	Sterile cytoplasm widely distributed in wild radish; occurrence of *Rf* allele is frequent in Europian and Chinese cultivars and *rf* in Japanese cultivars (Yamagishi and Terachi, 1996; Yamagishi, 1998), commercially utilized (Singh et al., 2001).

Note: The gms (†) and cms (*) lines exploited at commercial scale.

391

phytic system, expression of restorer allele is pollen specific, thus the pollen grains are the responding elements (e.g., S-cytoplasm in corn, abortive cytoplasm in rice, etc.). Therefore, a plant heterozygous for maintainer-restorer locus (*Rfrf*) produces 50% aborted (*rf*) and 50% normal (fertile) pollen (*Rf*). Pollen from such plant (*Rfrf*) crossed on to a sterile plant (*rfrf*), will again produce plants with 50% each of aborted and normal pollen. In contrast, all pollen are either fertile or sterile in sporophytic system, which is most common (Pearson, 1981). A heterozygous restorer line (*Rfrf*) in this system produces all fertile pollen and when crossed on to a sterile plant (*rfrf*), produces 50% absolute male sterile and 50% absolute male fertile plants. Cytoplasmic male sterility may originate from inter-generic or inter-specific crosses and may be artificially induced through mutagenesis or antibiotic effects on cytoplasmic genes (Kaul, 1988). Cytoplasmic male sterile plants have also been developed in several vegetables through protoplast fusion (Pelletier et al., 1995). In near future, genetically engineered cytoplasmic male sterility may be available after standardization of transformation technique for organelle genome.

Development of Cytoplasmic Male Sterility Line

The cms line (A line) is developed by back crossing of a selected maintainer line (B line) on to a already available A line for six to seven generations. This generates a pair of A and B line in the new genetic background. Restorer allele (*Rf*) is either introgressed into identified male parent (restorer breeding) or male parent is directly used for hybrid seed production on cms, if restorer (*Rf*) gene in homozygous state is already available in male parent or vegetative part is of economic importance in a crop like onion, cabbage, etc. Similarly, if maintainer (*rf*) gene is not available in the line, which is to be developed as A line, then maintainer breeding would be required to transfer recessive gene (*rf*) in the desirable line.

Utilization of Cytoplasmic Male Sterility

The cms system is the most commonly utilized male sterility to produce commercial hybrid seeds of several vegetables (Table 4). The cms based hybrid development is often term as three line method of hybrid breeding involving A line (male sterile; S-*rfrf*), B line (maintainer; N-*rfrf*) and C or R line (restorer; S-or N-*RfRf*). As mentioned, cms line without restorer male parent cannot be utilized in fruit producing vegetables (e.g., chilli), but it can be utilized in vegetables where vegetative part is of economic value (e.g., onion, cole vegetables, carrot, radish, leafy vegetables, etc.). The cms system though is the most commonly utilized, its utilization is restricted in specific species because of the following limitations:

- Non-availability of cms in many crops and their wild relatives.
- Need of fertility restorer allele in fruit producing vegetables.
- Undesirable pleiotropic effect of sterile cytoplasm on horticultural qualities.
- Highly unstable sterile cytoplasm in several cases.
- Poor cross pollination ability of flowers of plants with sterile cytoplasm due to altered morphology.
- Technical complexity involved in seed production and maintenance of parental lines.

Besides, vulnerability of sterile cytoplasm to specific diseases is a major risk due to monopolistic cultivation of hybrids derived from single source of sterile cytoplasm. The devastation of corn hybrids derived from T-cytoplasm by *Helminthosporium* blight in USA during 1970s (Levings, 1990), is a well-known example of such risk.

Transgenic Male Sterility Systems

From the beginning of 1990s, new genetic approaches have been proposed and implemented to develop male sterility systems through genetic transformation (Mariani et al., 1992). The ability to design new molecular strategies and their successful execution has been possible because of the isolation, cloning, and characterization of anther or pollen specific genes and promoter sequences. These genes are expressed in pollen themselves (gametophytic expression) or cells and tissues (sporophytic expression) that directly or indirectly support pollen development, such as tapetum, filament, anther wall. Williams et al. (1997) reviewed the reports on genetically engineered male sterility systems under dominant male sterility, recessive male sterility, targeted gamitocide, and dual method. However, based on mechanism of male sterility induction and fertility restoration, all transgenic male sterility systems developed so far can be described under five classes, viz., (i) abolition-restoration system, (ii) abolition-reversible system, (iii) constitutive-reversible system, (iv) complementary-gene system, and (v) gametocide-targeted system. Although in transgenic(s) developed within one system, mode of action of trans-gene(s) remains the same, there can be variations in trans-gene constructs including promoter, targeted site (depending upon the promoter used) and methodology adopted within one system. Since detail discussion of all the transgenic male sterile system developed so far is beyond the scope of this chapter, a comparison between these systems is given (Table 5). All the transgenic male sterile lines developed till date are gms, since they have been developed through transformation of male sterility causing gene construct(s) inside the nuclear genome.

TABLE 5. Comparison of transgenic male sterility systems for requirement of gene constructs and possible risk (due to system failure) involved (adapted from Kumar et al., 2000)

System	Requirements and risk involvement
Abolition-restoration	Two transgenics each with a distinct exogenous gene: (i) transgenic male sterile plant with disruptive gene under anther specific (usually tapetum) promoter and (ii) transgenic fertility restorer plant with inhibitor gene under same anther specific promoter. Risk will be involved in both the fields.
Abolition-reversible	One transgenic with two exogenous genes (co-transformed): reversible transgenic male sterile plant with (i) disruptive gene and (ii) inhibitor of disruptive gene under chemically inducible promoter. Risk will be involved in both the fields.
Constitutive-reversible	One transgenic with one exogenous gene: reversible transgenic male sterile plant with endogenous gene for male sterility (ms) and exogenous male fertility gene (mf) for same ms gene under inducible promoter. Risk will be involved only in the male sterile seed multiplication field.
Complementary-gene	Two transgenics, each having a complementary exogenous gene for male sterility: transgenic male fertile plant with (i) gene a and (ii) gene b, under the control of anther specific promoter (gene a & b are complementary). Risk will be involved in both the fields.
Gametocide-targeted	One transgenic with one exogenous gene: transgenic male fertile plant with disruptive gene under the control of pollen specific chemically inducible promoter. Risk will be involved only in the hybrid seed production field.

Regardless of the crop, all transgenic male sterility systems (except constitutive-reversible) with the same trans-gene construct(s) may be utilized to develop transgenic male sterile lines in those vegetables, where transformation and regeneration protocols have been standardized.

Male Sterility and Its Commercial Utilization

In tomato (*Lycopersicon esculentum*; $2n = 24$), approximately 19 male sterile based hybrids have been released, 17 of them are based on functional sterility controlled by gene positional sterility (*ps-2*). The use of *ps-2* mutant has become applicable at commercial scale in few countries like Czech Republic, Moldova and Bulgaria (Atanassova, 1999). The $ms-10^{35}$ gene is linked with a recessive marker gene (*a*) responsible for absence of anthocyanin. Hence, $ms-10^{35}$ sterile plant can be identified at seedling stage and fertile plant can be rouged out in the nursery itself (Georgiev, 1991). Atanassova and Georgiev (1986) suggested that genes $ms-10^{35}$, $ms-15^{26}$, and *ps-2* combined with short styles are most promising for hybrid development.

In pepper (*Capsicum annuum*; $2n = 24$), the induced male sterile gene in France (*mc-509*; Pochard, 1970; renamed *ms-10* by Daskalov and Poulos,

1994) was found allelic to *msk* allele isolated spontaneously in Korea (Shifriss, 1997). The *ms-2* line identified by Shifriss and Rylski (1972), was found non-allelic to *ms-1* isolated by Shifriss and Frankel (1969). The *ms-509* line (bell pepper type) of Pochard was introduced in India at Punjab Agricultural University (PAU) and *ms-10* was introgressed in three chilli, genotypes, viz., MS-12, MS-13, and MS-41 (Singh and Kaur, 1986). The MS-12 (*ms-10ms-10*) line is being utilized to produce hybrid seeds of CH-1 and CH-3 hybrids of hot pepper in Punjab state (Hundal, personal communication). In Hungary, male sterile lines possessing *ms-3* gene are being utilized to produce hybrid seeds (Kumar et al., 2000). In the recent past, development of stable cms lines of hot pepper has led to its increased utilization in hybrid development. In South Korea (Shifriss, 1997), China (Boaxi et al., 2000), and India, cms lines are being utilized to develop hybrids of hot pepper. However, non-availability of *Rf* genes in most of the sweet pepper genotypes is still a handicap in developing cms based commercial sweet pepper hybrids.

Among cole vegetables (*Brassica* spp. 2*n* = 18), although gms based experimental crosses have been developed, it has not been commercially utilized mainly because of the difficulty in multiplication of male sterile seeds and availability of self-incompatibility and cms systems. The cms cybrid plants developed by Pelletier and his associates were of normal flower morphology, with good nectar production and found highly stable. These promising cybrids, contained genomes resembling more to the *B. oleracea* type than the Ogura and being utilized by seed companies in France to develop hybrids of cabbage and cauliflower (Pelletier et al., 1995). Seed companies in India are also developing cms based hybrids of cabbage and cauliflower (Singh, 2002).

On the basis of floral morphology, radish (*Raphanus sativus*; 2*n* = 18) cms lines are generally classified into three types: (i) degenerative corolla, (ii) shrivelled stamen, and (iii) abortive pollen. Success has been achieved to identify maintainer of Ogura cytoplasm and transfer of sterile cytoplasm in the new genetic background (Nieuwhof, 1990; Hawaldar et al., 1997). Ogura cytoplasm has been found widely distributed among the wild Japanese radish plants and most of the early European radish populations, in which availability of mainainer allele is more frequent. Whereas, most of the Asian radish cultivars including Japanese cultivars possess normal cytoplasm except few cultivars from Tibet and Taiwan (Nieuwhof, 1990; Yamagishi and Terachi, 1994 a; b; 1996). In a study by Yamagishi (1998), restorer allele was found to be widely distributed in wild radish, European and Chinese cultivars, while occurrence of maintainer allele was more frequent in the Japanese cultivars. The male sterility in radish is being utilized by seed companies in Taiwan, China, Korea, and Japan. In India also seed companies are using cms system for hybrid development (Singh et al., 2001).

Carrot (*Daucus carota*; 2n = 18) is one of the few crops in which male sterility was documented very early, i.e., in the year 1885 (Kaul, 1988). Several other genic male sterile mutants have been described (Kaul, 1988), however, none of them have been utilized for commercial seed production due to the availability of more efficient cms system in carrot. Two types of sterile cytoplasm have been reported, namely, (i) petaloid and (ii) brown anther (Welch and Grimball, 1947; Morelock, 1974). Taki Seed Company of Japan developed the first F_1 hybrid variety in 1982 using cms (Pelletier et al., 1995). In USA, majority of hybrids are produced from one cytoplasm, i.e., Cornell cytoplasm. Considering the risk of disease vulnerability of hybrid variety due to the monopolistic use of Cornell sterile cytoplasm, USDA has released a petaloid type new cms line derived from sterile cytoplasm of Wisconsin (Morelock et al., 1996).

In onion (*Allium cepa*; 2n = 16), first male sterile plant was reported within the progenies of an onion cultivar Italian Red (Jones and Emsweller, 1936), which was cytoplasmically inherited and male sterility was under the control of single recessive nuclear restorer locus (Jones and Clarke, 1943). World wide more than 50% onion varieties currently cultivated are F_1 hybrids derived from S-cytoplasm (Pelletier et al., 1995). In India, public sector bred commercial hybrid has not been recommended so far, however, cms based hybrids are developed at several seed companies and institutes like as Indian Institute of Horticultural Research, Indian Agricultural Research Institutes, etc.

In muskmelon (*Cucumis melo*; 2n = 22), first recessive *ms* gene was reported by Bohn and Whitaker (1949), since then at least four additional male sterile recessive alleles, viz., *ms-2* (Bohn and Principe, 1962), *ms-3* (McCreight and Elmstrom, 1984), *ms-4* (Pilrat, 1990) and *ms-5* (Lecouviour, 1990) have been identified. The *ms-1* line has been successfully utilized in India, to develop first commercial hybrid (Punjab Hybrid-1) in vegetable crops, through the exploitation of male sterility at PAU, Ludhiana (Kumar et al., 2000).

SELF-INCOMPATIBILITY (SI)

Ever since the first discussion on self-incompatibility by Darwin (1877), the phenomenon has extensively studied in several plant families and now significant amount of information is available on genes and gene products involved in the expression of SI trait (Dodds et al., 1997). There are two types of SI, viz., gametophytic and sporophytic. In gametophytic system, SI reaction of pollen and stigma is determined by the genotype of the mother plant on which pollens are produced (e.g., tomato) while in sporophytic system, pollen phenotype (SI reaction) is determined by the genotype of the mother plant on which pollens are produced (e.g., cole vegetables). In Brassicacae, sporophytic self-incompatibility (SSI) has been best characterized and successfully utilized for the development of commercial hybrids (Pearson, 1983; McCubbin and Dickinson, 1997; Tripathy and Singh, 2000; Singh, 2000; Singh et al., 2001).

SPOROPHYTIC SELF-INCOMPATIBILITY (SSI)

Sporophytic self-incompatibility (SSI) was first observed in radish (Stout, 1920), and its inheritance pattern was first demonstrated (Bateman, 1955). The numbers of S allele at S-locus have been reported to be 34 in *Raphanus sativus* and 60 in *B. oleracea*.

Genetic and Molecular Basis of SSI

The complex genetics of SSI has been discussed by Dickson and Wallace (1986). The compatible or incompatible reaction of pollen and stigma is not only depends on the genotypes (homozygous/heterozygous) of male and female plant at S locus, but also on the existence of any of the following four levels of interactions between the two S alleles, which characterize the incompatibility/compatibility specificity of the pollen and stigma: dominance ($S1 < S2$), co-dominance ($S1 + S2$), mutual wakening (no action by either allele) and intermediate gradation (1-100% activity by each allele) (Dickson and Wallace, 1986). The analysis of sequence data of S-locus cDNA clones revealed that various S alleles exhibit high levels (up to 30%) of sequence divergence (McCubbin and Dickinson, 1997).

Assessment of SI

The first developed procedure for the quantification of the SI was based on the number of seed set after each specific self- or cross-pollination. The first disadvantage of this procedure is 60 days time (after pollination) required to harvest seeds in order to get the results. The second disadvantage is that seed counts at maturity stage often do not reflect degree of compatibility/incompatibility because before reaching maturity number of seeds may be reduced due to biotic or abiotic factors, which may lead to biased interpretation of the result. The fluorescent microscopic observations on pollen ability to penetrate style, has provided a more reliable method to directly measure incompatibility/compatibility reaction within 12-15 h after pollination (for detail see Dickson and Wallace, 1986).

Development of Homozygous SI Inbreds

This crucial step requires forced selfing of self-incompatible plants either by bud pollination or treatment with CO_2 gas (CO_2 enrichment) or sodium chloride, which break the self-incompatibility. The detailed procedure of developing self-incompatible inbred line development has been described by Pearson (1983).

Utilization

For hybrid seed production both the parental inbreds should have two different S alleles for strong self-incompatibility (in case of single cross hybrid). One SI inbred is used as female parent and a good pollinator (an open pollinated variety) as male to develop top cross hybrid, while four SI inbreds having altogether different S alleles are used to produce double cross hybrids. Among the cole vegetables like cabbage, cauliflower, broccoli, etc., self-incompatibility (sporophytic) mechanism is being utilized for hybrid seed production at several places including India (Singh, 2000). Usually in cauliflower SI is weak and SI reaction is breaks at high temperature, resulting into selfing and sibling (brother-sister mating) among the plants of female parent, thus deterioration in the genetic make up of F_1 seeds. Following strategies are utilized to overcome sibling problem: (i) use of parental lines with synchronized flowering, (ii) use of parental lines with similar morphology, (iii) pollination by stored pollen, and (iv) use of male sterility as an alternative to SI inbreds. Due to the inherent advantages, the last option, i.e., use of cms line is being getting more attention in the hybrid seed production of all cole vegetables.

GYNOECISM

Gynoecious sex expression has been reported in both pickling and slicing type of cucumber and is being used to develop cucumber hybrids. Further, this trait is linked with parthenocarpic genes (Kalloo, 1988). Hybrids developed utilizing gynoecious linesare having parthenocarpic gene and do not require pollination for fruit set. Use of silver nitrate ($AgNO_3$), 50-100 ppm or silver thiosulphate, 25-50 ppm on gynoecious plants at 2-3 leaf stage produces lots of staminate flower, which are used to multiply the seed of gynoecious plants. In several cucurbits populations with high frequency of pistillate, flowers are desirable because such population can be utilized as female parent and during hybrid seed production, removal of staminate flowers would be easy and required less labor, e.g., bitter gourd (Ram et al., 2002).

SEX REGULATION USING PLANT GROWTH REGULATORS

During the 1950, Laiback and Kribbeu (1950) first demonstrated that β-naphthalene acetic acid and β-indoleacetic acid could be used to increase proportion of pistillate flowers in cucumber. In monoecious cucurbits like cucumber, melons, bottlegourd, bitter gourd, etc., male sterility *per se* is not required for hybrid seed production, rather it is only necessary to prevent or disrupt development of staminate flowers. The application of ethylene releasing chemical, ethephon (ethrale) considerably enhances the development of pistillate flowers and delays the development of staminate flowers.

The ethephon has been most extensively utilized sex regulator to produce cost effective hybrid seeds in several monoecious cucurbits like summer squash (Shannon and Robinson, 1979), pumpkin (Hume and Lovell, 1981), muskmelon (Loy, 1978; Alvarez, 1988; Korzeniewska et al., 1995) and cucumber (Hunsperger et al., 1983). In muskmelon and watermelon, application of ethephon led to an absence of staminate flowers for 40 days (Andrasek, 1988). Interestingly, successful uses of chemicals to facilitate hybrid seed production have been achieved in those monoecious vegetables, having separate staminate and pistillate flowers.

AUXOTROPHY

Barabas (1991) at Central Research Institute, Szeged, Hungary demonstrated the exploitation of monogenic-recessive auxotrophic mutants (for nutritional requirement) in tomato and barley for the production of pure hybrid seeds. The F_1 hybrids between auxotroph and prototroph (normal gene for nutrition requirement) were phenotypically and physiologically normal. However, the self or sib pollinated auxotrophic parental lines produced lethal progenies. Three schemes, viz., female auxotrophy, male auxotrophy and complementary auxotrophy were proposed utilizing one or more auxotrophic mutants (Barabas, 1991). In tomato, utilizing thiamin dependent nutritional mutant (auxotroph; recessive homozygote; No-50-30W) as female parent and a prototroph (K509) as male parent (female auxotrophic scheme), it was demonstrated that pure F_1 seeds can be obtained even if selfing occurs in female parent, e.g., if the male sterility is imperfect. The thiamin dependent auxotroph mutant of tomato was successfully maintained through 1-2 treatments of thiamin treatments (@ 5ml/l). This scheme can work even where 50% of the female are fertile, e.g., gms lines because seedlings of selfed seeds (derived from male fertile female parent) will be prototroph, hence will not able to survive and eliminated in the hybrid nursery. Further, any normal (prototroph) line may be utilized as male parent because all will be able to restore prototrophy in the F_1s. Barabas (1991) mentioned that sowing of separate male and female strips as one of the disadvantages of female auxotrophy system, but such cultural practice would not be a handicap in tomato hybrid seed production. The utilization of auxotrophy mechanism does not need other marker because it is *per se* a genetic marker observable at seedling stage. Unfortunately, research on utilization of tomato auxotrophic mutant for hybrid seed production has been discontinued.

USE OF CHEMICAL HYBRIDIZING AGENTS (CHA)

Various terms have been used to describe chemicals that induce male sterility in plants such as gametocide or selective gametocide, pollen suppression, male sterilant, selective male sterilant, pollen suppressant, pollenocide, androcide, etc. As the ultimate goal of the application of chemicals is to produce hybrids, the term chemical hybridizing agent (CHA) is preferred (McRae, 1985). Potential of certain chemicals (maleic hydrazide) to induce selective male sterility was first demonstrated during 1950 in maize (Moore, 1950; Naylor, 1950). These discoveries stimulated interest in the possibility of using chemicals for hybrid seed production of vegetables (Table 6). The interest began to develop on chemical induction of male sterility in plants even though development of hybrids was achievable through cytoplasmic male sterility. It was recognized that despite of certain disadvantages there might be some advantages, especially in terms of time required to identify economically viable hybrids. This is because chemical methods for inducing male sterility can obviate the lengthy time frame required to obtain male sterile and restorer lines. Fairly large number of literature demonstrate the success of male sterility induction through chemical in many plant species, however, production of hybrid seed at commercial scale has not been demonstrated (Mc Rae, 1985; Cross and Schulz, 1997).

Site and Mode of Action of CHA

Most studies on the site and mode of action of CHA, have concentrated on observable micro and microscopic events that takes place in anthers. The most important general feature revealed from the literatures is that the earlier developed compounds (e.g., FW-450, ethephon, RH-531, PPX 3778) can induce a range of specific effects that are dependent upon treatment time and dosage interaction. Some general effects includes the followings (McRae, 1985):

- Meiosis is disrupted at early stage and anther development stage is arrested at an early stage.
- Exine formation is disrupted and microspores are thin walled, distorted in shape and non-viable.
- Microspores vacuole abnormalities, decreased starch deposition and persistent tapetum. Pollen is non-viable but anthers are normal.
- Pollen is present and viable but anthers either do not dehisce or show delayed dehiscence.

TABLE 6. Experimental demonstration of CHA application in hybrid development

Vegetable	Applied chemicals	Remark(s)	Reference
Tomato	GA₃	Treated plants produced separate stamens and split pistils	Chandra Sekhar and Sawhney, 1990
Tomato	Gibberlin synthesis CCC inhibitors	Selectively inhibited the development of stamen or suppressed pollen	Rastogi and Singh, 1988
Tomato	ABA (Abscisic acid)	Selectively inhibited the development of stamen or suppressed pollen	Rastogi and Singh, 1988
Tomato	FW-450 (Mendok)	Showed promise for commercial utilization	Moore, 1959
Tomato	Dalapon	Male sterility was induced	Brauer, 1959
Tomato	TIBA Triiodobenzoic acid	Some degree of male sterility was inducing	Rehm, 1952
Tomato	NAA	Induced male sterility	McRae, 1985
Eggplant	Ethephon	Induced selective male sterility	McRae, 1985
Pepper	GA₃	Pollen development was inhibited	Sawhney, 1981
Pepper	Dalapon and a-chloropropionate	These two chemicals were found to be most effective	Hirose and Fujime, 1973
Cole vegetables	GA	Reported promising for utilization	Van Der Meer and Van Dam, 1979
Cucumber	α NAA and β IAA	Increased the proportion of female flowers	Laiback and Kribben, 1950
Pea	Dalapon	Male sterility was induced	Brauer, 1959

These are the general effects and any of these is sufficient to confer CHA status on a chemical, providing female fertility is not adversely affected. However these types of effects are not the only ones that can make a chemical as useful CHA (McRae, 1985). Like cms and gms, the utilization of CHA is dependent on the followings:

- A source of viable pollen from a male parent that can outcross the male sterile/female parent.
- A female planting configuration that can maximize out crossing.
- An agency such as wind or insects to remove pollens from the male parent and transfer them to female parent.

- Female and male parents that are in synchrony, i.e., pollen from the male parent must shed when the male sterile (female) parent is receptive.
- An abundance of pollen from the male parent.
- Good flower opining characteristics in the female parent.

Hence irrespective of the mechanism involved for hybrid seed production, all these criteria are essential to maximize the efficiency of commercial hybrid seed production.

GENETIC MARKERS AND HYBRID DEVELOPMENT

Commercial hybrid seed production utilizing described genetic or non-genetic mechanisms always have certain degree of risk of getting selfed seeds in hybrid seeds lot. Recessive genetic markers (preferably seed or seedling markers) are available in several vegetables, which can be utilized to increase the purity of hybrid plant stand. Usually recessive phenotypic markers are incorporated in female parent and contaminated selfed/sib seeds (non-hybrid seeds) are differentiated at seed or seedling stage based on the presence of recessive marker depending upon the expression stage of the markers. Recessive or dominant seedling markers tightly linked with the nuclear male sterile (ms) gene have also been utilized in vegetables for the identification and removal of 50% male fertile plants at early stage from the hybrid seed production field (Table 7).

CONCLUSIONS

A number of mechanisms and methods have not been exploited for the development of commercial hybrids. The cms system is most widely utilized mechanism for the development of commercial hybrids in several vegetables, however, in many others, it has not been exploited mainly due to the non-availability of sterile cytoplasms (e.g., eggplant, melons, etc.). The use of gms lines has been restricted to only few crops and at few places, due the problem associated with the seed multiplication of male sterile lines. In the light of rapid advancement of bio-technology, it may be anticipated that transgenic male sterility systems will be increasingly utilized in near future, especially in those vegetables, where cms lines are not available. Development of practically feasible molecular markers may provide appropriate cost effective selection strategy to discard 50% male fertile sister plants at seedling stage, which may open the way to exploit monogenic recessive male sterile lines in several

TABLE 7. Utilization of morphological markers in hybrid development

Vegetable	Marker	Mechanism	Advantage
Tomato	Brown seed: *bs, bs2, bs3, bs4* genes (Bruzzone et al., 1983; Martiniello et al., 1983)	No emasculation + HP	Reduced cost on hybrid seed production; only 10% brown seeds (non-hybrid = selfed), which can be eliminated easily.
Tomato	Anthocyninless (aa) linked with *ms* gene (Georgive, 1991)	Male sterile line + HP	Planting of 100% male sterile plants in hybrid seed production field.
Tomato	Enzyme marker (*prx-2*) linked with ms-10 gene (Tanksley et al., 1984)		Proposed indirect use of marker in male sterile line development.
Tomato	Anthocyninless (*aa*) and exerted stigma (Andeev, 1986)	No emasculation + HP	Reduction in hybrid seed production cost and identification of selfed seeds at seedling satge.
Watermelon	Linkage of ms gene with delayed-green (dg) seedling marker gene (Zhang et al., 1996).	Male sterile line + HP	Planting of 100% male sterile plants in hybrid seed production field.
Onion	Brown seed (Fustos, 1986)	cms line + NP	Eliminate necessity of specific planting arrangement of female (cms) and male parents.
Chilli*	The *ms-10* gene is linked with taller plant height, erect growth and dark purple anther (Dash et al., 2001).	gms line + NP	Identification of male sterile plants at comparatively early stage in the hybrid seed production field.

* exploited at commercial scale

vegetables. Alternatively, phenotypic restoration of nuclear male sterile plants can be achieved using certain chemicals to over come maintenance problem, as already demonstrated in tomato. Nevertheless, search for fertilizing cytoplasm for specific nuclear *ms* gene also needs renewed attention. The utilization of auxotrophic mutants for hybrid development in tomato is an attractive proposition, which can be extended to other vegetable crops, where appropriate (i.e., monogenic-recessive) nutritional mutants are either available or can be induced. Apomixis has been reported in few *Allium* spp. like *A. tuberosum, A. nutans.* Considering the great importance of apomixes in fixing heterosis in hybrids, although researches on molecular apomixes are under progress at few international institutes, it would be imperative to continue studies on apomictic vegetable species.

REFERENCES

Alvarez J. 1988. Muskmelon hybrid seed production through ethephon induced feminization in andromonoecious cultivars. Proc. EUCARPIA Meet. Cucurbits Genet. Breed.. Avignon-Montavest, France, pp. 89-97.

Andeev Yu I. 1986. A method of breeding heterotic tomato hybrids. USSR Patent. As 1277930.

Andrasek K. 1988. Regulation in the hybrid seed production of hybrid F_1 of musk- and watermelon. Acta Horticulturae 220: 219-222.

Atanassova B and Georgiev H. 1986. Investigation on tomato male sterile lines in relation to hybrid seed production. Acta Horticulturae 190: 553-557.

Atanassova B. 1999. Functional male sterility (*ps-2*) in tomato (*Lycopersicon esculentum*) and its application in breeding and hybrid seed production. Euphytica 107: 13-21.

Barabas Z. 1991. Hybrid seed production using nutritional mutants. Euphytica 53: 67-72.

Barnes WC. 1961. A male sterile cucumber. Proc. Amer. Soc. Hort. Sci. 77: 415.

Bassett MJ. 1986. Breeding Vegetable Crops. AVI Pub. Com., Westport, Connecticut.

Bateman AJ. 1955. Self-incompatibility systems in angiosperms. Heredity 9: 52-68.

Bernninger E. 1965. Contribution a letude de la sterilite male de loifnon (*Allium cepa*). Ann. Amelior Plant 15: 183-199.

Boaxi Z, Sanwen H, Guimei Y and Jiazhen G. 2000. Two RAPD markers linked to a major fertility restorer gene in pepper. Euphytica 113:155-169.

Bohn GW and Principe JA. 1962. A second male-sterility gene in the muskmelon. J. Hered. 55: 211-215.

Bohn GW and Whitaker TW. 1949. A gene for male sterility in the muskmelon (*Cucumis melo* L.). Proc. Amer. Soc. Hort. Sci. 53: 309-314.

Brauer, H.O. 1959. Male sterility obtained with dalapon. Agr. Tech. Mex. 9: 6-8.

Bruzzone GB. Falavigna A and Soressi GP. 1983. Procedures for tomato hybrid seed production based on *bs* marker. Genmet. Agvar. 37: 149-150.

Chandra Sekhar KN and Sawhney VK. 1990. Regulation of the fusion of floral organs by temperature and gibbrellic acid in the normal and solanifolia mutant of tomato (*Lycopersicon esculentum*). Can. J. Bot. 68: 713-718.

Cho YH, Chung Ho and Kim BH 1985. Production of radish F_1 hybrid seed by means of combined use of male sterility and self-incompatibility. J. Korean Soc. Hortic. Sci. 26: 201-209.

Cross JW and Schulz PJ. 1997. Chemical induction of male sterility. In: Shivanna KR and Sawhney VK (eds.), Pollen Biotechnology for Crop Production and Improvement. Cambridge Univ. Press, pp. 218-236.

Darwin CR. 1877. The Different Forms of Flowers on the Same Species. John Murray, London.

Dash SS, Kumar Sanjeet and Singh JN. 2001. Cytomorphological characterization of a nuclear male sterile line of chilli pepper (*Capsicum annuum* L.). Cytologia 66: 365-371.

Daskalov S and Poulos JM. 1994. Updated *Capsicum* gene list. Capsicum and Eggplant Newsl. 13: 15-26.

Daskalov S. 1972. Male sterile pepper (*C. annuum* L.) mutants and their utilisation in heterosis breeding. Proc. Eucarpia Meet. Capsicum 7: 202-210.

Dickson MH and Wallace DH. 1986. Cabbage breeding. In: Bassett MJ (ed.), Breeding Vegetable Crops. AVI Pub. Com., Westport, Connecticut, pp. 395-432.

Dickson MH. 1970. A temperature sensitive male sterile gene in broccoli. *Brassica oleracea* L. var. *italica*. J. Amer. Soc. Hort. Sci. 95: 13-14.

Dodds PN, Clarke AE and Newbigin ED. 1997. Molecules involved in self-incompatibility in flowering plants. Plant Breed. Rev. 15: 19-42.

Dutta OP. 1983. Male sterility in okra (*Abelmoschus esculentus* (L.) Moench) and bottle gourd (*Lagenaria siceraria* (Mol.) Standl.) and its utilization in hybrid seed production. Thesis Abstracts, Univ. Agric. Sci., Bangalore, India, 9: 341-342.

Eisa HM and Munger MH (1968). Male sterility in *Cucurbita pepo*. Proc. Amer. Soc. Hort. Sci. 92: 473-479.

Fang ZY, Sun PT, Liu YM, Yang LM, Wang XW, Hou AF and Bian CS. 1997. A male sterile line with dominant gene (*Ms*) in cabbage (*Brassica oleracea* var. *capitata*) and its utilization for hybrid seed production. Euphytica 97: 265-268.

Fustos Z. 1986. Seed quality of brown seeds: A mutation in onion. Tagungsbuicht 239: 51-52.

Georgiev H. 1991. Heterosis in tomato breeding. In: Kalloo G (ed.), Genetic Improvement of Tomato. Monographs on Theor. Appl. Genet. 14, Springer-Verlag, Berlin, pp. 83-98.

Gorman SW, Banasiak D, Fairley C and McCormick S. 1996. A 610 Kb YAC clone harbors 7 CM of tomato (*Lycopersicon esculentum*) DNA that includes the male sterile 14 gene and a hotspot for recombination. Mol. Gen. Genet. 251: 52-59.

Hawaldar MSH, Mian MAK and Ali M. 1997. Identification of male sterility maintainer lines for Ogura radish (*Raphanus sativus* L.). Euphytica 96: 299-300.

Hirose T and Fujime Y. 1973. Studies of chemical emasculation in pepper. III. Effect of repeated application of 2, 2-dichloropionate (dalapon) and the genetocide action of dalapon and related compounds. J. Jap. Soc. Hort. Sci. 42: 235-240.

Hogenboom NG, Harten AM and Zeven AC. 1979. Broadening the gentic base of crops. In: Proc. Conference, Wageningen, Netherlands, pp. 299-309.

Horner HT and Palmer RG. 1995. Mechanisms of genic male sterility. Crop Sci. 35: 1527-1535.

Hume RJ and Lovell PH. 1981. Reduction of the cost involved in hybrid seed production of pumpkin (*Cucurbita maxima* Duchesne). New-Zealand J. Expri. Agric. 992: 209-210.

Hunsperger MH, Helsel DB and Baker LR. 1983. Silvar nitrate induction of staminate flowering in hermaphroditic pickling cucumber. HortScience 18: 347-349.

Jones HA and Clarke AE. 1943. Inheritance of male sterility in onion and the production of hybrid seed. Proc. Amer. Soc. Hort. Sci. 43: 189-194.

Jones HA and Emsweller SL. 1936. A male sterile onion. Proc. Amer. Soc. Hort. Sci. 34: 582-585.

Kalloo G and Bergh BO. 1993. Genetic Improvement of Vegetable Crops. Pergamon Press, U.K.

Kalloo G. 1988. Vegetable Breeding Vols. I, II & III. Panima Educational Agency, New Delhi.

Kaul MLH. 1988. Male Sterility in Higher Plants. Monographs on Theor. Appl. Genet. 10, Springer-Verlag, Berlin.

Korzeniewska A, Galecka T and Niemirowicz-Szczytt K. 1995. Effect of ethephon (2-chlorethylphosphonic acid) on monoecious muskmelon (*Cucumis melo* L.) F_1 hybrid seed production. Folia Horticulturae 7: 25-34.

Kumar Rajesh, Singh A, Kumar Sanjeet, Banerjee MK and Kalloo G. 2001. Introduction and characterization of male sterile and parthenocarpic lines of tomato (*Lycopersicon esculentum* Mill.). Veg. Sci. 28: 30-33.

Kumar Sanjeet, Banerjee MK and Kalloo G. 2000. Male sterility: mechanisms and current status on identification, characterization and utilization in vegetables. Veg. Sci. 27: 1-24.

Kumar Sanjeet, Kumar Sudhir, Rai SK, Kumar Virendra and Kumar Rajesh. 2002. Considerable amount of natural cross pollination on male sterile lines of chilli (*Capsicum annuum* L.). Capsicum and Eggplant Newsletter 21:48-51.

Laibach F and Kribben FJ. 1950. The effect of growth regulators on the flowering of cucumber. Naturwissenschaften 37: 114-115.

Lecouviour M, Pitrat M and Risser GA. 1990. Fifth gene for male sterility in *Cucumis melo*. Cucurbits Genet. Coop. Rep. 13: 34.

Levings CS. 1990. The Texas cytoplasm of maize: Cytoplasmic male sterility and disease susceptibility. Science 250: 942-947.

Liedle BE and Anderson NO. 1993. Reproductive barriers: Identification uses and circumvention. Plant Breed. Rev. 11: 11-154.

Loy B. 1978. Regulation of sex expression in gynoecious muskmelon for hybrid seed production. Cucurbit Genet. Coop. Rep.1: 18.

Mariani C, Gossele V, De Beuckeleer M, De Block M, Goldberg RB, De Greef W and Leemans J. 1992. A chimeric ribonuclease-inhibitor gene restores fertility to male sterile plants. Nature 357: 384-387.

Martiniello P, Falavigna A and Soressi SP. 1985. Influence of the tomato (*Lycopersicon esculentum* Mill.) brown seed gene (*bs*) on plant and fruit characteristics. Genet. Agraria 39: 417-422.

Mc Rae DH. 1985. Advances in chemical hybridization. Plant Breed. Rev. 3: 169-191.

McCreight JD and Elmstrom GW. 1984. A third male sterile gene in muskmelon. HortScience 19: 268-270.

McCreight JD, Nerson H and Grumet R. 1993. Melon. In: Kalloo G and Bergh BO (eds.), Genetic Improvement of Vegetable Crops. Pergamon Press, UK, pp. 287-294.

McCubbin A and Dickinson H. 1997. Self-incompatibilty. In: Shivanna KR and Sawhney VK (eds.), Pollen Biotechnology for Crop Production and Improvement. Cambridge Univ. Press, pp. 199-217.

Melchers G, Mohri Y, Watanabe K, Wakabayashi S and Harada K. 1992. One step generation of cytoplasmic male sterility by fusion of mitochondrial inactivated protoplasts with nuclear inactivated *Solanum* protoplasts. Proc. Nat. Acad. Sci. (USA) 89: 6832-6836.

Moore RH. 1950. Several effects of maleic hydrazide on plants. Science 112: 52-53.

Moore RH. 1959. Male sterility indeed in tomato by sodium 2,3-dichloroisobutyrate. Science 129: 1738-1740.

Morelock TE, Simon PW and Peterson CE. 1996. Wisconsin Wild: Another petaloid male-sterile cytoplasm for carrot. HortScience 31: 887-888.

Morelock TE. 1974. Influence of cytoplasm source on the expression of male sterility in carrot, *D. carota*. Ph.D. Thesis. Wisconsin University, USA.

Naylor AW. 1950. Observations on effects of maleic hydrazide on flowering of tobacco, maize and coclebut. Proc. Natl. Acad. Sci. 36: 230-232.

Nieuwhof M. 1961. Male sterility in some cole crops. Euphytica 10: 351-356.

Nieuwhof M. 1968. Effect of temperature on the expression of male sterility in Brussels sprouts (*Brassica oleracea* L. var. *gemmifera* DC.). Euphytica 17: 265-273.

Nishi S. 1967. F_1 seed production in Japan. Proc. XVIII Int. Hort. Cong. 3: 231-257.

Ogura H. 1968. Studies on the new male-sterility in Japanese radish with special reference to the utilisation of this sterility towards the practical raising of hybrid seed. Mem. Fac. Agric. Kagoshina Univ. 6: 39.

Pearson OH. 1933. Breeding plants of cabbage group. Calif. Agric. Exp. Stn. Bull. 532: 3-22.

Pearson OH. 1972. Cytoplasmically inherited male sterility characters and flavour components from the species cross *Brassica nigra* (L.) Koch × *B. oleracea* L. J. Amer. Soc. Hort. Sci. 97: 397-402.

Pearson OH. 1981. Nature and mechanism of cytoplasmic male sterility in plants: A review. HortScience 16: 482-487.

Pearson OH. 1983. Heterosis in vegetable crops. In: Frankel R (ed.), Heterosis, Monograph on Theor. Appl. Genet 6. Springer Berlag, Berlin, pp. 139-188.

Pekarkova TE. 1993. History of the discovery and utilization of in tomato (*Lycopersicon esculentum* Mill.). Zahradnictvi 20(4): 261-265.

Pelletier G, Ferault M, Lancelin D, Boulidard L, Dore C, Bonhomme S, Grelon M and Budar F. 1995. Engineering of cytoplasmic male sterility in vegetables by protoplast fusion. Acta Horticulturae 392: 11-17.

Peterson CE and Simon PW. 1986. Carrot breeding. In: Basset MJ (ed.), Breeding Vegetable Crops. AVI Publishing Co., West Port, Connecticut. pp. 321-356.

Peterson PA. 1958. Cytoplasmically inherited male sterility in *Capsicum*. Amer. Nat. 92: 111-119.

Petrova M, Yulkova Z, Gorinova N, Izhar S, Firon N, Jacquemin JM, Atanassov A and Stoeva P. 1999. Characterization of a cytoplasmic male sterile hybrid between *Lycopersicon peruvianum* Mill. × *Lycopersicon pennellii* Corr. and its crosses with tomato. Theor. Appl. Genet. 98.

Phatak SC and Jaworski CA. 1989. UGA 1-MS male-sterile eggplant germplasm. HortScience 24:1050.

Pilrat M. 1990. Gene list for *Cucumis melo*. L. Cucurbit & Genet. Coop. Rep. 13: 58.

Pochard E. 1970. Obtaining three new male sterile mutants of pepper (*C. annuum*) through application of mutagenes on monoploid material. Eucarpia, Versailles, France, pp. 93-95.

Ram D, Kumar Sanjeet, Banerjee MK and Kalloo G. 2002. Occurrence, identification and preliminary characterization of gynoecism in bitter gourd (*Momordica charantia*). Indian J. Agric. Sci.: 72:348-349.

Ranfort J, Saumitou LP, Cuguen J and Couvet D. 1995. Mitochondrial DNA diversity and male sterility in natural populations of *Daucus carota* ssp. carota. Theor. Appl. Genet. 91: 150-159.

Rastogi R and Swahney VK. 1988. Suppression of stamen development by CCC and ABA in tomato floral bud cultured in vitro. J. Plant Physiol. 9: 529-537.

Rehm S. 1952. Male sterile plants by chemical treatment. Nature 179: 38-39.

Rick CM. 1948. Genetics and development of nine male-sterile tomato mutants. Hilgardia 18: 599-633.

Rundfeldt H. 1961. Untersuchungen zur zuchtung des kopfkohhs (*B. oleracea* var *capitata*). Z Pflanzenzucht 44: 30-62.

Sakamoto Y, Ohnishi N, Hayashi M, Okamoto A, Mashiko T and Sanada M. 1991. Synthetic seed: The development of a botanical seed analogue. Chemical Regulation of Plants 26: 205-211.

Sampson DR. 1966. Linkage of genetic male sterility with a seedling marker and its use in producing F_1 hybrid seed of *Brassica oleracea* (cabbage, broccoli, kale etc.). Can. J. Plant Sci. 46: 703.

Sawhney VK. 1981. Abnormalities in pepper (*Capsicum annuum*) flowers induced by gibberellic acid. Can. J. Bot. 59: 8-16.

Sawhney VK. 1983. Temperature control of male sterility in a tomato mutant. J. Hered. 74: 51-54.

Sawhney VK. 1997. Genic male sterility. In: Shivanna KR and Sawhney VK (eds.), Pollen Biotechnology for Crop Production and Improvement. Cambridge Univ. Press, pp. 183-198.

Shannon S and Robinson RW. 1979. The use of ethephon to regulate sex expression of summer squash for hybrid seed production. J. Amer. Hort. Sci. 104: 674-677.

Shifriss C and Frankel R. 1969. A new male sterility gene in *Capsicum annuum*. J. Amer. Soc. Hort. Sci. 94: 385-387.

Shifriss C. 1997. Male sterility in pepper (*Capsicum annuum* L.). Euphytica 93: 83-88.

Sigareva MA and Earle ED. 1997. Direct transfer of a cold tolerant Ogura male sterile cytoplasm into cabbage (*Brassica oleracea* ssp. *capitata*) via protoplast fusion. Theor. Appl. Genet. 94: 213-220.

Singh J and Kaur S. 1986. Present status of hot pepper breeding for multiple resistance in Punjab. In: Proc. VIth Meet.Genet. Breed. Capsicum and Eggplant, Zaragoza, Spain, pp. 111-114.

Singh PK. 2000. Utilization and seed production of hybrid vegetable varieties in India. J. New Seeds. 2(4): 37-42.

Singh PK, Tripathi SK and Somani KV. 2001. Hybrid seed production of radish (*Raphanus sativus* L.). J. New Seeds. 3(4): 51-58.

Stout AB. 1920. Further experimental studies on self-incompatibility in hermaphrodite plants. J. Genet. 9: 85-129.

Tanksley SD, Rick CM and Vallejos CE. 1984. Tight linkage between a nuclear male sterile locus and an enzyme marker in tomato. Theor. Appl. Genet. 68: 109-113.

Tokumasu S. 1951. Male sterility in Japanese radish (*Raphanus sativus* L.). Sci. Bull. Fac. Agric. Kyushu Univ. 13: 83-89.

Tripathi SK and Singh PK. 2000. Hybrid seed production of cauliflower. J. New Seeds. 2(4): 43-49.

Van der Meer QP and Van Dam. 1979. Gibberellic acid as gametocide for cole crops. Euphytica 28: 717-722.

Wang X, Fang Z, Sun P, Liu Y and Yang L. 1998. Identification of a RAPD marker linked to a dominant male sterile gene in cabbage. Acta Horticulturae Sincia 25: 197-198.

Watts VM. 1962. A marked male sterile mutant in watermelon. Proc. Amer. Soc. Hort. Sci. 81: 498-505.

Welch JE and Grimball Jr EL. 1947. Male sterility in the carrot. Science 106: 594.

Williams ME, Lecmans J and Michiels F. 1997. Male sterility through recombinant DNA technology. In: Shivanna KR and Sawhney VK (eds), Pollen Biotechnology for Crop Production and Improvement. Cambridge Univ. Press, pp. 237-257.

Yamagishi H and Terachi T. 1994a. Molecular and biological studies on male-sterile cytoplasm in the Cruciferae. I. The origin and distribution of Ogura male sterile cytoplasm in Japanese wild radishes (*Raphanus sativus* L.) revealed by PCR-aided assay of their mitochondrial DNAs. Theor. Appl. Genet. 87: 996-1000.

Yamagishi H and Terachi T. 1994b. Molecular and biological studies on male sterile cytoplasm in the Cruciferae. II. The origin of Ogura male sterile cytoplasm inferred from the segregation pattern of male sterility in the F_1 progeny of wild and cultivated radishes (*Raphanus sativus* L.). Euphytica 80: 201-206.

Yamagishi H and Terachi T. 1996. Molecular and biological studies on male sterile cytoplasm in the Cruciferae. III. Distribution of Ogura-type cytoplasm among Japanese wild radishes and Asian radish cultivars. Theor. Appl. Genet. 93: 325-332.

Yamagishi H. 1998. Distribution and allelism of restorer genes for Ogura cytoplasmic male sterility in wild and cultivated radishes. Genes and Genetic Systems 73: 79-83.

Yordanov M. 1983. Heterosis in tomato. In: Frankel R (ed.), Heterosis. Monograph on Theor. Appl. Genet. 6. Springer Berlag, Berlin, pp. 189-219.

Zhang Q, Gabert AC and Baggett JR. 1994. Characterizing a cucumber pollen sterile mutant: inheritance, allelism, and response to chemical and environmental factors. J. Amer. Soc. Hort. Sci. 119: 804-807.

Zhang XP and Wang M. 1990. A genetic male sterile (*ms*) watermelon from China. Cucurbit Genet. Coop. Rep. 13: 45-46.

Zhang XP, Rhodes BB, Baird WV, Skorupska HT and Bridges WC. 1996. Development of genic male sterile watermelon lines with delayed-green seedling marker. HortScience 31: 123-126.

Zhang XP, Skorupska HT and Rhodes BB. 1994. Cytological expression in the male-sterile ms mutant in watermelon. J. Heredity 85: 279-285.

Transgenic Vegetable Crops

R. B. Ram

S. K. Dasgupta

SUMMARY. Modern gene transfer based on recombinant DNA technology is a rapidly growing subject and offers vast opportunity for manifesting the utility of this technology in economic terms. Various areas where there is an enormous scope for this technology application in vegetables include improved yield, altered nutrition quality, improved resistance to diseases, pests and herbicides and in food stuffs having therapeutic value. Accordingly the preferred target area of research in gene transfer, apart from the ones cited above, include tolerance to various abiotic stresses, plant productivity genes, genes affecting nitrogen fixation, usage of male sterility for production of hybrids, increased storage life, various aesthetic aspects of the product, etc. Keeping these advantages in view, there has been a drastic increase in the total cultivable area under transgenic crops globally. Among all vegetable crops potato, tomato, eggplant, peppers, cucumber, cauliflower, cabbage, and carrot has received the maximum attention particularly in the areas relating to insect pests, disease and herbicide resistance and quality improvement. The paper discusses on the specific technology adopted in these areas. It also discusses the various issues relating to biosafety issues, current constraints in transformation, etc. *[Article copies available for a fee from The Haworth Document Delivery Service: 1-800-HAWORTH. E-mail address: <docdelivery@haworthpress.com> Website: <http://www.HaworthPress.com> © 2004 by The Haworth Press, Inc. All rights reserved.]*

R. B. Ram is affiliated with the Division of Genetics, Indian Agriculture Research Institute, New Delhi 110012, India.

S. K. Dasgupta is affiliated with the Sungro Seeds Ltd., Shalimar Bagh, New Delhi 110088, India.

[Haworth co-indexing entry note]: "Transgenic Vegetable Crops." Ram, R. B., and S. K. Dasgupta. Co-published simultaneously in *Journal of New Seeds* (Food Products Press, an imprint of The Haworth Press, Inc.) Vol. 6, No. 4, 2004, pp. 411-431; and: *Hybrid Vegetable Development* (ed: P. K. Singh, S. K. Dasgupta, and S. K. Tripathi) Food Products Press, an imprint of The Haworth Press, Inc., 2004, pp. 411-431. Single or multiple copies of this article are available for a fee from The Haworth Document Delivery Service [1-800-HAWORTH, 9:00 a.m. - 5:00 p.m. (EST). E-mail address: docdelivery@haworthpress.com].

http://www.haworthpress.com/web/JNS
© 2004 by The Haworth Press, Inc. All rights reserved.
Digital Object Identifier: 10.1300/J153v06n04_06

KEYWORDS. Transgenics, vegetables, insect pest resistance, disease resistance, herbicide tolerance, biosafety issues

Modern gene transfer based on recombinant DNA technology is a rapidly growing subject and their intelligent use offers new opportunity for making directed changes to plant genotypes. Genetic transformation can introduce a gene for desired trait without disturbing the genetic architecture of the plant. Recent developments in molecular biology and genetic transformation have made it possible to identify, isolate and transfer of any desirable gene(s) from any living organism to crop plants and vice-versa. Agriculture biotechnology is no longer just an academic or basic research program but a commercial reality in terms of economic attributes. These attributes include plants with improved agronomic traits, resistance to insect-pest and diseases, tolerance to herbicides, enhanced quality, pharmaceutical products, chemicals, and biofuels.

OUTLINE OF METHODOLOGY

The major steps for obtaining transgenic plants in laboratory presented in a simplistic way as follows (Figure 1).

1. Identification and isolation of desired gene(s) or DNA fragment to be cloned.
2. Generation of suitable 'gene construct.'
3. Transfer of this gene construct with suitable protocol into plant cells/tissues *in vitro* (transformation).
4. Selection of transformed cells using suitable marker system.
5. Regeneration of plants from the transformed cells.
6. Characterization of transgene(s) for their stable integration, expression, and genetic behavior.
7. Evaluation of transgenic plants for desired trait.
8. Mass multiplication of transgenic plants for their commercial use.

SCOPE OF APPLICATION

1. Improvement of efficiency of the plants in terms of yield, nutritional quality or agronomic characteristics by specific metabolic pathway modification.
2. Useful to improve resistance to insect-pests and diseases and to correct some limiting factors such as toxins in plant products, intolerance to herbicide, and abiotic stresses.

3. By the use of metabolic engineering change the nature of harvested products, i.e., human foodstuff having therapeutic aesthetic value and commercial properties and industrial raw stock.

GOAL OF GENE TRANSFER TECHNOLOGY IN CROP PRODUCTION

1. Crop resistance to insect-pests and diseases.
2. Crop resistance to herbicides.
3. Tolerance to environmental stresses such as temperature, salt, moisture, cold, etc.
4. Reduction of photorespiration in C_3 plants.
5. Transfer of *nif* gene for atmospheric nitrogen fixation in crop plants.
6. Development of value added food products through enhancement of nutritional qualities of crop plants.
7. Increased productivity of food crops including horticultural crops.
8. Introduction of male sterility system for hybrid seed production.
9. Enhanced production of chemicals and enzymes used in pharmaceutical and food processing industries.
10. Transgenic plants as biorefineries for production of biomolecules (protein, carbohydrates lipids, etc.), edible vaccine and non-food materials, e.g., industrial oils and fatty acids and biodegradable plastics etc.
11. Prolonging the storage or shelf life of fruits and vegetables.

GROWTH OF TRANSGENIC CROPS–A GLOBAL PERSPECTIVE

In recent decades, there is increasing trend in total cultivated area, specific crop area and choice of specific traits under transgenic crops. The growth in global cultivated area under transgenic crops presented in Figure 2 showed drastic increase from 1996 to 1999. The global area with transgenic crops was 1.7 million hectares in 1996, 11.0 million hectares in 1997, 27.8 million hectares in 1998, and 39.9 million hectares in 1999.

The USA with 28.7 million hectare area (72%) of total area under transgenic crops rank first in the world followed by countries like Argentina (6.7 mha), Canada (4.0 mha), China (0.3 mha), and Australia and South Africa (0.1 mha). Mexico, Spain, France, Portugal, Romania, and Ukraine jointly contributed only 0.1 mha in transgenics (James, 1999 and Serageldin, 2000).

Specific crop area shown in Figure 3 under transgenic crops revealed that soybean is the major crop (51%) to be cultivated worldwide followed by cotton (28%), canola and maize (8%), and tobacco (3%). In concurrence with the

FIGURE 1. Schematic diagram for the procedure of gene transfer in crop plants.

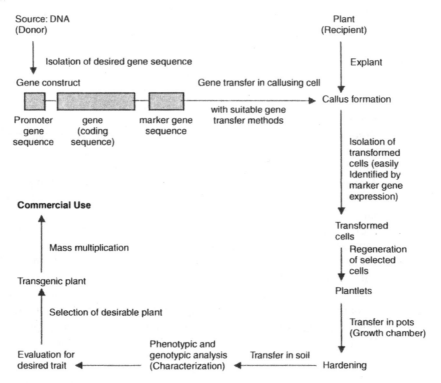

FIGURE 2. Growth in global cultivated area under transgenic crops (adapted from James, 1999)

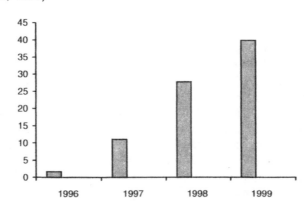

FIGURE 3. Specific crop area under transgenic crops worldwide (adapted from James, 1998).

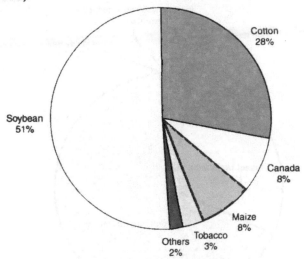

changing trend about 70 genetically modified varieties of crop plants are registered for commercial cultivation in more than 10 countries.

Choice of specific trait under transgenic crops (Figure 4) shows that the highest percentage of transgenic crops is for herbicide tolerance (67.3%) followed by insect resistance (27.9%), viral resistance (3.4%), insect resistance and herbicide tolerance (1.0%), and quality traits (0.3%).

Many transgenic varieties in vegetable crops have been approved for commercial cultivation in USA (Table 1). In tomato variety, Endless Summer™ was approved for commercial cultivation in USA during 1995 for delayed ripening, thicker skin and altered pectin content. In potato variety, New Leaf™ resistant to Colorado potato beetle and in squash variety Freedom™ resistant to viruses were also approved for commercial cultivation in 1995. Many other transgenic lines in potato, tomato, brinjal, cabbage, cauliflower, and other vegetable crops is under evaluation for resistant to insect-pests and diseases and for nutritional quality in many countries.

TRANSGENIC STUDY IN SOME VEGETABLE CROPS

Potato (Solanum tuberosum)

Transgenic potato plants for resistance to herbicide, viral, fungal, bacterial and insect-pests, tolerance to abiotic stresses, and improved quality as well as

FIGURE 4. Choice of specific trait: Cultivated area under transgenic crops (adapted from James, 1998)

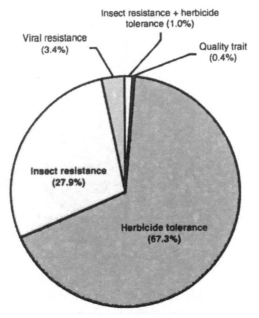

potato as a bioreactor for pharmaceuticals are the possibilities that can be explored.

Herbicide Resistance

The gene encoding EPSP (5-enolpyruvyl-shikimate-3 phosphate synthase) synthase has been cloned from petunia and introduced into potato and demonstrated to be resistant to glyphosate, a broad spectrum non-selective herbicide (Shah et al., 1986). Bromoxynil specific nitrilase gene was isolated from the soil bacterium *Klebsiella ozyaenae* and transformed into potato to make them resistant against bromoxynil (Stalker et al., 1988). The gene (*bar*) cloned from *Streptomyces hygroscopicus* was transferred into potato conferring resistance to biolophos, a non-selective herbicide that produces phosphonithricin (PPT) inside the plants and causes cell death (Thompson et al., 1987).

Resistance to Viral Diseases

Coat protein (*CP*)-mediated resistance is the most popular strategy to control viral diseases (Beachy et al., 1990). After multiplication, a virus moves to

TABLE 1. Transgenic crop approved in the USA for commercial use.

PRODUCTS	ALTERED TRAIT	COMPANY	YEAR	TRADE NAME
TOMATO	DELAYED RIPENING	CALGENE Inc.	1994	FLAVR SAVR
	DELAYED RIPENING	DNA Plant Tech	1995	ENDLESS SUMMER
	DELAYED RIPENING	MONSANTO Co.	1995	
	THICKER SKIN AND ALTERED PECTIN CONTENT	ZENECA/ PETOSEED	1995	
COTTON	Bt GENE	MONSANTO Co.	1995	BOLLGARD
	RESISTANT TO WEEDICIDE	CALGENE Inc.	1995	BXN COTTON
	RESISTANT TO WEEDICIDE	MONSANTO Co.	1996	ROUNDUP READY
	RESISTANT TO WEEDICIDE	DU PONT	1996	
SOYBEAN	RESISTANT TO WEEDICIDE	MONSANTO Co.	1995	ROUNDUP READY
POTATO	Bt GENE	MONSANTO Co.	1995	NEW LEAF
	Bt GENE	MONSANTO Co.	1996	
MAIZE	Bt GENE	CIBA-GEIGY	1995	MAXIMIZER
	RESISTANCE TO WEEDICIDES	DE KALAB GENETICS	1996	
	RESISTANCE TO WEEDICIDES	AGRO EVO	1996	LIBERTY LINK
	MALE STERILITY	PLANT GENETICS SYS	1996	
	Bt GENE	MONSANTO Co.	1996	YIELDGARD
	Bt GENE	NORTHRUP KING	1996	
RAPE SEED	ALTERED FATTY ACID COMP.	CALGENE Inc.	1995	LAURICAL
SQUASH	RESISTANT TO VIRUSES	ASHGROW SEEDS	1995	FREEDOM II
PAPAYA	RESISTANT TO VIRUSES	CORNELL UNIV.	1997	

the adjacent cells through plasmodesmata. Movement protein (mp) of virus is involved in this process. Transgenic plants expressing a defective mp would block all the sites of the viral genome and effectively minimize the binding of functional mp.

Resistance to Fungal Diseases

During fungal infection plants asccumulate a novel class of proteins called pathogenesis-related proteins or PR proteins in response to pathogen attack

(Van Loon, 1985). Information on host's active defense mechanism, i.e., gen-eration of reactive oxygen (like hydrogen peroxide) can be utilized to develop resistance to fungal attack. Wu et al. (1995) developed transgenic potato ex-pressing a glucose oxidase gene cloned from *Aspergillus niger* possessed re-sistance against leaf blight. Similarly osmotin gene encoding a class of PR-protein (*pr-5*) causing sporangial lysis has also been transferred into po-tato for improved resistance to late blight (Liu et al., 1994).

Resistance to Bacterial Diseases

Simple antimicrobial peptides produced by vertebrates, arthropods as well as by some plants in response to infection by different biotic agents have re-cently attracted attention of the biologists. Among various antimicrobial pep-tides, aeropins, attacins and lysozymes possess broad antimicrobial activity against both Gram-positive and Gram-negative bacteria. Cecropins are the most potent antibacterial agent and its analogue SB-37 and Shiva-1 was found to be more effective than the native molecule. Transgenic potato plants having genes encoding SB-37 and Shiva-1 showed reduced disease severity against *Ralstonia solanacearum* (Montanelli et al., 1995).

The lysozyme gene cloned from the bacteriophage T4 introduced into po-tato showed reduced rate of disease against soft rot caused by *Erwinia carotovora* (During, 1996). A gene from Southeast Asian horseshoe crab in-troduced into potato proved less susceptible to soft rot causing bacteria (Allefs et al., 1996).

Resistance to Insect-Pests

Use of insecticidal protein of a Gram-positive soil bacterium *Bacillus thuringiensis* is the most widely practiced transgenic technology strategy to impart resistance to insect-pests. Induced resistance through *Bt* is one of the first crop biotechnology applications where products have already reached the farmers field. The codon-modified *cry-V-Bt* gene from *B. thuringiensis* subsp. *Rurstaki* and potato Y poty virus Yo coat protein gene (*PV Yo cp*) co trans-formed into potato cv. spunta via *Agrobacterium tumefaciens* showed trans-genic lines to be more resistant to potato tuber moth (*Phthorimaea operculella*) and PV Yo infection than non-transgenic spunta (Lloid et al., 1999). At the Central Potato Research Institute, Shimla, India, several transgenic potato lines have been developed expressing a synthetic analogue of truncated *cry1*(Ab) gene and are presently under different stages of evaluation (Naik et al., 1997). Similarly, transgenic potato cultivar New Leaf™ having *cry* 3A gene has al-ready been released for commercial cultivation in USA by Monsanto Com-pany to impart resistance against Control Colorado potato beetle. Hoy (1999) reported that New Leaf™ transgenic Bt potatoes could provide substantial ecological and economic benefits to potato growers.

Aphids that feed on plant sap as well as responsible for spread of viruses has also been a subject of importance in transgenic development. Potato cultivar Desiree transformed with snowdrop lectin gene (*Galanthus nivalis* agglutinin, *gna*) showed significant reduction in fecundity of the potato aphid (*Aulacorthum solani*), which feed on plant sap and are responsible for the spread of viruses (Down et al., 1996) but it can also cause adverse effects to a predatory ladybird via aphid in its food chain. The significance of these potential ecological risks under field conditions needs to be further evaluated (Birch et al., 1999).

Tolerance to Abiotic Stresses

The reactive oxygen generated during normal photorespiration in plants is detoxified by an enzyme called superoxide dismutase (SOD). Constitutive expression of SOD activity was thought to improve the tolerance of plants to drought, frost, and sulphur dioxide and calcium deficiency. The chloroplastic and cytosolic Cu, Zn-SOD gene of tomato transferred into potato resulted in expression of this enzyme and transgenic potato clones were tolerant to a superoxide generating herbicide, paraquat (Perl et al., 1993). The transgenic strategy for drought tolerance came from analysis of plants that grow under desert (xerophytic) or extreme saline (halophytic) conditions. Transgenic potato producing fructans (water stress sugar) in leaves and tubers showed better drought tolerance (Rober et al., 1996).

Quality Improvement

Starch is the primary storage component of carbohydrate in potato tubers accounting up to 70% of tuber dry matter. Genetic engineering of starch biosynthesis in potato may lead to modification of starch granule shape, amylose to amylopectin chain length which can in turn alter significantly properties of starch like crystallinity, gelling properties, phosphorylation, lipid contents, etc., for use in both food and non-food industries.

The biosynthetic steps required for starch biosynthesis involve three enzymes, ADP glucose pyrophosphorylase (ADPGPPase), starch synthase (SS), and starch branching and debranching enzymes. ADPGPPase catalyses the syntheis of ADP glucose from glucose-1-phosphate. ADP glucose is the precursor for synthesis of both amylose and amylopectin. Therefore, the ADPGPPase regulation would determine the sink strength of tuber and its over expression would produce tubers with high starch content. Bacterium *Escherichia coli* gene glg C16 encoding bacterial ADPGP Pase when transferred into potato, the transgenic plant showed high starch content in the tubers

(Stark et al., 1992). Similarly high starch content in tubers of transgenic potato was found having ADPGPPase gene from barley. Granule bound starch synthase I (GBSSI) enzyme is responsible for regulating the amylose content of the starch. Transgenic potato was produced through antisense technology exhibited reduced level of amylose content in starch (Salehuzzaman et al., 1994). Two amylose free transgenic cultivars, released in the Netherlands during 1996, showed improved gel stability and paste clarity (Visser et al., 1997).

Essential amino acid content is an important constituent of quality protein. Gene transfer technology offers a novel approach to improve the nutritional quality of potato through modification of essential amino acid composition of plant protein. High essential amino acid encoding (*heaae*) gene was transferred to potato clones K-2 and K-7 and it showed increase in essential amino acids (Yang et al., 1989). Bovine growth hormone is important to boost the milk production and health of cattle. Keeping this in view a transgenic potato plant was developed by Zhao et al. (1995) having increased amount of bovine growth hormone in tubers. The human serum albumin has also been extracted from transgenic potato (Pen et al., 1993). Transgenic technology offers a unique opportunity to develop plants having antigenic epitope of the vaccine which can constitute an alternative for easy and economical delivery of oral vaccines (Goddijin and Pen, 1995). Successful immunity against diarrhoea was expressed in mice when transgenic potato were provided to them having a gene encoding a heat labile entertoxin (LT-B) from *E. coli* (Haq et al., 1995). Transgenic potato plants for hepatitis B surface antigen (Ehsani et al., 1998) and cholera toxin-B subunit (Arakawa et al., 1998) have also been developed for oral immunization. Boiling and frying of potato before consumption destroyed the antigenic property of recombinant vaccine and this is a major problem in vaccine production in potato. Therefore, transgenic potato can be a good source of oral immunization of poultry and cattle against major bacterial and viral diseases, i.e., capsid protein (VP1) gene for foot and mouth disease of cattle.

Tomato (Lycopersicon esculentum)

Tomato is an important vegetable because of its nutritional value and widespread use as a processed food. Its production and marketing is affected by various biotic and abiotic stresses and there is an urgent need to remove these limitations. The recent advancement in biotechnology may help to increase production as well as its nutritional and market value.

Disease Resistance

For the viral disease resistance, coat protein and satellite RNA have been extensively utilized to produce viral disease resistance plants in tomato. Nelson et al. (1988) produced transgenic tomato using coat protein of the U1 strain of tobacco mosaic virus and expressor plants were protected against symptom development after inoculation with TMV strain U1 or PV230 and TMV strain 2 and 22. *Agrobacterium*-mediated transformation of the coat protein gene of cucumber mosaic cucumovirus white leaf (CMV-WL) strain (Xue et al. 1994) and it was reported that transformed plants exhibited high level of resistance to systemic infection by virus strain CMV-WL and CMV-China. Dong et al. (1995) transformed cotyledon segments of tomato with *A. tumefaciens* having CMV sat RNA cDNA. Most of the transgenic plants had delayed symptom, lower disease index and slow development of CMV.

Insect Resistance

Transformation of tomato plants with two truncated genes from *Bacillus thuringiensis* Rurstaki strain HD1 via *Agrobacterium*, conferred tolerance to larvae of *Manduca sexta, Heliothis zea*, and *Heliothis virescens* and exhibited up to 100% mortality when fed on transformed plant leaves (Fischhoff et al., 1987). Rhim et al. (1995) constructed a chimaeric gene encoding coleopteron insect-specific toxin from *B. thuringiensis* and transferred into tomato via an *A. tumefaciens* binary vector system. Transgenic tomato plant expressed significant insecticidal activity against Colorado potato beetle (*Leptinotarsa decemlineata*) larvae.

Fruit Ripening

Vase life and enhanced storage ability is an important factor for the marketing and transportation of the tomato. Ethylene is a critical substance for the ripening of fruit. Enzyme ACC oxidase is responsible to control the synthesis of ethylene. By regulation of polygalacturonase (PG) using antisense technology the variety Flavr Savr™, of tomato developed possessed improved textural qualities and is being commercially exploited in USA. The Flavr Savr™ tomato is the first genetically-engineered food to be sold commercially (Kramer and Redenbaugh, 1994).

Eggplant (Solanum melongena)

Eggplant is a popular and normally self-fertilized perennial but grown as annual crop in the world. Like other solanaceous vegetables diseases and pests also affect it. Biotechnological tools provide a substantial opportunity to control these menace to increase the crop productivity.

Parthenocarpic Fruit Development

Heterologous gene *def*H9-*iaa* M was transformed in eggplant for parthenocarpic fruit development. Transgenic fruits showed increased level of IAA in the ovule. The fruit size was normal and yield was more than corresponding non-transgenic plants (Rotino et al., 1997).

Insect Resistance

The gene *Cry IIIb* from *Bacillus thuringiensis* was used to transform eggplant to study resistance against Colorado potato beetle (CPB) in transgenic plants. Transgenic plants did not show any detectable effect on larval mortality of CPB (Innacone et al., 1995). Similarly Chen et al. (1995) studied the *Cry IIIb* gene transgenic plants of eggplants and they did not demonstrate significant resistance to the first and second instar larvae of the CPB. Arpaia et al. (1997) transferred a modified gene of *B. thuringiensis* encoding coleopteran insect specific *Cry IIIB* toxin via *A. tumefaciens* to the eggplant and the toxic effect of *Bt*-transgene was reported on CPB. Rotino et al. (1998) developed an efficient protocol for genetic transformation of eggplant via *A. tumefaciens*. In transgenic lines having *Cry IIIb* gene from *B. thuringiensis* confirmed a good level of resistance to CPB and other insect-pests of eggplant.

Peppers (Capsicum sp.)

Capsicum is an important vegetable as well as spice crop of the world. Dong et al. (1995) transferred CMV sat-RNA gene via *A. tumefaciens* and observed about 10 days delay in expressing disease symptoms, slow CMV development and lower disease index in transgenic pepper plants. Kim et al. (1997) in hot pepper cv. Golden tower transferred satellite-RNA resistance against cucumber mosaic virus. PCR and RNA gel blot analyses showed stable inheritance and expression of cDNA of CMV satellite RNA in progenies. Zhang et al. (1994) transformed pepper plant with CMV-CP genes by co-culturing cotyledon explants of three cultivars with *A. tumefaciens*. Inoculation test with virus showed resistance to CMV in transgenic plants. To improve the fungal disease resistance ribosome inactivating proteins (RIPs) are known to be important because of their cytotoxic activity on eukaryotic cells.

Cucumber (Cucumis sativus)

Cucumber is an important vegetable crop among the cucurbits. Despite exploitation of heterosis through gynoecious lines and development of triploid having high level of heterosis, cucumber cultivation still faces many problems like, powdery mildew, mosaic, black rot, fruit fly and environmental stresses.

Recent development of gene transfer technology has paved the way for developing transgenic plants to transform the production scenario of cucumber.

Hyakawa et al. (1996) introduced CMV-O coat protein gene in cucumber using *Ti Agrobacterium* transformation system. The transgenic plant displayed strong resistance against cucumber mosaic cucumovirus. Tabei et al. (1998) using *Agrobacterium* mediated transformation introduced a rice chitinase cDNA (RCC2) into cucumber and tested transgenic plants against *Botrytis cinerea* (Gray mold) in controlled condition. Some 75% of regenerated shoots exhibited high level of resistance compared to the control.

Cauliflower (Brassica oleracea var. botrytis)

Cauliflower is an important winter season vegetable. It requires more care for successful cultivation than other vegetables. Its production is affected by large number of biotic and abiotic stress factors.

Transgenic plant having *barnase* gene of *Agrobacterium amyloquefaciens* expressed in tapetal cells of anthers may be an important strategy for sterility system useful for hybrid seed production in cauliflower. Passellgue and Kerlane (1996) using *Agrobacterium*-mediated transformation of capsid gene and the antisense gene VI of CaMV developed transgenic cauliflower plant. In transgenic plant multiple insertion of T-DNA and CaMV gene transcripts were detected but low amount of RNA transcribed in plant cells were observed.

Ding et al. (1998) developed insect resistant transgenic in Taiwan cauliflower (*Brassica oleracea* var. *botrytis*) against local insects to which control plants were susceptible. The transgenic plant was developed through *Agrobacterium*-mediated transformation of trypsin inhibitor gene isolated from *Ipomoea batata*. Similarly, Asokan (1999) produced transgenic plant showing toxicity against two pests of cauliflower namely, *Plutella xylostella* and *Crocidolomia binotalis* carrying insecticidal crystal proteins.

Cabbage (Brassica oleracea var. capitata)

Cabbage is one of the oldest and highly nutritive vegetable crop. As cauliflower its production is also affected by various biotic and abiotic stresses.

Resistance to Insects

Using *npt II* as selectable marker gene with CaMV 35 S promoter transgenic cabbage plants were developed having *Cry Ia* (c) gene of *Bacillus thuringiensis* (Bt). Transformants were found to confer tolerance to *Plutella xylostella* (Metz et al., 1995).

Root Formation and Auxin Synthesis

He et al. (1994) cloned two auxin synthesis genes in binary vector and introduced them into Chinese cabbage and cabbage using *Agrobacterium*. Both auxin genes initiated root formation on hypocotyl segments and cotyledonary explants.

Carrot (Daucus carota)

Carrot is an important vegetable used for human as well as animal consumption. It is a good source of vitamins and minerals. Diversified use and its nutritive value emphasized to enhance the yield potential of the carrot. Gene(s) of interest namely good flavor, tolerance to thermal stress, insect-pests, and pathogens can be transferred for their economic benefit.

Flavor

Dolgov et al. (1999a) transformed carrot with super sweet protein gene (thaumatin II) from *Thaumatococcus daniellii* using super virulent strain (BE2) of *Agrobacterium*. Transgenic plants expressed high amount of thaumatin II in callus tissues but low levels were observed in the leaves of greenhouse-grown plants.

Tolerance to Thermal Stress

Heat shock protein gene encoding Hsp 17.7 plays an important role in the ability of carrot cells and plants to survive in thermal stress condition (Malik et al., 1999). In transgenic plant *Hsp* gene were constitutively expressed (denoted as CaS lines). Thermo tolerance measurements showed that CaS lines were more tolerant than AH lines.

Insect Resistance

Murray et al. (1991) observed the expression of several insecticidal crystal protein genes of *B. thuringiensis* in electroporated protoplasts of carrot. Low levels of lepidopteran toxin *CryIA* (b) insecticidal crystal protein gene expression were detected due to RNA instability. Adang et al. (1993) designed a plant with *B. thuringiensis cry IIIA* delta-endotoxin and compared modified *Cry IIIA* gene expression with native gene expression by electroporation of protoplasts of carrot. *CryIIIA*-specific RNA and protein were detected in electroporated carrot protoplasts. There was good correlation between insect control and levels of delta-endotoxin RNA and protein. Plant expressed high levels of delta-endotoxin identified by their toxicity to more resistant third-instar larvae.

Resistance to Pathogens

Punja and Raharjo (1996) transformed carrot cultivars Nano and Golden State with two chitinase genes (from Petunia and tobacco) by detached petiole inoculation method. The transformants were evaluated against *Alternaria radicina, Botrytis cinerea, Rhizoctonia solani, Sclerotium rolfsii*, and *Thielaviopsis basicola*. The rate and final extent of lesion development after 7 days were significantly lower in transgenic plants inoculated with *Botrytis cinerea, Rhizoctonia solani*, and *Sclerotium rolfsii* expressing the tobacco chitinase gene, but not in plants expressing the chitinase gene. Dolgov et al. (1999) transferred plant defensin (pd) gene from *Raphanus sativus* via *Agrobacterium*-mediated transformation. Total 270 carrot lines were obtained and successfully rooted on high antibiotic and some of them showed high NPT activity in the greenhouse grown plants.

SOME RELEVANT QUESTIONS REGARDING TRANSGENIC TECHNOLOGY

1. How efficiently can we use biotechnology to feed increasing population in coming decades when the natural resources are diminishing?
2. How to further improve crop production and productivity when soils are becoming increasingly hungry?
3. How to control the emerging biotic and abiotic stresses that causes enormous losses before and after harvesting of crop?
4. What measures and code of conduct would be required to address key issues of biosafety, biodiversity, and genetic vulnerability?
5. Can modern revolution in biotechnology act as an effective instrument to fight poverty and hunger in developing countries?
6. Transgenic plant produces toxins against insect-pests and disease causing agents. Keeping this in view what is the limit of toxins in plant, which is not harmful to the human beings and animal nutrition?

CURRENT CONSTRAINS IN TRANSFORMATION

1. Gene transfer technology requires considerable technical skills, infrastructural requirements and financial assistance.

2. The mechanism of which foreign DNA is taken up and incorporated into the host genome remains largely unknown and sight of incorporation is difficult to target.
3. The randomness in insertion of transgene causes considerable limitations in biotechnological intervention. For instance, endogenous or native traits cannot readily be altered in transgenic plants.
4. *In vitro* culture may show somoclonal variation.
5. There is problem of gene silencing and the transgene has either become silent or has been lost in subsequent cycle of breeding.
6. There are some interactions between different transgenes in a transformed line, where one transgene construct interferes with expression of another transgene.
7. Most public programs working towards the betterment of agriculture in developing countries are not licensed to use these technologies such as patented genes of agronomic interest.

BIOSAFETY CONCERNS IN THE DEVELOPMENT AND COMMERCIALIZATION OF TRANSGENICS

Biosafety refers to policies and procedures adopted to ensure environmental safety during the course of the development and commercialization of the transgenic organisms.

1. Escape of engineered gene by the gene flow or gene disposal.
2. Non-target or ecological effects.
3. Invasiveness or weediness of transgenics. It means tendency of plant to spread beyond the field where first planted.
4. Creation of super-weeds and super-viruses.
5. Toxicity and allerginicity to human beings and animals.
6. Expression of undesirable phenotypic traits.
7. Genetic erosion of biological diversity.

BIOSAFETY OF ANTIBIOTIC RESISTANCE MARKERS

For distinction of transformed and non-transformed cells during regeneration of seedlings *in vitro*, some questions from the diverse sources have been raised to biosafety assessment of the marker gene are as follows:

1. Will the product of marker gene, such as npt II that confers Kanamycin resistance to the transgenic plant become toxic to human and animal health?

2. Can the plant or its products with antibiotic resistance be processed or eaten by human being and livestock without problem?
3. Will the spread of the marker genes from transgenic plants to other living organisms through gene flow or even bacteria through horizontal gene transfer cause unpredictable damage?
4. Will the possible transfer of antibiotic resistance genes from crop to the bacteria increase the level of such genes in the soil?
5. Will antibiotic resistance results due to an undesirable increase in the use of antibiotic?
6. Can marker genes from living organisms transferred to the plant genome cause any undesirable effect on human or animal bio-physiological processes?

FACTORS THAT MAY INFLUENCE GENE FLOW

1. Proximity of the transgenic with compatible wild relatives.
2. Sexual compatibility of crop plants with wild relatives.
3. Mode of pollination and mating system.
4. Synchronous flowering of crop plant and wild relatives.
5. Well viability of weed-crop hybrid.
6. Mode of seed dispersal.
7. Nature of transgenic character itself.

REFERENCES

Adang, M.J., Brody, M.S., Cardineau, G., Eagan, N., Roush, R.T., Shewmaker, C.K., Jones, A., Oakes, J.V. and McBride, K.E. (1993). The reconstruction and expression of a *Bacillus thuringiensis* cryIIIA gene in protoplasts and potato plants. Plant Mol. Biol., 21: 1131-1145.

Allefs, S.J.H.M., DeJong, E.R., Florack, D.E.A., Hoogendoorn, C. and Stiekma, W.J. (1996). Erwinia soft-rot resistance of potato cultivars expressing antimicrobial peptide Tachyplesin-I. Mol. Breed., 2: 97-105.

Arakawa, T., Chong, D.K.X. and Langride, W.H.R. (1998). Efficacy of a food plant - based oral cholera toxin B subunit vaccine. Nat. Biotech., 16: 292-297.

Arpaia, S., Mennella, G., Onofaro, V., Perri, E., Sunseri, F. and Rotino, G.L. (1997). Production of transgenic eggplant (*Solanum melongena* L.) resistance to Colorado potato beetle (*Leptinotarsa decemlineata* Say). Thero. Appl. Genet., 95: 329-334.

Asokan, R. (1999). A method of delivering *Bacillus thuringiensis* Berliner insecticidal crystal toxins for the management of diamond back moth on cabbage. Insect Environment, 4: 124.

Beachy, R.N., Laesch-Fries, S. and Tumor, N.E. (1990). Coat protein-mediated resistance against virus infection. Ann. Rev. Phytopathol., 28: 451-474.

Birch, A.N.E., Geoghegan, I.E., Majerus, M.E.N., McNicol, J.W., Hackett, C.A., Gatehouse, A.M.R. and Gatehouse, J.A. (1999). Tri-trophic interactions involving pest aphids, predatory 2-spot ladybirds and transgenic potatoes expressing snowdrop lectin for aphid resistance. Mol. Breed., 5: 75-83.

Chen, Q., Jelenkovic, G., Chin, C.K., Billings, S., Kberhardt, J., Goffreda, J.C. and Day, P. (1995). Transfer and transcriptional expression of coleopteral cryIIIB endotoxin gene of Bacillus thuringiensis in eggplant. J. Am. Soc. Hort. Sci., 120: 921-927.

Ding, L.C., Hu, C.Y., Yeh, K.W. and Wang, P.J. (1998). Development of insect-resistant transgenic cauliflower plants expressing the trypsin inhibitor gene isolated from local sweet potato. Plant Cell Rep., 17: 854-860.

Düring, K. (1996). Genetic engineering for resistance to bacteria in transgenic plants by introduction of foreign genes. Mol. Breeding, 2: 297-305.

Dolgov, S.V., Lebedev, V.G., Firsov, A.P., Taran, S.A. and Tjukavin, G.B. (1999a). Expression of thaumatinII gene in horticultural crops. In: Developments in plant breeding. Vol. 6 Genetics and breeding for crop quality and resistance. ed. S. Mugnozza, E. Porceddu and M. Pagnotta. Kluwer Ac. Publ. Dordrecht, Boston, London, pp. 165-172.

Dolgov, S.V., Lebedev, V.G., Firsov, A.P., Taron, S.A. and Tjukavin, G.B. (1999b). Phytopathogene resistance improvement of horticultural crops by plant-defensin gene introduction. In: Developments in plant breeding. Vol. 6. Genetic and breeding for crop quality and resistance, eds. S. Mugnozza, E. Porceddu and M. Pagnotta. Kluwer Ac. Publ. Dordrecht, Boston, London, pp. 111-118.

Dong, C.Z., Jiang, C.X., Feng, L.X., Li, S.D., Gao, Z.H., Guo, J.Z. and Zhu, De W. (1995). Transgenic tomato and pepper plants containing cmw sat-rna cdna. Acta Hortic., 402: 78-86.

Down, R.E., Gatehouse, A.M.R., Hamilton, W.D.O. and Gatehouse, J.A. (1996). Snowdrop lectin inhibits development and fecundity of the glasshouse potato aphid (Aulacorthum solani) when administered in vitro and via transgenic plants both in laboratory and glasshouse trials. J. Insect Physiol., 42: 11-12, 1035-1045.

Ehsani, P., Khabiri, A. and Domonsky, N.N. (1998). Polypeptide of HbsAg produced in trangenic potato. Gene, 190: 107-111.

Fischhoff, D.A., Bowdish, K.S., Perlak, F.J., Marrone, P.G., McCormick, S.M., Niedermeyer, J.G., Dean, D.A., Kusano-Kretzmer, K., Mayer, E.J., Rochester, D.E., Rogers, S.G. & Fraley, R.T. (1987). Insect tolerant transgenic tomato plants. Biotechnology, 5: 807-813.

Goddijin, O.J.M. and Pen, J. (1995). Plants as bioreactors. Trends Biotech., 13: 379-397.

Haq, T.A., Mason, H.S., Clements, J.D. and Arntzen, C.J. (1995). Oral immunization with a recombinant antigen produced in transgenic plants. Science, 268: 714-716.

He, Y.K., Wang, J.Y., Gong, Z.H., Wei, Z.M. and Xu, Z.H. (1994). Root development initiated by exogenous auxin genes in Brassica ssp. Plant Physiol. Biochem., 32: 493-500.

Hoy, C.W. (1999). Colorado beetle resistance management strategies for transgenic potatoes. Amer. J. Potato Res., 76: 215-219.

Innacone, R., Fiore, M.C., Macchi, A., Grieco, P.O., Arpaia, S., Perrone, D., Mennella, G., Sunseri, F., Cellini, F., Ratino, G.L. and Nishio, J. (1995). Genetic engineering of eggplant (*Solanum melongena* L.). Acta Hortic. 392: 227-233.

James, C. (1998). Global review of commercialized transgenic crops in 1998: ISAAA brief No. 8, Ithaca, NY.

James, C. (1999). Global review of commercialized transgenic crops in 1999: ISAAA brief No. 12, Ithaca, NY,

Kim, S.J., Lee, S.J., Kim, B.D. and Paek, K.H. (1997). Satellite-RNA mediated resistance to cucumber mosaic virus in transgenic plants of hot pepper (*Capsicum annuum* cv. Golden Tower). Plant Cell Rep. 16: 825-830.

Kramer, M.G. and Redenbaugh, K. (1994). Commercialization of a tomato with an antisense polygalacturonase gene: The FLAVR SAVR® tomato story. Euphytica, 79: 293-297.

Liu, D., Raghothama, K.G., Hasegawa, P.M. and Bressan, R.A. (1994). Osmotin overexpression in potato delays development of disease symptoms. Proc. Natl. Acad. Sci. USA. 91: 1888-1892.

Lloyd, J.R., Landschutze, V. and Kossmann, J. (1999). Simultaneous antisense inhibition of two starch synthase isoforms in potato tubers leads to accumulation of grossly modified amylopectin. Bioch. J., 338: 515-521.

Malik, M.K., Slovin, J.P., Hwang, C.H. and Zimmerman, J.L. (1999). Modified expression of a carrot small heat shock protein gene, Hsp17.7, results in increased or decreased thermotolerance. Plant J., 20; 89-99.

Metz, T.D., Dixit, R. and Earle, E.D. (1995). *Agrobacterium tumefaciens*-mediated transformation of broccoli (*Brassica oleracea* var. *italica*) and cabbage (*B. oleracea* var. *capitata*). Plant Cell Rep., 15: 287-292.

Montanelli, C., Stetanini, F.M., Chiari, A., Chiari, T. and Nascari, G. (1995). Variability in the response to *Pseudomonas solanacearum* of transgenic lines of potato carrying a cecropin gene analogue. Potato Res., 38: 371-378.

Murray, E.E., Rocheleau, T., Eberle, M., Stock, C., Sekar, V. and Adang, M.C. (1991). Analysis of unstable RNA transcripts of insecticidal protein genes of *Bacillus thuringiensis* in transgenic plants and electroporated protoplasts. Plant Mol. Biol., 16: 1035-1050.

Naik, P.S., Chakrabarti. S.K., Sarkar, D., Mandaokar. A.D., Chandla, V.K., Anand Kumar, P. and Sharma, R.P. (1997). An efficient protocol of *Agrobacterium tumefaciens*-mediated transformation and introduction of synthetic cry 1 Ab gene into potato In: Proc. Natl. seminar, IPA, OUAT, Bhubaneswar, Sept. 6, p. 19.

Nelson, R.S., S.M. McCormick, X. Delannay, P. Dubé, J. Layton, E.J. Anderson, M. Kaniewska, R.K. Proksch, R.B. Horsch, S.G. Rogers, R.T. Fraley, and R.N. Beachy. (1988). Virus tolerance, plant growth, and field performance of transgenic tomato plants expressing coat protein from tobacco mosaic virus. Bio/Technology 6:403-409.

Hayakawa, T., Kaneko, H., Nakajima, T., Nishibayashi, S., and Suzuki, M. (1996). CMV protection in transgenic cucumber plants with an introduced CMV-O CP gene. Theor. Appl. Genet., 93: 672-678.

Passelegue, E. and Kerlan, C. (1996). Transformation of cauliflower (*Brassica oleracea* var. *botrytis*) by transfer of cauliflower mosaic virus genes through combined cocultivation with virulent and avirulent strains of *Agrobacterium*. Plant Sci., 113: 79-88.

Pen, J., Sijmons, P.C., Van Ooijen, A.J.J. and Hoekema, A.C. (1993). Protein production in transgenic crops: Analysis of plant molecular farming. In: Transgenic plants: Fundamentals and application (Ed. Hiat, A.), Marcel Dekker Inc., New York, USA 239-254.

Perl, A., Perl-Trever, R., Galili, S., Aviv, D., Shalgi, E., Malkin, S. and Galum, E. (1993). Enhanced oxidative-stress defense in transgenic potato expressing tomato Cu, Zn superoxide dismutases. Theor. Appl. Genet., 85: 568-576.

Punja, Z.K. and Raharjo, S.H.T. (1996). Response of transgenic cucumber and carrot plants expressing different chitinase enzyme to inoculation with fungal pathogens. Plant Disease. 80: 999-1005.

Rober. M., Geider, K., Muller-Rober, B. and Willnitzer, L. (1996). Synthesis of fructans in tubers of transgenic starch-deficient potato plants does not result in an increased allocation of carbohydrates. Planta, 199: 528-536.

Rhim, S.L., Cho, H.J., Kim, B.D., Schnetter, W., Geider, K. (1995). Development of insect resistance in tomato plants expressing the d-endotoxin gene of *Bacillus thuringiensis* subsp. *tenebrionis*. Mol. Breed. 1: 229-236.

Rotino, G.L., Perri, E., Zottini, M., Sommer, H. and Spena, A. (1997). Genetic engineering of parthenocarpic plants. Nature Biotech., 15: 1398-1401.

Rotino, G.L., Spena, A., Ficcadenti, N., Arpaia, S., Aciarri, N., Mennella, G., Donzella, G., Sunseri, F. and Falavigna, A. (1998). Miglioramento della melanzana attraverso trasformazione genetica per i caratteri 'partenocarpia' e 'resistenza a dorifora'. Sementi Elette, 3-4: 63-68. (in Italian)

Serageldin, I. (2000). The challenge of poverty in the 21st century: The role of science. In: Agriculture biotecnology and the poor: Proceedings of an international conference. Washington, DC, 21-22 October, 1999 (Eds., G.J. Perseley and M.M. Lantin), Consultative Group on International Agricultural Research, Washington, DC, USA, 25-31.

Salehuzzaman, S.N.I.M., Jacobsen, E. and Visser, R.G.F. (1994). Expression patterns of two starch biosynthetic genes in in vitro cultured cassava plants and their induction by sugars. Plant Sci., 98: 53-62.

Shah, D., Horsch, R., Klee, H., Kishore, G., Winter, J., Turner, N., Hironaka, C., Sanders, P., Gasser, C., Aykent, S., Siegel, N., Rogers, S. and Fraley. R. (1986). Engineering herbicide tolerance in transgenic plants. Science, 233: 470-481.

Stalker, D.M., Malyj, L.D. and McBride, K.E. (1988). Purification and properties of a nitrilase specific for the herbicide bromoxynil and corresponding nucleotide sequence analysis of the bxn gene J. Biol. Chem., 263: 6310-6314.

Stark. D.M., Timmerman. K.P., Barry, K.F., Preiss, J. and Kishore. G.M. (1992). Regulation of the amount of starch in plant tissues by ADP glucose pyrophosphorylase. Science, 258: 287-292.

Tabei, Y., Kitade, S., Nishizawa, Y., Kikuchi, N., Kayano, T., Hibi, T. and Akutsu, K. (1998). Transgenic cucumber plants harboring a rice chitinase gene exhibit enhanced resistance to gray mold (*Botrytis cinerea*). Plant Cell Rep., 17: 159-164.

Thompson, C.J., Movva, N.R., Tizard, R., Crameri, R., Davies, J.E., Lauwereys, M. and Botterman, J. (1987). Characterization of the herbicide-resistance gene bar from *Streptomyces hygroscopicus*. EMBO J., 6: 2519-2523.

Van, Loon L.C. (1985). Pathogenesis related proteins. Plant Mol. Biol., 4: 111-116.
Visser, R.G.F., Suurs, L.C.J.M., Steneken, P.A.M. and Jacobsen, E. (1997). Some physico-chemical properties of amylose-free potato starch. Starch, 49: 443-448.
Wu, G., Short, B.J., Lawrence, E.B., Levine, E.B., Fitzsimmons, K.C. and Shah, D.M. (1995). Disease resistance conferred by expression of a gene encoding H_2O_2-generating glucose oxidase in transgenic potato plants. The plant cell, 7: 1357-1368.
Xue, B., Gonsalves, C., Provvidenti, R., Slightom, J.L., Fusch, M. and Gonsalves, D. (1994). Development of transgenic tomato expressing a high level of resistance to cucumber mosaic virus strains of subgroup I and II. Plant Disease, 78: 1038-1041.
Yang, M.S., Espinoza, N.O., Nagpala, P.G., Dodds, J.H. White, F.F., Schnor, K.L. and Jaynes, J. (1989). Expression of a synthetic gene for improved protein quality in transformed potato plants. Plant Sci, 64: 99-111.
Zhao, Q., Ao, G.M., Liu, S. and Han, B.W. (1995). Expression of bovine growth hormone gene in transgenic potato. Acta Botanica Sinica, 37: 842-847.
Zhang, Z.J., Zhou, Z.X., Liu, Y.J., Jiang, Q.Y., You, M., Liu, G.M. and Mi, J.J. (1994). CMVcp gene transformation into pepper and expression in the offspring of the transgenic plants. Acta Agriculturae Boreali-Sinica, 9: 67-71.

Index

T - #0356 - 101024 - C16 - 216/152/26 - PB - 9781560221197 - Gloss Lamination